ANNALS OF THE NEW YORK ACADEMY OF SCIENCES

Volume 968

EDITORIAL STAFF

Executive Editor
BARBARA M. GOLDMAN

Managing Editor
JUSTINE CULLINAN

Associate Editor
JOYCE HITCHCOCK
RICHARD STIEFEL

The New York Academy of Sciences
2 East 63rd Street
New York, New York 10021

THE NEW YORK ACADEMY OF SCIENCES
(Founded in 1817)

BOARD OF GOVERNORS, September 2001 – September 2002

TORSTEN N. WIESEL, *Chairman of the Board*
JOHN F. NIBLACK, *Vice Chairman of the Board*
BILL GREEN, *Past Chairman*

Honorary Life Governors
WILLIAM T. GOLDEN JOSHUA LEDERBERG
JOHN T. MORGAN, *Treasurer*

Governors

ELEANOR BAUM	D. ALLAN BROMLEY	KAREN E. BURKE
LAWRENCE B. BUTTENWIESER	PRAVEEN CHAUDHARI	
JOHN H. GIBBONS	MICHAEL GOLDEN	RONALD L. GRAHAM
JACQUELINE LEO	SANDRA PANEM	RICHARD A. RIFKIND
JOHN J. ROCHE		SARA LEE SCHUPF
JAMES H. SIMONS		LEE VANCE

HELENE L. KAPLAN, *Counsel* [ex officio]

PROTEIN KINASE A AND HUMAN DISEASE

ANNALS OF THE NEW YORK ACADEMY OF SCIENCES
Volume 968

PROTEIN KINASE A AND HUMAN DISEASE

Edited by Constantine A. Stratakis and Yoon S. Cho-Chung

The New York Academy of Sciences
New York, New York
2002

Copyright © 2002 by the New York Academy of Sciences. All rights reserved. Under the provisions of the United States Copyright Act of 1976, individual readers of the Annals are permitted to make fair use of the material in them for teaching or research. Permission is granted to quote from the Annals provided that the customary acknowledgment is made of the source. Material in the Annals may be republished only by permission of the Academy. Address inquiries to the Permissions Department (permissions@nyas.org) at the New York Academy of Sciences.

Copying fees: For each copy of an article made beyond the free copying permitted under Section 107 or 108 of the 1976 Copyright Act, a fee should be paid through the Copyright Clearance Center, Inc., 222 Rosewood Drive, Danvers, MA 01923 (www.copyright.com).

☉ The paper used in this publication meets the minimum requirements of the American National Standard for Information Sciences—Permanence of Paper for Printed Library Materials, ANSI Z39.48-1984.

Library of Congress Cataloging-in-Publication Data

Protein kinase a and human disease / edited by Constantine A. Stratakis and Yoon S. Cho-Chung.
  p. ; cm. — (Annals of the New York Academy of Sciences ; ISSN 0077-8923 ; v. 968)
Includes bibliographical references and index.
  ISBN 1-57331-412-9 (cloth : alk. paper) — ISBN 1-57331-413-7 (paper : alk. paper)
  1. Protein kinases—Pathophysiology—Congresses. 2. Cyclic adenylic acid—Pathophysiology—Congresses. 3. Cellular signal transduction—Congresses. 4. Pathology, Molecular—Congresses. I. Stratakis, Constantine A. II. Cho-Chung, Yoon S. III. Series.
  Q11 .N5 vol. 968
  [RB113]
  500 s—dc21
  [571.9'48]   2002005816

GYAT/PCP
Printed in the United States of America
ISBN 1-57331- 412-9 (cloth)
ISBN 1-57331-413-7 (paper)
ISSN 0077-8923

ANNALS OF THE NEW YORK ACADEMY OF SCIENCES

Volume 968
June 2002

PROTEIN KINASE A AND HUMAN DISEASE

Editors and Conference Organizers
CONSTANTINE A. STRATAKIS AND YOON S. CHO-CHUNG

This volume is the result of a conference entitled Protein Kinase A and Human Disease, which was sponsored by the National Institute of Child Health and Human Development, National Institutes of Health, and held on September 10, 2001 in Bethesda, Maryland.

CONTENTS

Preface. By CONSTANTINE A. STRATAKIS vii

Introduction. By MICHAEL M. GOTTESMAN 1

Part I. Protein Kinase A: Signaling, Model Systems, Pathophysiology, and Genetics

Mutations of the Gene Encoding the Protein Kinase A Type I-α Regulatory Subunit (*PRKAR1A*) in Patients with the "Complex of Spotty Skin Pigmentation, Myxomas, Endocrine Overactivity, and Schwannomas" (Carney Complex). By CONSTANTINE A. STRATAKIS 3

Dissecting the Circuitry of Protein Kinase A and cAMP Signaling in Cancer Genesis: Antisense, Microarray, Gene Overexpression, and Transcription Factor Decoy. By YOON S. CHO-CHUNG, MARIA NESTEROVA, KEVIN G. BECKER, RAKESH SRIVASTAVA, YUN GYU PARK, YOUL NAM LEE, YEE SOOK CHO, MEYOUNG-KIN KIM, CATHERINE NEARY, AND CHRIS CHEADLE 22

Regulatory Subunits of PKA and Breast Cancer. By W. R. MILLER 37

Reinventing the Wheel of Cyclic AMP: Novel Mechanisms of cAMP Signaling. By KHEW-VOON CHIN, WENG-LANG YANG, ROALD RAVATN, TSUNEKAZU KITA, ELENA REITMAN, DAVID VETTORI, MARY ELLEN CVIJIC, MICHAEL SHIN, AND LISA IACONO 49

Role of the PKA-Regulated Transcription Factor CREB in Development and Tumorigenesis of Endocrine Tissues. *By* D. ROSENBERG, L. GROUSSIN, E. JULLIAN, K. PERLEMOINE, X. BERTAGNA, AND J. BERTHERAT 65

The Essential Role of RIα in the Maintenance of Regulated PKA Activity. *By* PAUL S. AMIEUX AND G. STANLEY MCKNIGHT 75

Deficient Protein Kinase A in Systemic Lupus Erythematosus: A Disorder of T Lymphocyte Signal Transduction. *By* GARY M. KAMMER 96

The Role of Cyclic AMP and Its Effect on Protein Kinase A in the Mitogenic Action of Thyrotropin on the Thyroid Cell. *By* S. DREMIER, K. COULONVAL, S. PERPETE, F. VANDEPUT, N. FORTEMAISON, A. VAN KEYMEULEN, S. DELEU, C. LEDENT, S. CLÉMENT, S. SCHURMANS, J. E. DUMONT, F. LAMY, P. P. ROGER, AND C. MAENHAU 106

Cyclic AMP and the Reverse Transformation Reaction. *By* THEODORE T. PUCK, PATRICIA WEBB, AND ROBERT JOHNSON 122

Protein Kinase A as Target for Novel Integrated Strategies of Cancer Therapy. *By* GIAMPAOLO TORTORA AND FORTUNATO CIARDIELLO 139

Protein Kinase A and Chromosomal Stability. *By* LUDMILA MATYAKHINA, SARA M. LENHERR, AND CONSTANTINE A. STRATAKIS 148

Protein Kinase A: Regulation and Receptor-Mediated Delivery of Antisense Oligonucleotides and Cytotoxic Drugs. *By* P. G. SVESHNIKOV, I. D. GROZDOVA, M. V. NESTEROVA, AND E. S. SEVERIN 158

Part II. Other Diseases, Signaling Molecules, and Pathways: Interactions with PKA

Gs_α Mutations and Imprinting Defects in Human Disease. *By* LEE S. WEINSTEIN, MIN CHEN, AND JIE LIU 173

Regulation of Phospholipase D and Secretion in Mast Cells by Protein Kinase A and Other Protein Kinases. *By* WAHN SOO CHOI, AHMED CHAHDI, YOUNG MI KIM, PAUL F. FRAUNDORFER, AND MICHAEL A. BEAVEN ... 198

Role of PTEN, a Lipid Phosphatase Upstream Effector of Protein Kinase B, in Epithelial Thyroid Carcinogenesis. *By* CHARIS ENG 213

Signaling Pathways in Adrenocortical Cancer. *By* LAWRENCE S. KIRSCHNER . 222

Cyclic AMP–Dependent Signaling Aberrations in Macronodular Adrenal Disease. *By* ISABELLE BOURDEAU AND CONSTANTINE A. STRATAKIS .. 240

Protein Kinase A Signaling: "Cross-Talk" with Other Pathways in Endocrine Cells. *By* AUDREY ROBINSON-WHITE AND CONSTANTINE A. STRATAKIS 256

Index of Contributors 271

The New York Academy of Sciences believes it has a responsibility to provide an open forum for discussion of scientific questions. The positions taken by the participants in the reported conferences are their own and not necessarily those of the Academy. The Academy has no intent to influence legislation by providing such forums.

Preface

It was said by Hippocrates: "...ἡ δέ ἰητρική νῦν τε καί αὐτίκα, οὐ τό αὐτό ποιέει..." ("...medicine does not do the same thing at this moment and the next..."; Hippocrates, *Places in Man*, edited by P. Potter, Harvard University Press, 1995, p. 81). This is true in every field of medicine, but it is particularly applicable in the case of investigations related to the protein kinase A (PKA) system, an important pathway that mediates cyclic nucleotide signaling and appears to integrate many messages from a variety of senders and centers. The central position of PKA in cellular function qualifies it as a "hub" through which many lines pass, get finessed and even re-directed; signals get amplified, diverted, or deleted; and downstream effectors are released, recruited, or put in a "standby" mode. It is no surprise that what we thought of PKA in the 1970s, 1980s, and 1990s may not be the same today; in the era of post-genomic, neo-new, translational medicine, PKA and its regulated genes constitute a huge "transcriptosome" that current investigators are trying to elucidate—hence the cover of this volume from Dr. Cho-Chung's pioneering work. As for the volume's contents, they are, in essence, the proceedings of a meeting that was held at the Warren Magnuson National Institutes of Health Clinical Center in Bethesda, Maryland on September 10th, 2001—the day before the terrorist attacks in the US. Because of the latter, many of our speakers had to spend several days in Bethesda, much to their dismay, but also to our benefit: they had plenty of time to write wonderful, illuminating chapters on the many aspects of PKA function and its role in human disease. It is our hope that the present volume will not only add to our understanding of this important signaling pathway by concentrating for the first time in one text all relevant material written by some of the pioneers of the field, but also lead to more clinical applications of this basic knowledge!

—CONSTANTINE A. STRATAKIS

Acknowledgments

This conference was funded by a grant from NICHD, NIH, with additional support from Eli Lilly and company. Special thanks to Caroline Sandrini (Santa Catarina, Brazil), who designed the meeting poster and supervised the art of many papers in the volume. Additional thanks to Ruth Maraio and Sue Perdue from NICHD administration for their hard work in making this conference possible. We also thank the NICHD Scientific Director, Dr. Owen Rennert, and the NICHD Director, Dr. Duane Alexander, for their continuing support of this program and our research efforts on PKA and endocrine tumors.

Introduction

MICHAEL M. GOTTESMAN

Laboratory of Cell Biology, National Cancer Institute, National Institutes of Health, Bethesda, Maryland 20892, USA

Many commentators have noted that science revisits previous styles and trends from time to time, using current technology to build on past knowledge and create a more and more precise view of the world. A recent one-day meeting at the National Institutes of Health on **Protein Kinase A and Human Disease** held on September 10, 2001 illustrates this paradigm perfectly. The role of cAMP in endocrinology and signal transduction resulted in the awarding of three Nobel prizes for the basic physiology and biochemistry underlying transduction of signals from hormones, through G proteins, to adenylate cyclase, to cAMP and its activation of cAMP-dependent protein kinase (protein kinase A). These discoveries, though seminal, had to await the application of molecular techniques and, in particular, genomic approaches such as positional cloning, before they could be applied to an understanding of specific human disorders of this system, including inherited endocrine disorders and human cancers.

There is much still to be learned from studying the biology of cAMP. As illustrated in this volume, disorders resulting from mutations in Gs_α (e.g., McCune-Albright syndrome and Albright hereditary osteodystrophy) owe their pathophysiology to complex inheritance of Gs_α including maternal imprinting and tissue-specific imprinting. Loss of one allele of the type I regulatory subunit of protein kinase A (RIα) produces a multiple tumor syndrome known as Carney complex (CNC), whereas in mice, loss of both alleles produces an embryonic lethal defect due to failure of formation of the heart. Although the precise molecular basis of each of these defects is not known, they illustrate the broad biological consequences of disruption of the cAMP signaling system. Evidence for cAMP signaling through RIα independent of the catalytic subunit of protein kinase A and alterations in RI subunits in the immunological disorder systemic lupus erythematosis (SLE) reported in these pages adds to the mystique of cAMP.

Given that inherited disruption of cAMP signaling results in cancer, it is not surprising that researchers have turned their attention to manipulation of this system to improve the treatment of cancer. Ample correlative evidence suggests defects in protein kinase A in many kinds of cancers, including breast cancer, and the genetic data have begun to point to protein kinase A and its subunits as targets for treatment of cancer. Several papers in this volume explore this approach to cancer treatment.

Address for correspondence: Dr. Michael M. Gottesman, Chief, Laboratory of Cell Biology, National Cancer Institute, National Institutes of Health, Bethesda, Maryland 20892.

This collection of papers should serve as a landmark along the route from the discovery of cAMP as a second messenger in the cell to a more complete understanding of the role of cAMP and protein kinase A in health and disease. It is a satisfying source of solutions to old problems and poses enough new quandries to keep us busy for some time to come.

Mutations of the Gene Encoding the Protein Kinase A Type I-α Regulatory Subunit (*PRKAR1A*) in Patients with the "Complex of Spotty Skin Pigmentation, Myxomas, Endocrine Overactivity, and Schwannomas" (Carney Complex)

CONSTANTINE A. STRATAKIS

Unit on Genetics & Endocrinology (UGEN), Developmental Endocrinology Branch, (DEB), National Institute of Child Health and Human Development (NICHD), National Institutes of Health (NIH), Bethesda, Maryland 20892, USA

ABSTRACT: Carney complex (CNC) is a familial multiple neoplasia syndrome associated with abnormal skin and mucosal pigmentation. The complex has features overlapping those of McCune-Albright syndrome (MAS) and the other multiple endocrine neoplasias (MENs). CNC is inherited as an autosomal dominant trait, and the responsible genes have been mapped by linkage analysis to loci at 2p16 and 17q22-24. Because of its unusual biochemical features (e.g., paradoxical responses to various endocrine signals) and its clinical similarities to MAS, genes implicated in cyclic nucleotide–dependent signaling, including *GNAS1* (which is responsible for MAS), had been considered likely candidates for causing CNC. The gene encoding the protein kinase A (PKA) type I-α regulatory subunit (RIα), *PRKAR1A*, had been mapped to 17q22-24; loss-of-heterozygosity (LOH) analysis using polymorphic markers from this region revealed consistent changes in tumors from patients with CNC, including those from one family previously mapped to 17q22-24. Investigation of a polymorphic site within the 5′ of the *PRKAR1A* gene showed segregation with the disease and retention of the allele bearing the disease gene in CNC tumors. Mutations of the *PRKAR1A* gene were also found to have occurred *de novo* in sporadic cases of CNC; no mutations were found in kindreds mapping to 2p16. Thus, genetic heterogeneity in CNC was confirmed; in total, 41% of all patients with CNC had mutations in the *PRKAR1A* gene. All mutations were frame-shifts, insertions, and deletions that led to nonsense mRNA and premature termination of the predicted peptide product. Functional studies in CNC tumors suggested that inactivating mutations of the *PRKAR1A* gene led to nonsense mRNA decay (the mutant peptide product was not present) and were associated with dysregulated PKA activity, increased responsiveness to cAMP, and excess of type-II PKA activity. We conclude that the *PRKAR1A* gene, coding for the RIα subunit of PKA, a critical cellular component of a number of cyclic

Address for correspondence: Constantine A. Stratakis, M.D., D.Sc., Chief, Unit on Genetics and Endocrinology, DEB, NICHD, NIH, Building 10, Room 10N262, 10 Center Dr. MSC1862, Bethesda, Maryland 20892-1862.Voice: 301-496-4686/402-1998; fax 301-435-4358.
stratakc@cc1.nichd.nih.gov

Ann. N.Y. Acad. Sci. 968: 3–21 (2002). © 2002 New York Academy of Sciences.

nucleotide–dependent signaling pathways, is mutated in a subset of patients with CNC. In their tumors, there is LOH of the normal allele, suggesting that normal RI-α may have tumor suppression function in the tissues affected by CNC. An excess of type-II PKA activity was present in affected tissues, which may be responsible for the apparent tumorigenicity of PRKAR1A mutations in endocrine tissues.

KEYWORDS: protein kinase A; regulatory subunit; Carney complex; chromosome 17; amplification; tumor suppressor gene

CARNEY COMPLEX: AN INTRODUCTION

The complex of "spotty skin pigmentation, myxomas, endocrine overactivity, and schwannomas" or Carney complex (CNC) [Mendelian Inheritance in Man (MIM) catalog number 160980] is an autosomal-dominant, multiple neoplasia syndrome that was initially described in 1985 as "the complex of myxomas, spotty pigmentation, and endocrine overactivity."[1] Isolated patients with some components of the complex—in particular, cardiac myxomas and pigmentation anomalies—had previously been described under the acronyms NAME (nevi, atrial myxomas, and ephelides) and LAMB (lentigines, atrial myxomas, and blue nevi); today, it is accepted that most, if not all, of these patients had CNC.[2–5]

CNC may be viewed as a form of multiple endocrine neoplasia (MEN), because affected patients often have tumors of two or more endocrine glands, including primary pigmented nodular adrenocortical disease (PPNAD), growth hormone and prolactin-producing pituitary adenoma, thyroid adenoma or carcinoma, testicular neoplasms [primarily large-cell calcifying Sertoli cell tumor (LCCSCT)], and ovarian cysts. Additional unusual manifestations include psammomatous melanotic schwannoma (PMS), breast ductal adenoma, and, probably, a rare bone tumor, osteochondromyxoma.

EPIDEMIOLOGY AND INHERITANCE

Three hundred thirty-eight patients with CNC have so far been identified: 144 (43%) male and 194 (57%) female, including Caucasians, African-Americans, and Asians, from all continents [North and South America, Europe, Asia (Japan, China, India), Australia, and New Zealand]. Most of the patients (70%) belonged to 67 families; 88 were sporadic cases, whereas the origin of the disease could not be definitely determined in 12.

Previous estimates had indicated that only approximately half of the cases of CNC were familial. The increased number of familial cases we observed reflects the application of a rigorous screening protocol for all first-degree relatives of affected patients. Careful history taking often identified ancestors of an affected patient with cutaneous pigmented spots or obvious signs of endocrine disease. The detailed family data also demonstrated significant variability in clinical manifestations between patients, including members of the same family. This clinical variability was responsible for the apparent "skip" of a generation in extended CNC pedigrees, and renders designation of a case as "sporadic" doubtful, unless careful clinical, imaging, and biochemical screening of all first-degree relatives has been obtained.

Transmission of CNC occurred through a female affected parent in 43 cases and from a male in only 9. CNC may have non-Mendelian features in some aspects of its inheritance, not unlike MEN 2 and perhaps other familial cancer syndromes. However, it should be noted that LCCSCT, a frequent component of CNC in male patients, causes displacement and obstruction of seminiferous tubules and may also impair fertility by inappropriate hormone production or aromatization; in addition, several patients with CNC had undergone orchiectomy.

Although there were many families with CNC, the number of affected members in the majority of these was small. The maximum number of affected generations in a family was five. The small size of most CNC families precludes the use of genetic linkage studies in counseling kindreds that do not have *PRKAR1A* mutations.

CNC is a developmental disorder. Diagnosis of the disease was made at birth in at least five patients. The median age at detection among 235 cases was 20 years. Although abnormal skin pigmentation may be present at birth, the lentigines usually do not assume their characteristic distribution, density, and intensity until around and shortly after puberty. Unlike other pigmented lesions affecting the aging skin, lentigines associated with CNC tend to fade after the fourth decade of life.

Other pigmented lesions, including blue and other nevi, café-au-lait spots, and depigmented lesions may also be present at birth and are referred to as "birthmarks"; more frequently, however, these lesions develop in the early childhood years.[2] The café-au-lait spots in CNC are usually smaller and less pigmented than those in McCune-Albright syndrome (MAS); they also tend to fade with time. Their shape was more reminiscent of the neurofibromatosis (NF) syndromes.

During infancy, heart and cutaneous myxomas and PPNAD are the most common tumors encountered.[6] LCCSCT and thyroid nodules (appearing as microcalcifications and multiple, small, hypoechoic lesions upon testicular and thyroid ultrasonography, respectively) appear often within the first 10 years of life. The earliest detection of LCCSCT (by ultrasonography) was made in a two-year-old boy.

There seems to be a bimodal age distribution of PPNAD among CNC patients: a minority of patients present in the first 2–3 years, whereas the majority manifest in the second and third decade of life. Acromegaly usually is observed during the third and fourth decade of life; gigantism is rare, as in the case of other familial forms of acromegaly. Heart myxomas, on the other hand, are fairly equally distributed among the ages.

CNC COMPONENTS AND THEIR RELATIVE FREQUENCY

The criteria used for the diagnosis of CNC are presented in TABLE 1;[6] at least two of the major criteria need to be present for the establishment of the diagnosis. As has already been mentioned, spotty skin pigmentation is the most common clinical manifestation of CNC (FIG. 1), although it is not invariably present. Additional pigmentary abnormalities in our patients included usual and epithelioid blue nevi, combined nevi, café-au-lait spots, and depigmented lesions.

Heart myxomas occurred at a young age, multicentrically, in any or all cardiac chambers. Fifty-one patients had two or more operations for recurrent tumor. Classic sites for skin myxomas included the eyelid, external ear canal, and nipple. Breast myxomas, often bilateral, were present in 34 female and 2 male patients. Occasional

TABLE 1. Diagnostic criteria for Carney complex

Major criteria

1. Spotty skin pigmentation with a typical distribution (buccal mucosa, lips, conjuctiva and inner or outer canthi, vaginal and penile mucosa)
2. Myxoma (cutaneous and mucosal)^a
3. Cardiac myxoma^a
4. Breast myxomatosis ^{a} *or* detection by fat-suppressed magnetic resonance imaging
5. PPNAD ^{a} *or* paradoxical positive response of urinary glucocorticosteroids to dexamethasone administration during Liddle's test
6. Acromegaly due to GH-producing adenoma ^{a}
7. LCCSCT^a *or* characteristic calcification on testicular ultrasonography
8. Thyroid carcinoma^a *or* characteristic multiple, hypoechoic nodules on thyroid ultrasonography
9. Psammomatous melanotic schwannoma^a
10. Blue nevus, epithelioid nevus (multiple)^a
11. Breast ductal adenoma (multiple)^a
12. Osteochondromyxoma of the bone^a
13. First-degree relative with established diagnosis of Carney complex
14. Inactivating mutation of the *PRKAR1A* gene

Minor criteria

1. Intense freckling (without darkly pigmented spots or typical distribution)
2. Blue nevus, usual type (if multiple)
3. Café-au-lait spots or other "birthmarks"
4. Elevated IGF-1 levels, abnormal oGTT, or paradoxical GH responses to TRH testing *in the absence of clinical acromegaly*
5. Cardiomyopathy
6. Pilonidal sinus
7. History of Cushing syndrome, acromegaly, or sudden death in extended family

Other conditions that may be seen in patients with Carney complex but are not diagnostic of the disease

1. Multiple skin tags and other skin lesions; lipomas
2. Colonic polyps (usually in association with acromegaly)
3. Hyperprolactinemia (usually mild and almost always in association with clinical or subclinical acromegaly)
4. Single, benign thyroid nodule in a young patient; multiple thyroid nodules in an older patient (detected by ultrasonography)
5. Family history of carcinoma, in particular of the thyroid, colon, pancreas, and the ovary; other multiple benign or malignant tumors

^{a}With histologic confirmation.

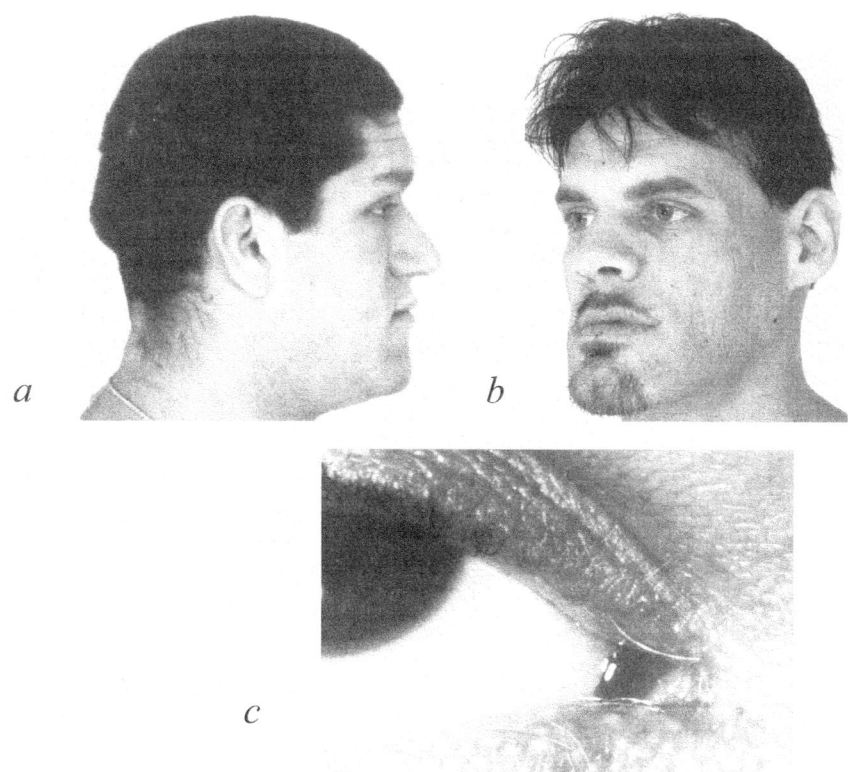

FIGURE 1. Clinical manifestations of CNC. **a.** The proband of family CAR01 (YC01 in ref. 9) (CAR01.08 in FIG. 3c) has a few pigmented spots and clinical stigmata of acromegaly; his growth hormone– and prolactin-producing pituitary adenoma, which also had loss-of-heterozygosity for an 8 cM-long region around *PRKAR1A* (FIG. 3a), is shown in FIGURE 3d. **b.** The proband of family CAR20 (CAR20.III.1 in FIG. 2), who presented with a cardiac myxoma, is shown here. He also had acromegaly and facial pigmentation, including pigmented spots on his lips. **c.** Pigmentation at the inner canthus of the eye in patient CAR01.08 was also present in several other patients with CNC; this clinical sign is characteristic for CNC.

other sites for myxomas were: oropharynx (tongue, 3; hard palate, 3; pharynx, 2); the female genital tract (uterus, 2; cervix, 1; vagina, 1), and the female pelvis (1).

Among the endocrine tumors, PPNAD was the most frequent manifestation of the disease, having occurred in about one quarter of the patients. This estimate, however, is likely to be low because biochemical screening by a dexamethasone-stimulation test, as suggested by Stratakis *et al.*, has been shown to detect patients with PPNAD-associated subclinical, atypical, or periodic Cushing syndrome. Furthermore, histological evidence of PPNAD has been found at autopsy in almost every patient with the complex.

LCCSCT was often multicentric and bilateral. Ultrasonography, however, identified testicular microcalcifications in most examined affected adult patients with CNC. Thus, LCCSCT is very prevalent, and ultrasonography is an effective and inexpensive screening technique for its detection. LCCSCT almost always is benign; one case of malignancy has occurred in a 62-year-old patient. Testicular ultrasonography has also detected other tumors in CNC patients, including Leydig cell (2 patients) and (pigmented nodular) adrenocortical rest tumors (3 patients). In all the latter patients, LCCSCT was also present; one patient had all three testicular tumors.

LCCSCT in CNC, as in Peutz-Jeghers syndrome (PJS), may be hormone producing; it has caused gynecomastia in prepubertal and peripubertal boys (5 patients in our series) due to increased P-450 aromatase expression. The gynecomastia, unlike that due to familial aromatase excess, in which medical treatment with blockers of aromatization appears to be effective, usually requires orchiectomy to avoid premature epiphyseal fusion and induction of central precocious puberty.

Clinically evident acromegaly is a relatively infrequent manifestation of CNC. However, asymptomatic GH and elevation of insulin-like growth factor type 1 (IGF-1) levels and subtle hyperprolactinemia may be present in up to 75% of the patients. Biochemical acromegaly is often unmasked by abnormal results of oral glucose tolerance test (oGTT) or paradoxical responses to thyrotropin-releasing hormone (TRH) administration. Somatomammotropic hyperplasia, a putative precursor of GH-producing adenoma, may explain the insidious and protracted period of establishment of clinical acromegaly in CNC patients.

Up to 75% of patients with CNC may have multiple thyroid nodules detected as small, hypoechoic lesions on ultrasonography. Most of these nodules were follicular adenomas, which were confirmed histologically in six patients (hyperfunctioning in 2). Five thyroid carcinomas occurred among CNC patients, three papillary and two follicular, including one that developed in a patient with a long history of multiple adenomas. Thyroid ultrasonography is recommended as a satisfactory, cost-effective method for determining thyroid involvement in pediatric and young adult patients with CNC; its value, however, is questionable in older patients.

Psammomatous melanotic schwannoma (PMS), a very rare tumor, occurred in 33 patients. In 6, the tumor was malignant. PMS may occur anywhere in the central and peripheral nervous system, although its most frequent site has been in the gastrointestinal tract (esophagus, liver, stomach, rectum) and paraspinal sympathetic chain. CNC is the only genetic condition other than the NF syndromes and isolated familial schwannomatosis that includes schwannomas. The particular schwannoma in CNC is distinctive because of its heavy pigmentation (melanin), frequent calcification, and multicentricity. Imaging of the brain, spine, chest, abdomen (in particular the retroperitoneum), and the pelvis may be necessary for the detection of PMS, if there are suggestive symptoms.

Breast ductal adenoma, an unusual mammary tumor akin to intraductal papilloma, was detected in 6 women with CNC, bilateral in 3. Other conditions probably associated with CNC are presented in TABLE 1; among them, only osteochondromyxoma of the bone, at present, is considered a likely candidate component of the disorder. Parotid mixed tumor, marfanoid habitus, bronchogenic cyst, and hepatocellular adenoma each occurred in 1 patient and are thought unlikely to be related to CNC. Overall, lifespan was decreased in patients with CNC. Fifty-one patients in the series (15%) are deceased, 29 due to heart-related causes (57% of the deaths).

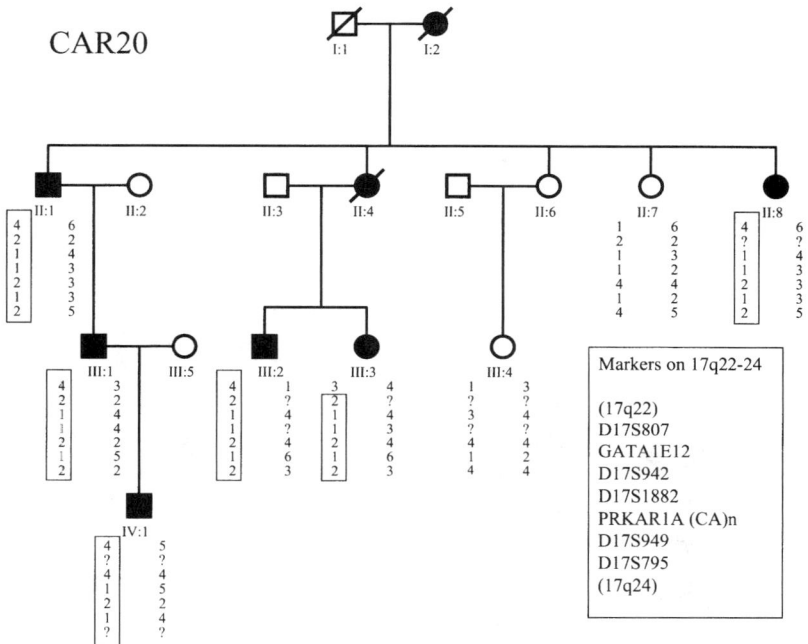

FIGURE 2. Pedigree and chromosome 17 genotyping of family CAR20. Polymorphic markers in the 17q22-24 region were examined in members of the family that had been evaluated at the National Institutes of Health and carefully phenotyped. The boxed alleles indicate the affected chromosome in this family; the list of the markers is in chromosomal order according to the on-line databases (see METHODS).

MOLECULAR GENETICS OF CNC

An initial linkage study of families with CNC demonstrated a genetic locus on 2p16 with an aggregate logarithm-of-odds (LOD) score of 5.97 (θ = 0.03), although no single family had a LOD score greater than 1.8 for the locus.[7] Additional genetic studies uncovered families, in whom CNC did not segregate with 2p16 markers. A genome-wide screen among the latter demonstrated linkage to a locus on 17q22-24.

Following the identification of genetic changes at 17q22-24 in a pituitary tumor from a patient with CNC belonging to a family that was mapped to that region,[9,10] we looked for additional families mapping to 17q22-24 and for genetic alterations in their tumors. Polymorphic markers from the 17q22-24 area proximal to the *PRKAR1A* gene[11-13] showed complete segregation with the disease in the CAR01 family (published as YC01 in ref. 8), a large pedigree with primary pigmented nodular adrenocortical disease (PPNAD), the adrenal tumor characteristic of CNC, and other manifestations of the complex, that was first reported in the literature in 1987.[14] The disease also segregated with markers from 17q in a large, previously unreported family, CAR20, that was screened for manifestations of CNC recently (FIG. 2).

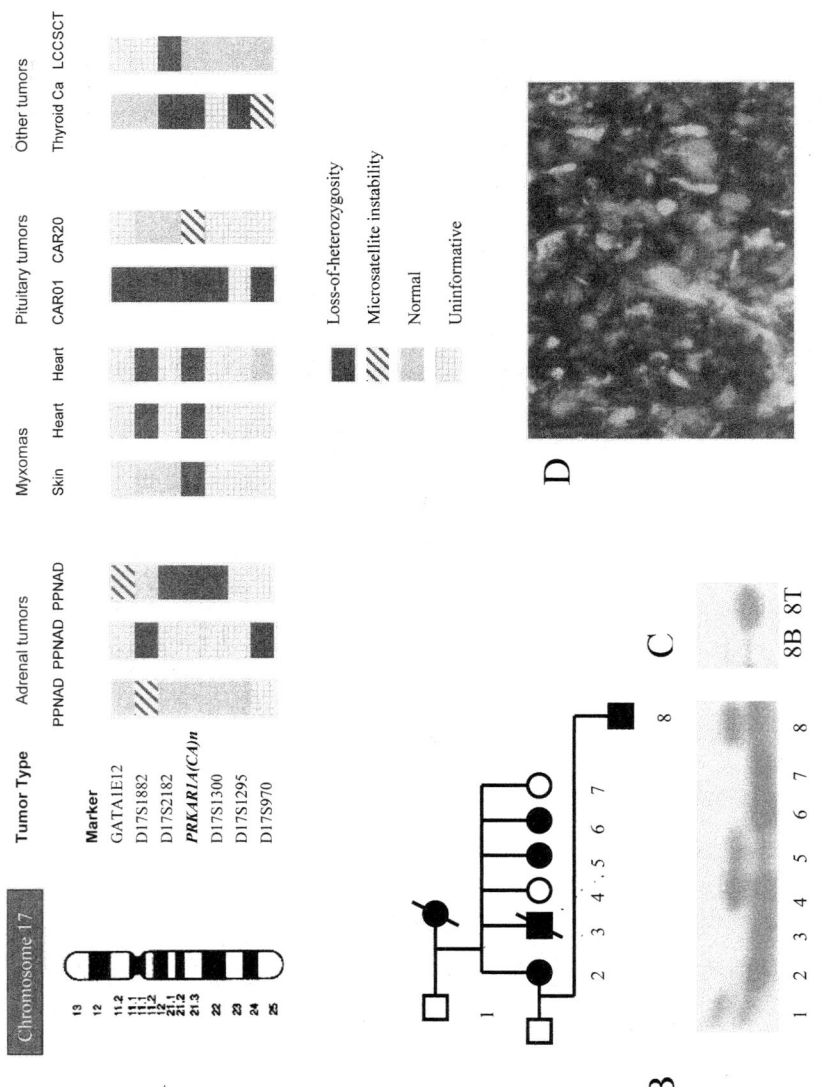

FIGURE 3. See following page for legend.

Loss-of-heterozygosity (LOH) analysis was then performed in DNA obtained from three tumors of members of the two large families with CNC (pituitary tumors from individuals CAR01.08 and CAR20.III.1 and a thyroid carcinoma from individual CAR01.10) and six sporadic cases; all samples showed areas of LOH around or within the *PRKAR1A* gene, in an 8 cM-long area of 17q22-24 (FIG. 3a). The allele segregating with the disease was retained in all the tumor samples that were informative for LOH (FIG. 3b and c). The tumor with the most extensive LOH changes was an aggressive growth hormone– and prolactin-producing pituitary adenoma (FIG. 3d) from individual CAR01.08 (FIG. 1a).

To evaluate the *PRKAR1A* gene as a candidate for CNC, denaturing high-performance liquid chromatography (DHPLC) of its exons was performed. This identified heteroduplex formation in all affected members of families CAR01 and CAR20. Sequencing of the heteroduplexes identified the same 2–base pair deletion at position 578 of the *PRKAR1A* gene in affected patients from both families (FIG. 4). The 578delTG mutation, which is located in exon 4B (FIG. 5) and corresponds to position 163 of the protein, results in a truncated RIα. The two families did not share 17q22-24 haplotypes of the disease-bearing allele, making it unlikely that they carried the same chromosome 17 (data not shown). Furthermore, to confirm the identity of *PRKAR1A* with the CNC-causing gene, we sought mutations in the gene in other families and sporadic cases. Patients with a sporadic form of the disease also had the 578delTG mutation; this mutation appears to have occurred *de novo* because it was absent in parental peripheral blood DNA (FIG. 6). The mechanism responsible for mutagenesis at this site remains unclear; DNA-polymerase stuttering may play a role because of the TGTG pattern present at the deletion site,[15] although other mechanisms cannot be excluded. Screening of additional patients revealed additional mu-

---

FIGURE 3. Loss-of-heterozygosity (LOH) studies of the chromosome 17 *PRKAR1A* locus. **a.** LOH analysis was performed using paired samples of blood and tumors using the markers indicated. With the exception of one testicular tumor (LCCSCT), all informative loci demonstrated LOH at the *PRKAR1A* locus. Abbreviations: Ca = carcinoma; LCCSCT = large-cell calcifying Sertoli cell tumor; PPNAD = primary pigmented nodular adrenocortical disease. **b.** The pedigree and linkage analysis of family CAR01 demonstrate segregation of this marker with the CNC phenotype. Note that the upper allele is derived from the unaffected father, and the affected mother can be inferred to be, most likely, a homozygote for the lower allele. **c.** LOH analysis of the pituitary tumor of CAR01.08, demonstrating loss of the "unaffected" (*upper*) allele in the tumor and retention of the mutant allele of the GATA1E12 marker, 8B = peripheral blood lymphocytes DNA, 8T = pituitary tumor DNA. **d.** The pituitary tumor of CAR01.08 (the patient is shown in FIG. 1a) stained for both growth-hormone and prolactin (× 40); this was an aggressive tumor with bilateral cavernous sinus extension that demonstrated LOH for the entire *PRKAR1A* region (shown in 3**a**) and had a number of other genetic changes by comparative genomic hybridization.

FIGURE 4. Detection of a frameshift mutation in CAR01 in *PRKAR1A* exon 4B. **a.** Six siblings from CAR01 were analyzed for the presence of heteroduplexes in exon 4B using denaturing HPLC (DHPLC). Abbreviations: Het = heterozygote; Ho = homozygote. **b.** Sequence traces from a homozygote (CAR01.04) and a heterozygote (CAR01.10) are shown, indicating the presence of a sequence alteration after the sequence GACTG. **c.** Heteroduplex DNA from CAR01.10 was subsequently TA-cloned for clarification of the sequencing defect; the colonies were selected and sequenced, revealing the presence of a 2-bp deletion in the mutant form of the sequence (578delGT). Abbreviations: mut = mutant; WT = wild type.

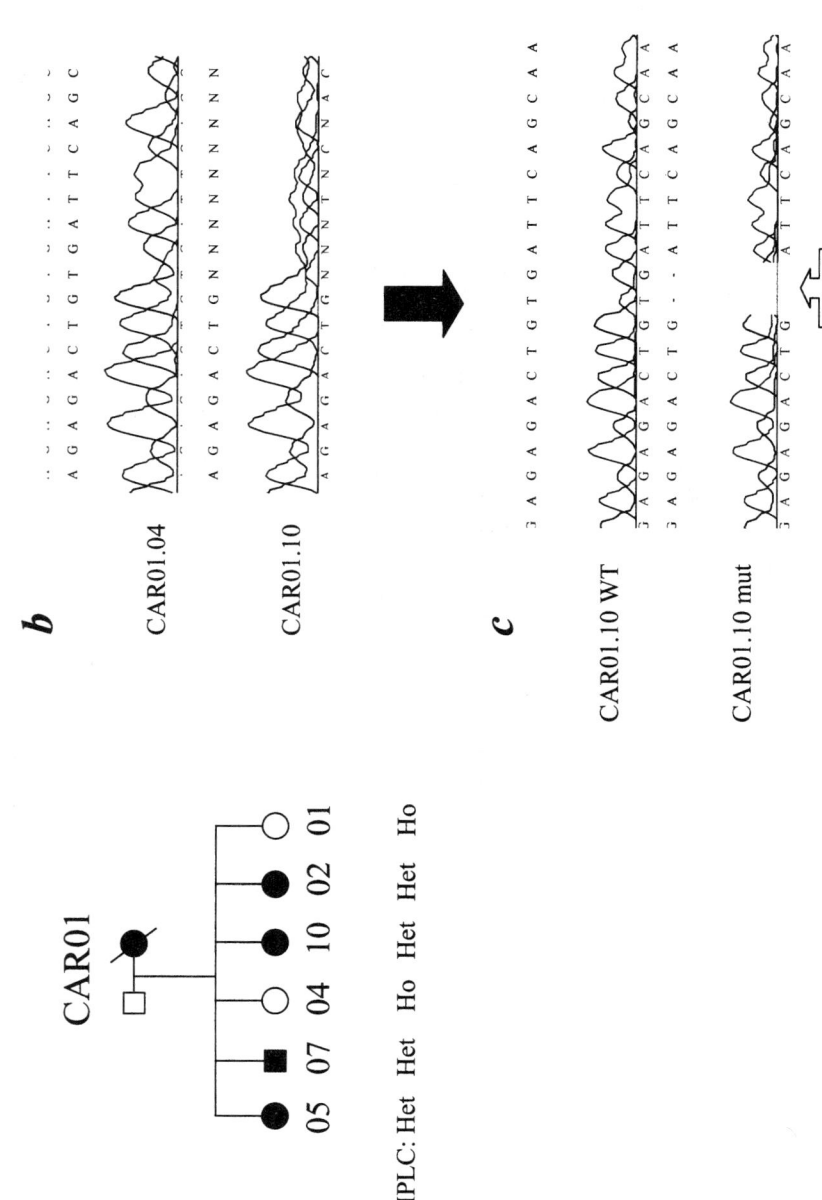

FIGURE 4. *See previous page for legend.*

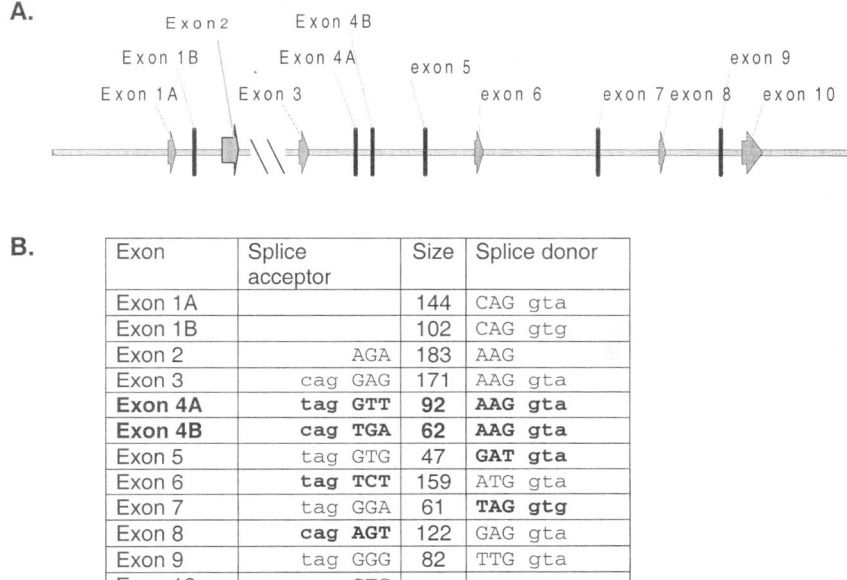

FIGURE 5. Gene structure of PRKAR1A. **Bold text** indicates changes from the intron-exon structure of Solberg *et al.*[12] Note that exon 4 in that paper has been divided into two smaller exons (4A and 4B).

tations of the *PRKAR1A* gene in families mapping to 17q22-24 and approximately half of the sporadic cases: a total of 53 kindreds for CNC were screened for mutations of the *PRKAR1A* gene. The samples were first analyzed for heterozygosity. This analysis detected the presence of mutations in 15 kindreds, in addition to the original 7.[16] No mutations were seen in families mapping to chromosome 2, confirming heterogeneity in the syndrome.[17] In 10 of the newly identified kindreds, the *PRKAR1A* mutation segregated with CNC in more than one family member; in 5 kindreds the mutation was found only in the proband (sporadic cases). TABLE 2 contains a comprehensive list of all PRKAR1A mutations identified to date in our kindreds and their relative frequency.

In addition to the 578delTG mutation, the only other *PRKAR1A* sequence change that appeared to have arisen *de novo* in more than one kindred with CNC were:[16] (1) a nonsense point mutation in exon 2 (C211T) that was found in 2 kindreds, and (2) a G-to-C transversion in the 5′ splice site of intron 3 of the PRKAR1A gene (exon 3 IVS +1G>C) that was also found in 2 kindreds. Of the remaining 12 mutations, 3 that were reported in ref. 25 and 9 newly reported here, each represents a unique mutation. These mutations have been detected throughout the coding region of the

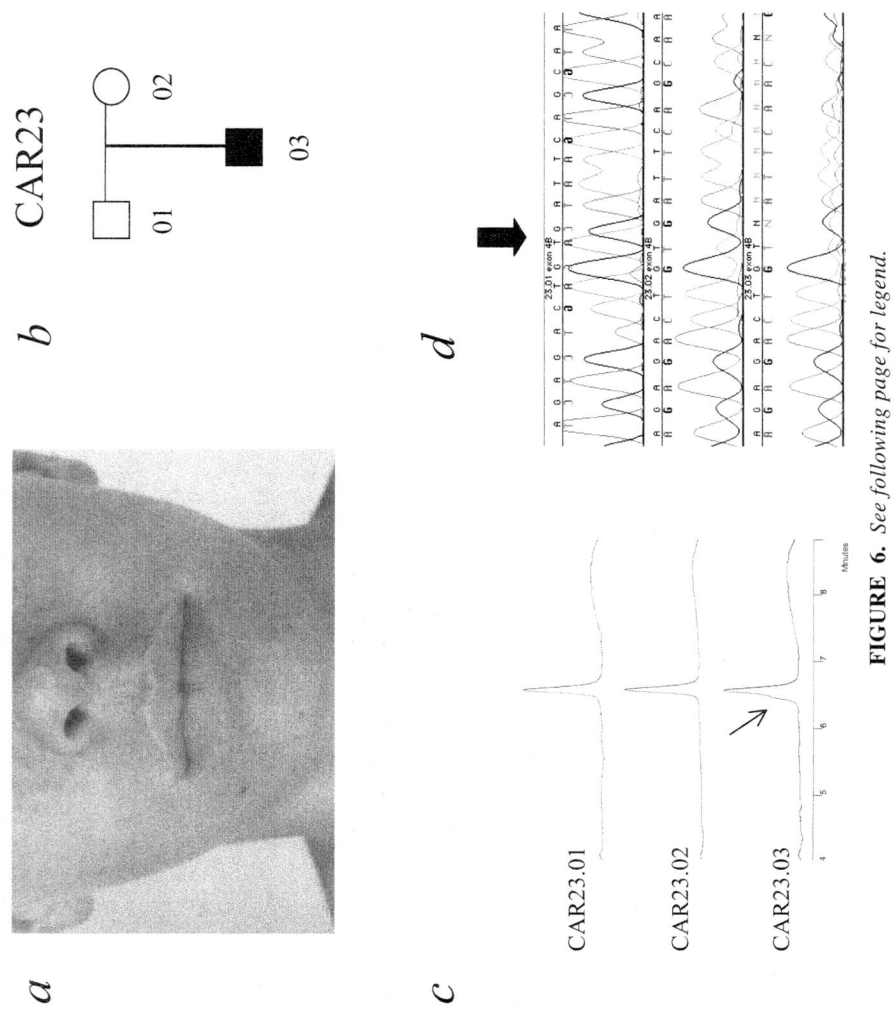

FIGURE 6. *See following page for legend.*

TABLE 3. Mutations of the PRKAR1A gene in patients with Carney complex

	Exon	No. of alleles	Effect	Frequency
Point mutations				
88A>G	2	1	abolishes initiator ATG	4.5
169C>T	2	1	nonsense	4.5
211C>T	2	2	nonsense	9.1
769C>T	6	1	nonsense	4.5
873GG>CT	8	1	nonsense	4.5
Frameshift mutations				
188delCTATT	2	1	frameshift/stop	4.5
578delGT	4B	6	frameshift/stop	27.3
618delTGAT	5	1	frameshift/stop	4.5
653AA>CAC	6	1	frameshift/stop	4.5
781insT	6	1	frameshift/stop	4.5
799insAA	7	1	frameshift/stop	4.5
Splice site mutations				
exon2 IVS −2A>G		1	unsure	4.5
exon3 IVS +1G>C		2	unsure	9.1
exon6 IVS del(−9>−2)		1	unsure	4.5
exon8 IVS +3G>A		1	pending	4.5
Total		22		

PRKAR1A cDNA and involve every exon except 4A, 9 and 10. It is interesting to note that *PRKAR1A* mutations tend to cluster in certain exons, with the majority of mutations in exons 2 ($N = 5$) and 6 ($N = 4$). On the other hand, 4B is the most commonly mutated exon of the *PRKAR1A* gene, albeit with a single genetic defect (578delTG) accounting for all mutations ($N = 6$) identified to date in that exon. It is not surprising, perhaps, that exons 2 and 6 harbor many of the mutations, as these are the largest and third-largest exons of *PRKAR1A*, respectively. On the other hand, it was unexpected that the second largest exon, exon 3, has not been found to be mutated in any families.

FIGURE 6. Detection of the exon 4B frameshift mutation shown in FIGURE 4 in a sporadic case of CNC (CAR23.03). **a.** Typical CNC facial pigmentation in individual CAR23.03, an 11-year old boy with a testicular and adrenal tumors. **b.** Partial pedigree of this patient. **c.** DHPLC analysis of *PRKAR1A* exon 4B in the individuals shown in panel b. The *arrow* indicates an alteration in peak shape in CAR23.03 indicative of heteroduplex formation. **d.** Sequence analysis of the father (*top*), mother (*middle*), and affected son (*bottom*) showing the *de novo* generation of the same sequence alteration in CAR23.03 as was seen in in CAR01 family members (FIG. 4, panel b); cloning and sequencing identified the same 578delGT mutation.

With the exception of the 88A>G mutation, which changes the initiator ATG to a GTG codon, each of the mutations detected in this study is predicted to lead to a premature termination codon.[16] Each of the five point mutations (including the 2 bp change in CAR13) encodes a nonsense mutation. The frameshift mutations also lead to the generation of missense residues, followed by an early termination codon. The effects of the four splice site mutations on the protein coding potential of their respective messages have not yet been fully elucidated, although a similar phenomenon is expected. It should be mentioned that the presence of a PRKAR1A pseudogene on chromosome 1 did not interfere with mutational analysis of the functional PRKAR1A gene on chromosome 17, because its sequence could not be detected by the methods used above (data not shown).

Overall, 10 families fully mapped to the chromosome 17 locus.[16] With the exception of one family, for whom we did not have sufficient amounts of DNA to complete the analysis, each of these families was found to have a mutation of *PRKAR1A* gene. In contrast, none of the families exhibiting recombination with 17q was found to have a mutation, suggesting that genetic linkage analysis was able to correctly identify the families in whom a mutation search was worthwhile. This observation suggests that genetic linkage will be helpful in identifying families to screen for *PRKAR1A*, such as the large family described by Casey *et al.*[8]

In total, 36 familial cases of CNC were evaluated, and 16 were found to carry *PRKAR1A* mutations (44.4%). Similarly, 6/17 (35.3%) of sporadic CNC patients were found to have mutations in that gene. Assuming that 5–10% of mutations may be undetectable by our current means of screening, this would suggest that *PRKAR1A* mutations would be expected to account for up to half of the patients with CNC, although the real number may be closer to 35–40%.

Penetrance of CNC in kindreds carrying *PRKAR1A* mutations appears to be over 97%, because only one of 48 mutation carriers (2.08%) (one individual in family CAR108) did not fully meet the diagnostic criteria (TABLE 1). However, penetrance of the disease per age group remains to be accurately estimated after extended family members are screened for *PRKAR1A* mutations, a process that is ongoing in our laboratory.

Western blot experiments on samples from patients with known truncating mutations failed to detect the presence of foreshortened forms of the protein (FIG. 7c).[9,16] To investigate this phenomenon, we analyzed mRNA in transformed lymphocytic and other tissue cell lines established from CNC patients. Amplification of full-length or shorter forms of the *PRKAR1A* cDNA from kindreds with the 578delTG and exon 8 IVS +3G>A mutations and subsequent sequence analysis demonstrated only the presence of normal *PRKAR1A* mRNA. This suggested that the mutant mRNA was not widely present in the lymphocytes and other cells of these patients, perhaps due to nonsense-mediated mRNA decay (NMD). NMD is a mechanism by which cells can degrade mRNAs containing premature stop codons before they reach the translational machinery. This mechanism has been shown to prevent translation of mutant mRNAs containing premature stop codons in a variety of human diseases, including the Ehlers-Danlos and Stickler syndromes. Treatment of cells with translation inhibitors such as cycloheximide (CHX) or puromycin (PUR) can abrogate this phenomenon.

Indeed, treatment of transformed lymphocytes from the CAR01 family that carried the 578delTG mutation with CHX, followed by cloning of its cDNA and sequencing revealed that this treatment led to the production of mRNA containing the expected 2-bp deletion, as evidenced by the disruption of the sequence trace. Similarly, treatment of transformed lymphocytes and a tissue cell line from family CAR25 that carried the exon 8 IVS +3G>A mutation with CHX led to the induction of a readily visible splice variant cDNA, which was the result of a cryptic splice donor site produced by mutation of the cognate site. Subsequent sequence analysis also showed that this led to the presence of a premature stop codon in the mRNA (data not shown).

In contrast to this, the A88G mutation, which abolishes the translational start codon but does not introduce a premature stop codon, would not be predicted to be subject to NMD. Analysis of mRNA from a patient carrying this mutation (family CAR19) demonstrated that the mRNA population was a 50:50 mixture of wild-type and mutant mRNA (data not shown). In Western blotting studies from this patient, no shortened forms of *PRKAR1A* were detected. Analysis of the primary sequence of the mRNA revealed that after the cognate ATG, there are two out-of-frame ATG sequences, neither of which is part of a Kozak initiation consensus. There was an in-frame ATG located within a reasonably good match sequence (at least equivalent to the cognate), which would be predicted to produce a protein lacking the 46 N terminal amino acids of the protein (data not shown). This protein was not detected in patients' cells, although the mechanism of suppression of translation of this particular transcript has not yet been elucidated.

GENOTYPE-PHENOTYPE CORRELATION IN CNC

Because all of the CNC mutations are predicted to lead to premature peptide chain termination and NMD was shown to operate in these cells, it appears that all of the CNC mutations detected to date are functionally equivalent to null alleles.[6,9,16] As such, one would predict that any patient with a *PRKAR1A* mutation has the same defect at the molecular level. This analysis suggests that differences in patient presentations may be due to contributions from disease-modifying loci located outside of the *PRAKR1A* locus. There is perhaps no more convincing argument about this statement from the comparison of the two larger families that shared the 578delTG *PRKAR1A* mutation but had a dramatically different clinical phenotype: Family CAR01 had many individuals with PPNAD, whereas this manifestation is completely absent from CAR20. Acromegaly and cardiac myxomas were strongly featured in CAR20, but were almost absent from CAR01. Many members of CAR20 (although not all) were highly pigmented, whereas this feature was much less prominent in CAR01. Thus, the protean manifestations of CNC appeared to be present or absent regardless of genotype, suggesting that a genotype-phenotype correlation does not exist. Similarly, there did not appear to be a difference between families that map to chromosome 17 and those that map elsewhere.

Finally, the absence of expression of the abnormal *PRKAR1A* protein in CNC cells, along with the previously identified LOH of the normal PRKAR1A allele, suggest that, in tissues affected by CNC, tumorigenesis is caused by abolition of type I PKA activity (data not shown); PKA activity is preserved, albeit it appears to be dysregulated (FIG. 7) (see below).

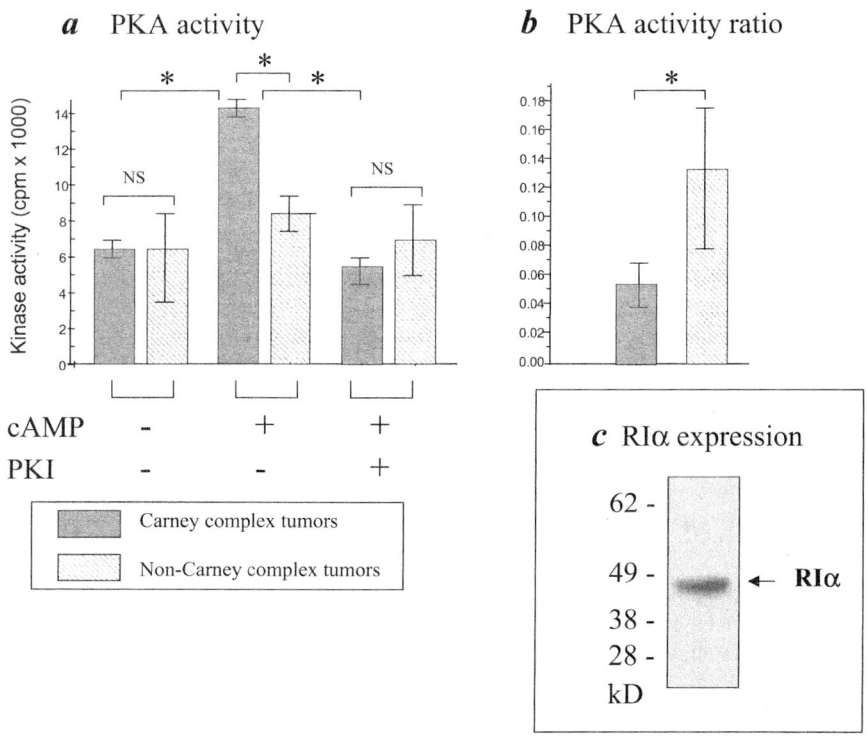

FIGURE 7. PKA activity and expression of PRKAR1A in CNC tumors and cell lines. **a.** Total PKA activity was not different between 4 CNC steroid-producing tumors from families CAR01, CAR20 and CAR25 and 3 control tumors from non-CNC patients at baseline; PKA activity increased in response to cAMP in CNC tumors, and that increase was blocked by PKI (for both $P < 0.001$). These responses were greater in CNC tumors than the corresponding changes in non-CNC tumors ($P < 0.05$); in addition, the peak kinase activity value in response to cAMP was higher in CNC tumors vs. non-CNC tumors ($P < 0.001$). All *error bars* represent standard error of the mean; NS = not significant; * = significant at $P < 0.05$; +, − indicate the presence or absence of cAMP or PKI, respectively. **b.** PKA activity ratio, calculated as the ratio between free (basal) PKA and total PKA, was lower in 6 CNC tumors bearing inactivating *PRKAR1A* mutations than that in 4 non-CNC samples ($P < 0.001$). **c.** Western blotting of protein lysates using a PRKAR1α-specific antibody: only the normal protein is detected because the predicted truncated product of the mutant allele is not synthesized due to NMD of the mutant mRNA (see text for details).

STUDIES OF PKA ACTIVITY IN CNC TUMORS

To determine the functional consequences of *PRKAR1A* mutations, we measured the cAMP responsiveness and PKA activity ratio in tumor cell extracts from CNC families and compared them with tumor extracts from non-CNC patients (FIG. 7, a and b). Measurement of nonstimulated PKA activity was not significantly different in CNC tumors vs. non-CNC tumors (FIG. 7a, left). Addition of cAMP led to stimulation in kinase activity in CNC tumors (kinase activity at baseline vs. following stimulation with cAMP, $P < 0.001$) (FIG. 7a). This response was PKA specific, as demonstrated by its inhibition by PKI, a specific PKA inhibitor[18] (kinase activity in response to cAMP vs. cAMP and PKI, $P < 0.001$), and it was not different between tumors with the 578delTG genotype (2 tumors from family CAR01 and 2 from family CAR20) and those with a different *PRKAR1A* mutation. Both the stimulation of kinase activity with cAMP and the inhibition of that stimulation by PKI were greater in CNC tumors than those in the control samples ($P < 0.05$). PKA activity ratio, a measure of how much PKA is in its active form (free/basal PKA),[18] was decreased in CNC tumors vs. the control samples ($P < 0.001$), as expected from RIα inactivating mutations (FIG. 7b).

More recent experiments in our laboratory suggested that total PKA activity in CNC tissues consists mostly of type-II PKA (data not shown).

SUMMARY

CNC is inherited as an autosomal dominant trait, and the responsible genes have been mapped by linkage analysis to loci at 2p16 and 17q22-24; other loci are also possible for some families. Mutations of the *PRKAR1A* gene at 17q22-24 have been identified in approximately half of the CNC patients. A defective cyclic nucleotide–dependent pathway has long been considered a candidate mechanism for the various manifestations of CNC, including tumors similar to those of MAS and paradoxical responses to hormonal stimuli. However, measurements of baseline and poststimulation intracellular cAMP levels in cultured tumors from patients with CNC (C.A. Stratakis, unpublished) and mutation analysis of the *GNAS1* gene[19] gave negative results. Thus the defect in CNC was placed downstream from cAMP activation; the PKA complex, a critical step in cAMP-dependent signaling,[20] was a likely candidate for the identification of mutations in patients with CNC. The present study identified LOH for the *PRKAR1A* gene in CNC tumors, suggesting (but far from proving) a tumor suppression function for the normal RIα.[21] Might haplo-insufficiency of the RIα, a gene that has been linked to cancer through its overexpression (see other articles in this volume), lead to an increase in PKA signaling and explain the multiple tumor formation and tendency for carcinogenesis that are seen in patients with CNC? It appears that RIα deficiency leads to overexpression of the genes coding for the other regulatory subunits of the PKA complex (data not shown), as has been demonstrated in other contexts.[22,23] After all, it is dysregulation of the PKA tetramer, rather than the levels of a particular subunit, that is the mediator of tumorigenic signals. The CNC mutations are unique experiments of nature that can help us understand the complex interactions and functions of the PKA signaling pathway.

ACKNOWLEDGMENTS

The author wishes to thank Dr. J. Aidan Carney (Mayo Clinic, Rochester, MN) for his contributions to both the text and clinical material presented in this paper; his clinical advice and academic guidance have been immensely helpful in the success of all endeavors related to the research on the syndrome that bears his name!

REFERENCES

1. CARNEY, J.A. & W.F. YOUNG. 1992. Primary pigmented nodular adrenocortical disease and its associated conditions. Endocrinologist **2:** 6–21.
2. CARNEY, J.A. 1995 Carney complex: the complex of myxomas, spotty pigmentation, endocrine overactivity, and schwannomas. Semin. Dermatol. **14:** 90–98.
3. STRATAKIS, C.A., L.S. KIRSCHNER & J.A. CARNEY. 1998 Carney complex: diagnosis and management of the complex of spotty skin pigmentation, myxomas, endocrine overactivity and schwannomas. Am. J. Med. Genet. **80:** 183–185.
4. CARNEY, J.A., L.S. HRUSKA, G.D. BEAUCHAMP & H. GORDON. 1985 Dominant inheritance of the complex of myxomas, spotty pigmentation and endocrine overactivity. Mayo Clin. Proc. **61:** 165–172.
5. STRATAKIS, C.A. 2000 Genetics of Carney complex and related familial lentiginoses, and other multiple tumor syndromes. Front. Biosci. **5:** D353–D366.
6. STRATAKIS, C.A., L.S. KIRSCHNER & J.A. CARNEY. 2001. Clinical and molecular features of the Carney complex: diagnostic criteria and recommendations for patient evaluation. J. Clin. Endocrinol. Metab. **86:** 4041–4046.
7. STRATAKIS, C.A. *et al.* 1996. Carney complex, a familial multiple neoplasia and lentiginosis syndrome: analysis of 11 kindreds and linkage to the short arm of chromosome 2. J. Clin. Invest. **97:** 699–705.
8. CASEY, M. *et al.* 1998. Identification of a novel genetic locus for familial cardiac myxomas and Carney complex. Circulation **98:** 2560–2566.
9. KIRSCHNER, L.S. *et al.* 2000. Mutations of the gene encoding the protein kinase A type I-alpha regulatory subunit in patients with the Carney complex. Nat. Genet. **26:** 89–92.
10. GIATZAKIS, C., S.E. TAYMANS & C.A. STRATAKIS. 1999. Expression profile of the adrenal gland from patients with primary bilateral adrenocortical diseases: primary pigmented adrenocortical disease (PPNAD) (associated with Carney complex) and massive macronodular adrenocortical disease (MMAD). Am. J. Hum. Genet. **65:** A127.
11. DELOUKAS, P. *et al.* 1998. A physical map of 30,000 human genes. Science **282:** 744–746.
12. SOLBERG, R. *et al.* 1997. The human gene for the regulatory subunit RI alpha of cyclic adenosine 3′, 5′-monophosphate-dependent protein kinase: two distinct promoters provide differential regulation of alternately spliced messenger ribonucleic acids. Endocrinology **138:** 169–181.
13. SCHOENBERG-FEJZO, M., *et al.* 1999. Integrated map of chromosome 17q critical region in multiple sclerosis. Am. J. Hum. Genet. **65:** A442.
14. DANOFF, A., S. JORMARK, D. LORBER & N. FLEISCHER. 1987. Adrenocortical micronodular dysplasia, cardiac myxomas, lentigines and spindle cell tumors: report of a kindred. Arch. Intern. Med. **143:** 443–448.
15. TAUTZ, D. 1989. Hypervariability of simple sequences as a general source for polymorphic DNA markers. Nucleic Acids Res. **17:** 6463–6471.
16. KIRSCHNER, L.S. *et al.* 2000. Genetic heterogeneity and spectrum of mutations of the PRKAR1A gene in patients with the carney complex. Hum. Mol. Genet. **9:** 3037–3046.
17. STRATAKIS, C.A., L.S. KIRSCHNER, S.E. TAYMANS, *et al.* 1999. Genetic heterogeneity in Carney complex (OMIM 160980): contributions of loci at chromosomes 2 and 17 in its genetics. Am. J. Hum. Genet. **65:** A447.
18. LEE, G.R. *et al.* 1999. Ala99ser mutation in RI alpha regulatory subunit of protein kinase A causes reduced kinase activation by cAMP and arrest of hormone-dependent breast cancer cell growth. Mol. Cell. Biochem. **195:** 77–86.

19. DEMARCO, L. *et al.* 1996. Sporadic cardiac myxomas and tumors from patients with Carney complex are not associated with activating mutations of the Gsα gene. Hum. Genet. **98:** 185–188.
20. BOSHART, M., F. WEIH, M. NICHOLS & G. SCHUTZ. 1991 The tissue specific extinquisher locus TSE1 encodes a regulatory subunit of cAMP-dependent protein kinase. Cell **66:** 849–859.
21. SCOTT, J.D. 1991. Cyclic nucleotide-dependent protein kinases. Pharmacol. Ther. **50:** 123-145.
22. CUMMINGS, D.E. *et al.* 1996. Genetically lean mice result from targeted disruption of the RIIβ subunit of protein kinase A. Nature **382:** 622–626.
23. AMIEUX, P.S. *et al.* 1997. Compensatory regulation of RIα protein levels in protein kinase A mutant mice. J. Biol. Chem. **272:** 3993–3998.

Dissecting the Circuitry of Protein Kinase A and cAMP Signaling in Cancer Genesis

Antisense, Microarray, Gene Overexpression, and Transcription Factor Decoy

YOON S. CHO-CHUNG,[a] MARIA NESTEROVA,[a] KEVIN G. BECKER,[b] RAKESH SRIVASTAVA,[a] YUN GYU PARK,[a] YOUL NAM LEE,[a] YEE SOOK CHO,[a] MEYOUNG-KIN KIM,[a] CATHERINE NEARY,[a] AND CHRIS CHEADLE[b]

[a]*Cellular Biochemistry Section, BRL, CCR, National Cancer Institute, National Institutes of Health, Bethesda, Maryland 20892-1750, USA*

[b]*DNA Array Unit, National Institute on Aging, National Institutes of Health, Baltimore, Maryland 21224-6820, USA*

> ABSTRACT: Expression of the RIα subunit of the cAMP-dependent protein kinase type I (PKA-I) is enhanced in human cancer cell lines, in primary tumors, in transformed cells, and in cells upon stimulation of growth. Signaling via the cAMP pathway may be complex, and the biological effects of the pathway in normal cells may depend upon the physiological state of the cells. However, results of different experimental approaches such as antisense exposure, 8-Cl-cAMP treatment, and gene overexpression have shown that the inhibition of RIα/PKA-I exerts antitumor activity in a wide variety of tumor-derived cell lines examined *in vitro* and *in vivo*. cDNA microarrays have further shown that in a sequence-specific manner, RIα antisense induces alterations in the gene expression profile of cancer cells and tumors. The cluster of genes that define the "proliferation-transformation" signature are down-regulated, and those that define the "differentiation-reverse transformation" signature are up-regulated in antisense-treated cancer cells and tumors, but not in host livers, exhibiting the molecular portrait of the reverted (flat) phenotype of tumor cells. These results reveal a remarkable cellular regulation, elicited by the antisense RIα, superimposed on the regulation arising from the Watson-Crick base-pairing mechanism of action. Importantly, the blockade of both the PKA and PKC signaling pathways achieved with the CRE-transcription factor decoy inhibits tumor cell growth without harming normal cell growth. Thus, a complex circuitry of cAMP signaling comprises cAMP growth regulatory function, and deregulation of the effector molecule by this circuitry may underlie cancer genesis and tumor progression.
>
> KEYWORDS: antisense; protein kinase A; cancer; growth inhibition; cDNA microarrays; transcription factor decoy

Address for correspondence: Dr. Yoon S. Cho-Chung, Chief, Cellular Biochemistry Section, Basic Research Laboratories, Center for Cancer Research, National Cancer Institute, National Institutes of Health, Bethesda, MD 20892-1750. Voice: 301496-4020; fax: 301-496-2443.
yc12b@nih.gov

INTRODUCTION

cAMP-dependent protein kinase (PKA), the primary mediator of cAMP action in mammalian cells,[1] exists in two isoforms, type 1 (PKA-I) and type 2 (PKA-II). These isoforms are distinguished by their cAMP-binding regulatory (R) subunits, RI versus RII, and they share an identical catalytic (C) subunit.[2] Four different regulatory subunits (RIα, RIβ, RIIα, and RIIβ) and three distinct C subunits (Cα, Cβ, and Cγ) have been identified,[3,4] but preferential coexpression of any one of these C subunits with either RI or RII subunit has not been found.[5,6]

RI and RII contain two tandem cAMP-binding domains at the carboxyl terminus, which is highly conserved. These subunits differ significantly in the amino terminus at a proteolytically sensitive hinge region that occupies the peptide substrate binding site of the C subunit in the holoenzyme complex.[7] RI and RII also differ in molecular weight, isoelectric point, and immunological characteristics. Other differences among the types of R subunits, such as affinity for cAMP, subcellular localization, expression in development, expression in cell cycle, and expression in transformation and differentiation, suggest that the regulatory role of the PKA isozymes differs under varying physiological conditions.[2,8,9] This difference in the roles of PKA isozymes may dictate the regulatory function of cAMP.

This review describes the effects of modulating PKA R subunit expression and cAMP response element (CRE)–directed transcription on cAMP signaling in cancer. The experimental approaches described here, such as antisense, 8-Cl-cAMP, gene transfer, transcription factor decoy, and cDNA microarrays, may not only provide the molecular tools to critically assess cAMP signaling in cancer genesis and progression, but they also provide opportunities to develop target-based drugs, which are more specific in their action and have fewer side effects, for therapeutic and preventive intervention against cancer.

ANTISENSE INHIBITION OF RIα

Nucleic acid therapeutics represent a direct genetic approach to cancer treatment. Such an approach takes advantage of genes known to confer a growth advantage to neoplastic cells and mechanisms that activate these genes.[10] Moreover, the ability to block these genes allows the exploration of normal growth regulation in addition to its therapeutic value.

Results from an antisense approach provided the first direct evidence that the cAMP receptor RIα receptor is a positive effector of cancer cell growth. A synthetic RIα antisense oligodeoxynucleotide (ODN) corresponding to the N terminal seven codons of human RIα (15–30 μM) produced growth inhibition in breast (MCF-7), colon (LS-174T), and gastric carcinoma (TMK-1) and neuroblastoma (SK-N-SH) cells,[11] as well as HL-60 leukemia cells,[12] with no sign of cytotoxicity. Furthermore, treatment with an RIα antisense phosphorothioate oligodeoxynucleotide (PS-ODN) (6 μM) brings about a marked reduction in RIα levels with a concomitant increase in RIIβ levels.[11] Strikingly, a single injection of an RIα antisense PS-ODN targeted against codons 8–13 of human RIα results in reduction of RIα expression and inhibition of tumor growth.[13] Tumor cells behave like untransformed cells by making less PKA-I.[13] An antisense PS-ODN directed to the N terminal codons of RIIβ brings about converse effects.[14] Thus RIα and RIIβ antisense ODNs each exert a specific effect on their respective target mRNAs and promote the compensatory, enhanced expression of the other R subunit.

To address the issue of nonspecific toxicity and side effects associated with antisense ODNs, the polyanionic nature[15] of the antisense PS-ODN (targeted against 8–13 codons of RIα) was minimized and the immunostimulatory GCGT motif[16] blocked with the placement of four 2′-O-methyl-ribonucleosides (RNA) at both the 5′ and 3′ ends.[17] An RNA-DNA mixed backbone (MBO) ODN has improved antisense activity over the PS-ODN,[18,19] is more resistant to nucleases, forms more stable duplexes with RNA than the parental PS-ODN,[18,20] and retains the capability to induce RNase H.[18] Thus, in addition to reducing nonspecific effects, the MBO RIα antisense ODN facilitates the exploration of sequence-specific antisense effects.[17] This antisense ODN (100–200 nM) markedly inhibits RIα expression in cancer cells, both *in vivo* and *in vitro*,[17,21–26] and further modulates the cAMP signaling pathway.[17] This modulation ultimately inhibits growth and induces apoptosis in various cancer cell lines and in tumors in nude mice.[17,21–26]

The target specificity of RIα antisense has been thoroughly addressed. Pulse-chase experiments have revealed that RIα has a relatively short half-life: 17 h in control cells and 13 h in antisense-treated cells (i.e., LS-174 colon carcinoma).[25] The short half-life of RIα, along with its message down-regulation, is consistent with the rapid RIα down-regulation observed in antisense-treated tumors.[13] In addition, levels of RIIβ protein increase because of the resultant longer half-life (about 5.5-fold),[25] leading to a decrease in the PKA-I–to–PKA-II ratio in tumor cells. The half-lives of RIIα and Cα remain unchanged in antisense-treated cells. The RIα antisense-induced stabilization of the RIIβ protein is consistent with results in RIβ and RIIβ knockout mice, in which compensatory stabilization-induced elevation of the RIα protein appears in tissues that normally express the β isoforms of the R subunit.[27] These results show a clear correlation between growth inhibition induced by RIα antisense and the target-specific antisense effect—namely, RIα down-regulation.

Multiple ODN sequences were also used to demonstrate sequence-specific antisense effects of RIα.[17] Four sequences that encode N terminal RIα codons and target-specific regions of the exposed nucleotide sequences (i.e., hairpin loop) are effective, whereas three sequences that target the 3′ UTR are ineffective. All of the sequences tested contain one or more CpG motifs or a contiguous G triplet; thus, the presence of these motifs does not affect the ability of the antisense to inhibit RIα expression or cancer cell growth. Finally, the sequence-specific antisense effects observed do not occur with other proteins, including the related isoform RIIα.

The effects of RIα antisense MBO ODN on the cAMP signaling cascade are dependent on the expression of PKA-I/PKA-II in the cell. In LS-174T colon cancer cells and in LNCaP prostate cancer cells, in which both PKA-I and PKA-II are expressed,[28] the antisense-directed loss of RIα results in the expected compensatory stabilization of the RIIβ protein, again because RIIβ's half-life is lengthened.[25] The antisense also triggered an increase in the activity of PDE4,[17] a cAMP-responsive enzyme[29,30] and nuclear translocation of the PKA Cα subunit[31] in the absence of an increase of cellular cAMP. Thus, the loss of RIα activates cAMP signaling by activating PKA-I and bypassing adenylate cyclase. However, in the case of HCT-15 MDR colon carcinoma cells, in which PKA-I is primarily expressed,[32] the antisense-directed loss of RIα decreases Cα subunit stability because of Cα's shortened half-present in the cell. Modulation of the R subunit expression using antisense strategy, 8-Cl-cAMP, or retroviral vector–mediated gene transfer has shown that two isoforms of cAMP receptor proteins, the RIα and RIIβ regulatory subunits of PKA, have opposite roles in cancer cell growth. RIα appears to promote cell growth, and RIIβ inhibits growth and induces differentiation.

RIIβ rapidly responds to the loss of RIα through protein stabilization into the PKA-II holoenzyme complex. This compensatory stabilization of RIIβ protein may represent an important biochemical mechanism of RIα antisense that ensures depletion of PKA-I and ultimately inhibits tumor cell growth. The RIα antisense-induced blockade of cancer growth induces apoptosis and differentiation, the terminal end point of cancer cell survival.

cDNA microarrays have revealed a specific subset of genes in cancer cells that are coordinately regulated by antisense RIα in a sequence-specific manner. Importantly, the differentiation and proliferation expression signatures were specifically up- and down-regulated, respectively, in tumor cells; these signatures were quiescent and unaltered in the host livers of antisense-treated animals. Examining global gene expression in cancer cells exposed to antisense ODN can further refine the antisense mechanism with respect to its sequence and target specificity.

The inhibition of cancer cell growth, but not normal cell growth, in response to a block in cAMP-responsive gene expression is an intriguing observation.[70] This observation clearly indicates separate and distinct cAMP signaling pathways that regulate growth for normal cells versus cancer cells. For example, germline mutations in the RIα subunit of PKA are responsible for a multiple neoplasia syndrome characteristic of Carney complex.[85] This suggests a role for RIα as a tumor suppressor in normal embryonic cell growth regulation.[85] However, an oncogene, RET/ptc2, is the product of a papillary thyroid carcinoma translocation event and consists of the c-*ret* protooncogene fused with the RIα subunit of PKA;[86] this suggests a role for RIα as a positive regulator of tumor cell growth.

Thus, the diversity and complexity of the cAMP signaling are highly dependent on different stages of normal cellular development and differentiation, and such signaling is disrupted in an abnormal physiology such as cancer. An intervention targeting cAMP signaling may therefore provide a more selective and effective method of restraining tumor cell growth without affecting normal cell growth.

ACKNOWLEDGMENT

We thank Dr. Frances McFarland of Palladian Partners, Inc., who provided editorial support under contract number NO2-BC-76212/C2700212 with the National Cancer Institute.

REFERENCES

1. KREBS, E.G. & J.A. BEAVO. 1979. Phosphorylation-dephosphorylation of enzymes. Annu. Rev. Biochem. **48:** 923–939.
2. BEEBE, S.J. & J.D. CORBIN. 1986. Cyclic nucleotide-dependent protein kinases. *In* The Enzymes: Control by Phosphorylation. Ed.: 43–111. Academic Press. New York.

3. MCKNIGHT, G.S., C.H. CLEGG, M.D. UHLER, *et al.* 1988. Analysis of the cAMP-dependent protein kinase system using molecular genetic approaches. Recent Prog. Horm. Res. **44:** 307–335.
4. LEVY, F.O., O. OYEN, M. SANDBERG, *et al.* 1988. Molecular cloning, complementary deoxyribonucleic acid structure and predicted full-length amino acid sequence of the hormone-inducible regulatory subunit of 3′,5′-cyclic adenosine monophosphate-dependent protein kinase from human testis. Mol. Endocrinol. **2:** 1364–1373.
5. SHOWERS, M.O. & R.A. MAURER. 1986. A cloned bovine cDNA encodes an alternate form of the catalytic subunit of cAMP-dependent protein kinase. J. Biol. Chem. **261:** 16288–16291.
6. BEEBE, S.J., O. OYEN, M. SANDBERG, *et al.* 1990. Molecular cloning of a unique tissue-specific protein kinase Cγ from human testis—representing a third isoform for the catalytic subunit of the cAMP-dependent protein kinase. Mol. Endocrinol. **4:** 465–475.
7. TAYLOR, S.S., J. BUBIS, J. TONER-WEBB, *et al.* 1988. cAMP-dependent protein kinase: prototype for a family of enzymes. FASEB J. **2:** 2677–2685.
8. CHO-CHUNG, Y.S., T. CLAIR, G. TORTORA, *et al.* 1991. Role of site-selective cAMP analogs in the control and reversal of malignancy. Pharmacol. Ther. **50:** 1–33.
9. CHO-CHUNG, Y.S., S. PEPE, T. CLAIR, *et al.* 1995. cAMP-dependent protein kinase: role in normal and malignant growth. Crit. Rev. Oncol. Hematol. **21:** 33–61.
10. ZAMECNIK, P. & M. STEPHENSON. 1978. Inhibition of Rous sarcoma virus replication and cell transformation by a specific oligodeoxynucleotide. Proc. Natl. Acad. Sci. USA **75:** 280–284.
11. YOKOZAKI, H., A. BUDILLON, G. TORTORA, *et al.* 1993. An antisense oligodeoxynucleotide that depletes RIα subunit of cyclic AMP-dependent protein kinase induces growth inhibition in human cancer cells. Cancer Res. **53:** 868–872.
12. TORTORA, G., H. YOKOZAKI, S. PEPE, *et al.* 1991. Differentiation of HL-60 leukemia cells by type I regulatory subunit antisense oligodeoxynucleotide of cAMP-dependent protein kinase. Proc. Natl. Acad. Sci. USA **88:** 2011–2015.
13. NESTEROVA, M. & Y.S. CHO-CHUNG. 1995. A single-injection protein kinase A-directed antisense treatment to inhibit tumor growth. Nat. Med. **1:** 528–633.
14. TORTORA, G., T. CLAIR & Y.S. CHO-CHUNG. 1990. An antisense oligodeoxynucleotide targeted against the type RIIβ regulatory subunit mRNA of protein kinase inhibits cAMP-induced differentiation in HL-60 leukemia cells without affecting phorbol ester effects. Proc. Natl. Acad. Sci. USA **87:** 705–708.
15. AGRAWAL, S. & Q. ZHAO. 1998. Mixed backbone oligonucleotdes: improvement in oligonucleotide-induced toxicity in vivo. Antisense Nucleic Acid Drug Dev. **8:** 135–139.
16. KRIEG, A.M., A.K. YI, S. MATSON, *et al.* 1995. CpG motifs in bacterial DNA trigger direct B-cell activation. Nature **374:** 546–549.
17. NESTEROVA, M. & Y.S. CHO-CHUNG. 2000. Oligonucleotide sequence-specific inhibition of gene expression, tumor growth inhibition, and modulation of cAMP signaling by an RNA-DNA hybrid antisense targeted to protein kinase A RIα subunit. Antisense Nucleic Acid Drug Dev. **10:** 423–433.
18. METELEV, V., J. LISZLEWICZ & S. AGRAWAL. 1994. Study of antisense oligonucleotide phosphorothioates containing segments of oligodeoxynucleotides and 2′-O-methyloligoribonucleotides. Bioorg. Med. Chem. Lett. **4:** 2929–2934.
19. MONIA, B.P., E.A. LESNIK, C. GONZALEZ, *et al.* 1993. Evaluation of 2′-modified oligonucleotides containing 2′-deoxygaps as antisense inhibitors of gene expression. J. Biol. Chem. **268:** 14514–14522.
20. SHIBAHARA, S., S. MUKAI, H. MORISAWA, *et al.* 1989. Inhibition of human immunodeficiency virus (HIV-1) replication by synthetic oligo-RNA derivatives. Nucleic Acids Res. **17:** 239–252.
21. CHO-CHUNG, Y.S., M. NESTEROVA, A. KONDRASHIN, *et al.* 1997. Antisense-protein kinase A: a single-gene-based therapeutic approach. Antisense Nucleic Acid Drug Dev. **7:** 217–223.
22. SRIVASTAVA, R.K., A.R. SRIVASTAVA, P. SETH, *et al.* 1999. Growth arrest and induction of apoptosis in breast cancer cells by antisense depletion of protein kinase A-RI alpha subunit: p53-independent mechanism of action. Mol. Cell. Biochem. **195:** 25–36.

23. SRIVASTAVA, R.K., A.R. SRIVASTAVA, Y.G. PARK, et al. 1998. Antisense depletion of RI-alpha subunit of protein kinase A induces apoptosis and growth arrest in human breast cancer cells. Breast Cancer Res. Treat. **49:** 97–107.
24. ALPER, O., N.F. HACKER & Y.S. CHO-CHUNG. 1999. Protein kinase A-Ialpha subunit-directed antisense inhibition of ovarian cancer cell growth: crosstalk with tyrosine kinase signaling pathway. Oncogene **18:** 4999–5004.
25. NESTEROVA, M., K. NOGUCHI, Y.G. PARK, et al. 2000. Compensatory stabilization of RIIβ protein, cell cycle deregulation, and growth arrest in colon and prostate carcinoma cells by antisense-directed down-regulation of protein kinase A RIα protein. Clin. Cancer Res. **6:** 3434–3441.
26. CHO-CHUNG, Y.S., M. NESTEROVA, S. PEPE, et al. 1999. Antisense DNA-targeting protein kinase A-RIA subunit: a novel approach to cancer treatment. Front. Biosci. **4:** D898–907.
27. AMIEUX, P.S., D.E. CUMMINGS, K. MOTAMED, et al. 1997. Compensatory regulation of RI-alpha protein levels in protein kinase A mutant mice. J. Biol. Chem. **272:** 3993–3998.
28. NESTEROVA, M.V., H. YOKOZAKI, L. MCDUFFIE, et al. 1996. Overexpression of RIIβ regulatory subunit of protein kinase A in human colon carcinoma cell induces growth arrest and phenotypic changes that are abolished by site-directed mutation of RIIβ. Eur. J. Cancer **253:** 486–494.
29. BEAVO, J.A. & D.H. REIFSNYDER. 1990. Primary sequence of cyclic nucleotide phosphodiesterase isozymes and the design of selective inhibitors. Trends Pharmacol. Sci. **11:** 150–155.
30. CONTI, M., S.L. JIN, L. MONACO, et al. 1991. Hormonal regulation of cyclic nucleotide phosphodiesterases. Endocr. Rev. **12:** 218–234.
31. NEARY, C. & Y.S. CHO-CHUNG. Nuclear translocation of the catalytic subunit of protein kinase A induced by an antisense oligonucleotide directed against the RIα regulatory subunit. Oncogene. In press.
32. NESTEROVA, M.V.&Y.S. CHO-CHUNG. Unpublished observations.
33. STEINBERG, R.A. & D.A. AGARD. 1981. Turnover of regulatory subunit of cyclic AMP-dependent protein kinase in S49 mouse lymphoma cells. Regulation by catalytic subunit and analogs of cyclic AMP. J. Biol. Chem. **256:** 10731–10734.
34. MONTMINY, M.R. & L.M. BILEZIKJIAN. 1987. Binding of a nuclear protein to the cyclic-AMP response element of the somatostatin gene. Nature **328:** 175–178.
35. GONZALEZ, G.A., W. BIGGS, III, W.W. VALE, et al. 1989. A cluster of phosphorylation sites on the cyclic AMP-regulated nucleoar factor CREB predicated by its sequence. Nature **337:** 749–752.
36. GINTY, D.D., A. BONNI & M.E. GREENBERG. 1994. Nerve growth factor activates a ras-dependent protein kinase that stimulates c-*fos* transcription via phosphorylation of CREB. Cell **77:** 713–725.
37. ROHLFF, C. & R.I. GLAZER. 1995. Regulation of multidrug resistance through the cAMP and EGF signaling pathways. Cell. Signal **7:** 431–443.
38. WANG, H., Q. CAI, X. ZENG, et al. 1999. Antitumor activity and pharmacokenetics of a mixed-backbone antisense oligonucleotide targeted to the RIα subunit of protein kinase A after oral administration. Proc. Natl. Acad. Sci. USA **96:** 13989–13994.
39. CHEN, H.X., J.L. MARSHALL, E. NESS, et al. 2000. A safety and pharmacokinetic study of a mixed-backbone oligonucleotide (GEM 231) targeting the type I protein kinase A by 2-hour infusions in patients with refractory solid tumors. Clin. Cancer Res. **6:** 1259–1266.
40. ROBINSON-STEINER, A.M. & J.D. CORBIN. 1983. Probable involvement of both interchain cAMP binding sites in activation of protein kinase. J. Biol. Chem. **258:** 1032–1040.
41. ØGREID, D., R. EKANGER, R.H. SUVA, et al. 1985. Activation of protein kinase isozymes by cyclic nucleotide analygs used singly or in combination. Eur. J. Biochem. **150:** 219–227.
42. CHO-CHUNG, Y.S. 1990. Role of cyclic AMP receptor proteins in growth, differentiation, and suppression of malignancy: new approaches to therapy. Cancer Res. **50:** 7093–7100.
43. ROHLFF, C., T. CLAIR & Y.S. CHO-CHUNG. 1993. 8-Cl-cAMP induces truncation and down-regulation of the RIα subunit and up-regulation of the RIIβ subunit of cAMP-

dependent protein kinase leading to type II holoenzyme-dependent growth inhibition and differentiation of HL-60 leukemia cells. J. Biol. Chem. **268:** 5774–5782.
44. ALLY, S., G. TORTORA, T. CLAIR, *et al.* 1988. Selective modulation of protein kinase isozymes by the site-selective analog 8-chloroadenosine 3′,5′-cyclic monophosphate provides a biological means for control of human colon cancer cell growth. Proc. Natl. Acad. Sci. USA **85:** 6319–6322.
45. CHO-CHUNG, Y.S., T. CLAIR, P. TAGLIAFERRI, *et al.* 1989. Site-selective cyclic AMP analogs as new biological tools in growth control, differentiation and proto-oncogene regulation. Cancer Invest. **7:** 161–177.
46. KATSAROS, D., G. TORTORA, P. TAGLIAFERRI, *et al.* 1987. Site-selective cyclic AMP analogs provide a new approach in the control of cancer cell growth. FEBS Lett. **223:** 97–103.
47. TAGLIAFERRI, P., D. KATSAROS, T. CLAIR, *et al.* 1988. Synergistic inhibition of growth of breast and colon human cancer cell lines by site-selective cyclic AMP analogues. Cancer Res. **48:** 1642–1650.
48. TAGLIAFERRI, P., D. KATSAROS, T. CLAIR, *et al.* 1988. Reverse transformation of Harvey murine sarcoma virus-transformed NIH/3T3 cells by site-selective cyclic AMP analogs. J. Biol. Chem. **263:** 409–416.
49. TORTORA, G., P. TAGLIAFERRI, T. CLAIR, *et al.* 1988. Site-selective cAMP analogs at micromolar concentrations induce growth arrest and differentiation of acute promyelocytic, chronic myelocytic, and acute lymphocytic human leukemia cell lines. Blood **71:** 230–233.
50. CHO-CHUNG, Y.S. 1989. Site-selective 8-chloro-cyclic adenosine 3′,5′-monophosphate as a biologic modulator of cancer: restoration of normal control mechanisms. J. Natl. Cancer Inst. **81:** 982–987.
51. RAMAGE, A D., S.P. LANGDON, A.A. RITCHIE, *et al.* 1995. Growth inhibition by 8-chloro cyclic AMP of human HT29 colorectal and ZR-75-1 breast carcinoma xenografts is associated with selective modulation of protein kinase A isoenzymes. Eur. J. Cancer **31A:** 969–973.
52. TORTORA, G., A. BUDILLON, H. YOKOZAKI, *et al.* 1994. Retroviral vector-mediated overexpression of the RIIβ subunit of the cAMP-dependent protein kinase induces differentiation in human leukemia cells and reverts the transformed phenotype of mouse fibroblasts. Cell Growth Differ. **5:** 753–759.
53. SRIVASTAVA, R.K., A.R. SRIVASTAVA & Y. S. CHO-CHUNG. 1998. Synergistic effects of 8-chlorocyclic-AMP and retinoic acid on induction of apoptosis in Ewing's sarcoma CHP-100 cells. Clin. Cancer Res. **4:** 755–761.
54. KIM, S.N., S.G. KIM, J.H. PARK, *et al.* 2000. Dual anticancer activity of 8-Cl-cAMP: inhibition of cell proliferation and induction of apoptotic cell death. Biochem. Biophys. Res. Commun. **273:** 404–410.
55. TORTORA, G., F. CIARDIELLO, S. PEPE, *et al.* 1995. Phase I clinical study with 8-chloro-cAMP and evaluation of immunological effects in cancer patients. Clin. Cancer Res. **4:** 377–384.
56. BUDILLON, A., A. CERESETO, A. KONDRASHIN, *et al.* 1995. Point mutation of the autophosphorylation site or in the nuclear location signal causes protein kinase A RIIβ regulatory subunit to lose its ability to revert transformed fibroblasts. Proc. Natl. Acad. Sci. USA **92:** 10634–10638.
57. HARADA, H., B. BECKNELL, M. WILM, *et al.* 1999. Phosphorylation and inactivation of BAD by mitochondria-anchored protein kinase A. Mol. Cell **3:** 413–22.
58. LEE, G.R., S.N. KIM, K. NOGUCHI, *et al.* 1999. Ala99ser mutation in RI alpha regulatory subunit of protein kinase A causes reduced kinase activation by cAMP and arrest of hormone- dependent breast cancer cell growth. Mol. Cell. Biochem. **195:** 77–86.
59. CHO, Y.S. & Y.S. CHO-CHUNG. Unpublished observations.
60. OTTEN, A.D. & G.S. MCKNIGHT. 1989. Overexpression of the type II regulatory subunit of the cAMP-dependent protein kinase eliminates the type I holoenzyme in mouse cells. J. Biol. Chem. **264:** 20255–20260.

61. AMBESI-IMPIOMBATO, F.S., L.A. PARKS & H.G. COON. 1980. Culture of hormone-dependent functional epithelial cells from rat thyroids. Proc. Natl. Acad. Sci. USA **77:** 3455–3459.
62. TORTORA, G., S. PEPE, C. BIANCO, et al. 1994. The RIα subunit of protein kinase A controls serum dependency and entry into cell cycle of human mammary epithelial cells. Oncogene **9:** 3233–3240.
63. TORTORA, G., S. PEPE, A.M. CIRAFICI, et al. 1993. Thyroid-stimulating hormone-regulated growth and cell cycle distribution of thyroid cells involve type I isozyme of cyclic AMP-dependent protein kinase. Cell Growth Differ. **4:** 359–365.
64. TORTORA, G., S. PEPE, C. BIANCO, et al. 1994. Differential effects of protein kinase A sub-units on Chinese-hamster-ovary cell cycle and proliferation. Int. J. Cancer **59:** 712–716.
65. SCHENA, M., D. SHALON, R.W. DAVIS, et al. 1995. Quantitative monitoring of gene expression patterns with a complementary DNA microarray. Science **270:** 467–470.
66. CHO, Y.S., M.K. KIM, C. CHEADLE, et al. 2001. Antisense DNAs as multisite genomic modulators identified by DNA microarray. Proc. Natl. Acad. Sci. USA **98:** 9819–9823.
67. BIELINSKA, A., R.A. SHIVDASANI, L.Q. ZHANG, et al. 1990. Regulation of gene expression with double-stranded phosphorothioate oligonucleotides. Science **250:** 997–1000.
68. MORISHITA, R., G.H. GIBBONS, M. HORIUCHI, et al. 1995. A gene therapy strategy using a transcription factor decoy of the E2F binding site inhibits smooth muscle proliferation in vivo. Proc. Natl. Acad. Sci. USA **92:** 5855–5859.
69. ROESLER, W.J., G.R. VANDENBARK & R.W. HANSON. 1988. Cyclic AMP and the induction of eukaryotic gene transcription. J. Biol. Chem. **263:** 9063–9066.
70. PARK, Y.G., M. NESTEROVA, S. AGRAWAL, et al. 1999. Dual blockade of cyclic AMP response element- (CRE) and AP-1-directed transcription by CRE-transcription factor decoy oligonucleotide. Gene-specific inhibition of tumor growth. J. Biol. Chem. **274:** 1573–1580.
71. HABENER, J.F. 1990. Cyclic AMP response element binding proteins: a cornucopia of transcription factors. Mol. Endocrinol. **4:** 1087–1094.
72. SUN, P., H. ENSLEN, P.S. MYUNG, et al. 1994. Differential activation of CREB by Ca2+/calmodulin-dependent protein kinases type II and type IV involves phosphorylation of a site that negatively regulates activity. Genes Dev **8:** 2527–2539.
73. XING, J., D.D. GINTY & M.E. GREENBERG. 1996. Coupling of the RAS-MAPK pathway to gene activation by RSK2, a growth factor-regulated CREB kinase. Science **273:** 959–963.
74. SHENG, M., M.A. THOMPSON & M.E. GREENBERG. 1991. CREB: a Ca(2+)-regulated transcription factor phosphorylated by calmodulin-dependent kinases. Science **252:** 1427–1430.
75. TAN, Y., J. ROUSE, A. ZHANG, et al. 1996. FGF and stress regulate CREB and ATF-1 via a pathway involving p38 MAP kinase and MAPKAP kinase-2. EMBO J. **15:** 4629–4642.
76. CHRIVIA, J.C., R.P. KWOK, N. LAMB, et al. 1993. Phosphorylated CREB binds specifically to the nuclear protein CBP. Nature **365:** 855–859.
77. ARIAS, J., A.S. ALBERTS, P. BRINDLE, et al. 1994. Activation of cAMP and mitogen responsive genes relies on a common nuclear factor. Nature **370:** 226–229.
78. GOODMAN, R.H. & S. SMOLIK. 2000. CBP/p300 in cell growth, transformation, and development. Genes Dev. **14:** 1553–1577.
79. LEE, Y.N., Y.G. PARK, Y.H. CHOI, et al. 2000. CRE-transcription factor decoy oligonucleotide inhibition of MCF-7 breast cancer cells: cross-talk with p53 signaling pathway. Biochemistry **39:** 4863–4868.
80. PARK, Y.G., S. PARK, S.O. LIM, et al. 2001. Reduction in cyclin D1/Cdk4/retinoblastoma protein signaling by CRE-decoy oligonucleotide. Biochem. Biophys. Res. Commun. **281:** 1213–1219.
81. WALTON, K.M., R.P. REHFUSS, J.C. CHRIVIA, et al. 1992. A dominant repressor of cyclic adenosine 3′,5′-monophosphate (cAMP)-regulated enhancer-binding protein activity inhibits the cAMP-mediated induction of the somatostatin promoter in vivo. Mol. Endocrinol. **6:** 647–655.

82. CHO, Y.S., Y.G. PARK, Y.N. LEE, *et al.* 2000. Extracellular protein kinase A as a cancer biomarker: its expression by tumor cells and reversal by a myristate-lacking C alpha and RII beta subunit overexpression. Proc. Natl. Acad. Sci. USA **97:** 835–840.
83. CVIJIC, M.E., T. KITA, W. SHIH, *et al.* 2000. Extracellular catalytic subunit activity of the cAMP-dependent protein kinase in prostate cancer. Clin. Cancer Res. **6:** 2309–2317.
84. CHO, Y.S., Y.N. LEE & Y.S. CHO-CHUNG. 2000. Biochemical characterization of extracellular cAMP-dependent protein kinase as a tumor marker. Biochem. Biophys. Res. Commun. **278:** 679–684.
85. KIRSCHNER, L., J. CARNEY, D. SVETLANA, *et al.* 2000. Mutations of the gene encoding the protein kinase A type I-α regulatory subunit in patients with the Carney complex. Nat. Genet. **26:** 69.
86. LANZI, C., M.G. BORRELLO, I. BONGARZONE, *et al.* 1992. Identification of the product of two oncogenic rearranged forms of the RET proto-oncogene in papillary thyroid carcinomas. Oncogene **7:** 2189–2194.

Regulatory Subunits of PKA and Breast Cancer

W. R. MILLER

*Breast Unit Research Group, University of Edinburgh,
Western General Hospital, Edinburgh EH4 2XU, UK*

ABSTRACT: Overexpression of the R subunits of PKA (in particular, RI) is associated with high proliferation in normal breast, malignant transformation in the breast, poor prognosis in established breast cancer, and resistance to antiestrogens. These data, together with the observation that successful antiestrogen therapy is associated with reduced expression of RI mRNA, suggest that targeting R subunits is an appropriate therapeutic strategy for breast cancer. Initial experimental results, using antisense RI oligonucleotides, are promising in terms of reducing the growth rate of breast cancer cells and xenografts. While clinical trials designed to target RI subunits have yet to be established (and interventions as preventative measures are even more distant), the concept of these approaches to prevent and treat breast cancer should be developed and exploited.

KEYWORDS: protein kinase A; regulatory subunits; breast cancer; tumor

BREAST CANCER

Breast cancer is a common disease affecting about one in ten women in the UK and the USA.[1,2] The cause(s) of the malignancy is/are largely unknown, making implementation of preventative measures difficult. Because of this, it is important to define the molecular changes that occur during the transformation of normal breast cells into those with the malignant phenotype. Most women with established breast cancer present with "early" disease in that the disease appears restricted to the breast.[3] However, clinical experience indicates that many patients will already have occult metastatic disease at distant sites.[4] The behavior of this disease is very variable: some patients survive several decades with the minimum of treatment, while others die of cancer within months despite intensive therapy.[5,6] Because of this variable outlook and the toxicity of many therapies, there is a need to identify biological markers that can predict prognosis. In this way, patients with tumors that confer a favorable outcome may be spared the side effects associated with the more intensive treatment regimes, whereas aggressive tumors may be treated with correspondingly aggressive treatment. Breast cancers may respond to one or more treatment modali-

---

Address for correspondence: Professor W. R. Miller, Breast Unit Research Group, University of Edinburgh, Western General Hospital, Edinburgh EH4 2XU, UK. Voice: 44-131-537-2501, ext. 2401; fax: 44-537-332-2449.
w.r.miller@ed.ac.uk

ties, including hormone deprivation therapy.[1,7] The latter has advantages over cytotoxic chemotherapy, most notably a relative lack of toxicity. The problems with endocrine therapy are (i) only the minority of tumors respond, and (ii) initial responses are frequently followed by acquired resistance. Consequently there is a need to identify both predictive markers of endocrine sensitivity and the mechanism by which resistance occurs. The present manuscript summarizes the evidence that regulatory subunits of protein kinase A (PKA) are (i) increased in proliferating normal breast, (ii) overexpressed in breast cancers, (iii) higher in tumors with poor prognosis, (iv) predictive of endocrine sensitivity, (v) raised in tumors resistance to antiestrogens, (vi) an appropriate target for treating breast cancer.

REGULATORY SUBUNITS (R) AND NORMAL BREAST

The functional units of the normal breast (in which most breast cancers arise[8]) are the terminal ductal lobular units. These elements have been dissected from breasts in a variety of physiological conditions, including resting state, pregnancy, lactation, postlactational involution, and prolonged involution.[9] Levels and the proportion of different types of R units of PKA varied widely between individual breasts and physiological groups. However, within the nulliparous resting breast the proportion of RI was significantly higher in the second half of the menstrual cycle compared with the first (interestingly, cellular proliferation was also higher in the second half of the cycle[10]). This provided the impetus for a systematic evaluation of proliferative activity in these breasts so that a correlation could be made with R subunit expression.

These results showed highly significant positive associations between level of epithelial proliferation and both total subunit and RI expression, irrespective of whether proliferation was represented as a continuous variable (unpublished results) or subdivided into groups with high and low values[9] (TABLE 1). Whether this association is causal or casual is unknown. There is, however, a potential relevance for susceptibility to breast cancer. Highly proliferative lesions within the breasts of women may be associated with increased risk of breast cancer,[11] and there is a general concept that abnormally high rates of cell division are likely to create genetic mistakes that, if left uncorrected, may lead to malignant transformation and ultimately to overt cancer.[12,13]

TABLE 1. Relationships between level and type of R subunit expression and proliferation in normal breast

	Total subunits (% binding)	Type I (% of total)
High proliferation (21) (>2% cells)	1.05	63 (19)
Low proliferation (26) (<2% cells)	0.33	35 (25)
	$P < 0.015$	$P < 0.002$

NOTE: Values are medians. Figures in parenthesis are number of cases. P values by Wilcoxon rank test.

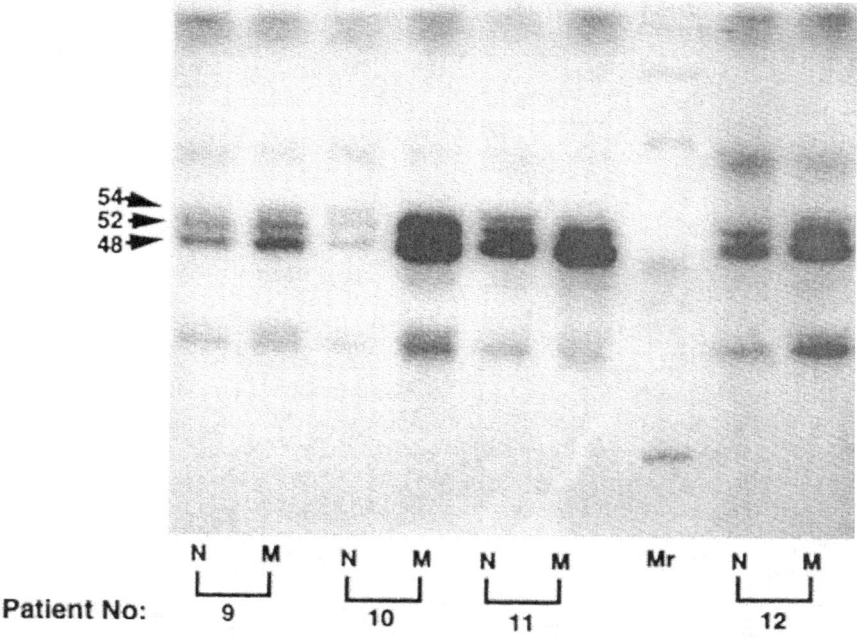

FIGURE 1. PKA R subunits in paired samples from normal (N) and malignant (M) breast. Typical autoradiograph of tissue cytosols photoaffinity labeled with [^{32}P]98-ozidocyclic AMPO. *Arrows* indicate position of cAMP binding proteins of molecular mass 48,000 Da (RI subunit), 52,000 Da, and 54,000 Da (RII subunits) as determined by comparison with ^{14}C methylated molecular weight markers (Mr).

COMPARISON OF THE NORMAL AND MALIGNANT BREAST

Measurements of regulatory subunits in paired samples of normal and malignant tissues from the breasts of 13 women with breast cancer have been reported.[14] Illustrative examples are shown in FIGURE 1. Expression of RI and RII PKA regulatory subunits were higher in malignant tissues from 12 ($P = 0.0005$ by paired Wilcoxon Rank Test) and 9 ($P = 0.01$) of the pairs, respectively (TABLE 2). However, the degree of RI subunit overexpression in malignant tissue was greater than that of the RII subunit, as illustrated by an increase in the RI/RII unit ratio in 10 of the 13 paired samples ($P = 0.017$). Interestingly, we have reported similar findings when comparing regulatory subunits in colorectal cancer and related mucosa.[15] It may be that the differences between the normal and malignant phenotype reflect the accompanying changes of increased cellular proliferation and dedifferentiation that are generally associated with malignant transformation. This would be compatible with the proposition that RI subunit expression programs for cell division, whereas RII expression is associated with a differentiated phenotype.[16] Clearly the hypothesis would be dependent upon an antagonistic interaction between subunits since expression of RII subunits also increases in most breast cancers, while the ratio with RI decreases. In

TABLE 2. Comparative expression of R subunits in paired samples of normal and malignant breast tissue

	Normal breast (arbitrary units)	Tumor (arbitrary units)	
RI	31.6 (15.3–43.1)	122.2 (36.0–373.5)	$P = 0.0005$
RII	34.7 (22.4–50.6)	71.7 (14.8–176.9)	$P = 0.01$
RI/RII	0.97 (0.40–1.56)	1.74 (0.92–3.03)	$P = 0.017$

NOTE: Values are means. Figures in parenthesis are ranges. P values by paired Wilcoxon rank test.

order to define further the role of R subunits in the process of malignant transformation, it would have been useful to identify and analyze premalignant lesions or atypias within the breast. To the author's knowledge this study has not been performed. However, a limited examination of preinvasive cancer (ductal carcinoma *in situ*) indicates the increased expression of R subunits (especially RI) is apparent at this stage (unpublished results), an observation that would again be consistent with the high proliferative rate and invasive potential of these lesions.

R SUBUNITS AND BREAST CANCERS

A feature of breast cancer is the variation in level and size of R subunits expressed by different tumors (FIG. 2). These differences in patterns have been reviewed in a large cohort of breast cancers.[17] Thus, following photoaffinity labeling, subunits with molecular masses of 48 kDa and 52–56 kDa were detected in all tumor cytosols and have been characterized as being RI and RII species, whereas subunits with molecular masses of 39 kDa and 37 kDa, which occur in most but not all tumors, have been suggested to be proteolytic fragments of parent RI and RII.[18] Interestingly, a positive relationship was found between the 48-kDa species and total R subunit expression and inverse relationships between the 52–56 kDa species and total expression and that of 48 kDa. This again confirms that in breast cancers overexpression of R subunits is associated primarily with increased expression of RI and that there is a yin-yang phenomenon between RI and RII.

PROGNOSIS AND BREAST CANCER

Because of the variation in expression of R subunits between different breast cancers, it was of interest to determine whether these related to other tumor features and particularly the outcome of patients. It was shown that, while levels of R subunits were higher in estrogen receptor–negative tumors, there were no significant relationships between R subunits and tumor histological grade and patient lymph node positivity.[19] However, in two separate cohorts of breast cancer patients (one studied retrospectively and one prospectively) presenting with "early" disease, high levels of R expression were associated with poor prognosis in terms of both disease recurrence and overall survival.[19,20] This association was independent of known estab-

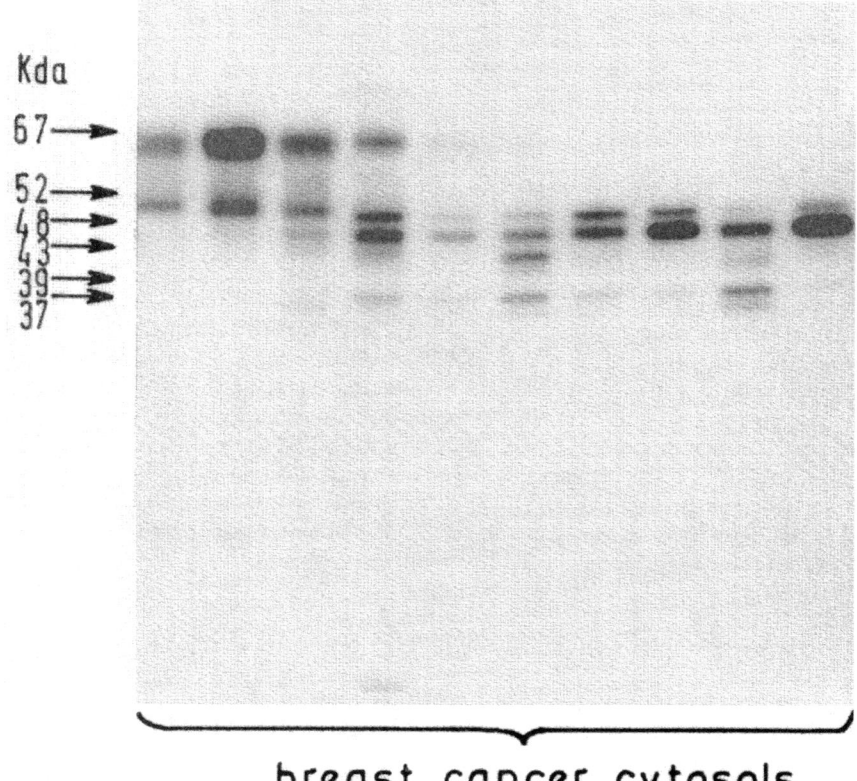

FIGURE 2. Autoradiogram of 10 different breast cancer cytosols photoaffinity labeled with [^{32}P] 8-azido cyclic AMP.

lished prognostic factors and allowed the identification of a small subgroup of patients whose poor outlook warranted implementation of aggressive systemic therapy. However, a deficiency in both these investigations was that the study population represented only a small proportion of the patients presenting with operable breast cancer and had a selection bias towards larger tumors. We therefore undertook a more comprehensive study.[21] This failed to confirm earlier findings, and total R unit expression had no prognostic significance. It has been important to identify the reasons for this. A possibility is the impact of adjuvant endocrine therapy. Thus, while our initial and earlier study included substantial numbers of patients who had not received adjuvant therapy, more recent practice is to give women systemic therapy following their surgery (most notably tamoxifen). As is indicated below, R subunit expression interacts with estrogen receptors to identify patients more or less likely to respond to endocrine therapy, and tamoxifen itself may modulate RI levels. Furthermore, patient follow-up was comparatively short, and expression of individual subunits was not considered. Because of this a further study was performed in which

TABLE 3. Prognostic significance of RI subunit in breast cancer

	Recurrence	Disease free
Majorative	29	20
Other	38	98

NOTE: RI was either the major component of the integrated densitometry measurement of an autoradiograph lane following photoaffinity labeling of a cytosol (majorative) or not (other). P value by chi-square. $\chi^2 = 13.9$, $P < 0.002$.

patient follow-up was at least five years and photoaffinity labeling of tumor cytosols was performed. After PAGE, autoradiograms were subject to densitometric scanning; when the band corresponding to RI constituted >50% of the total, they were classified as majorative. Results are shown in TABLE 3 and indicate that when RI was the major subunit in tumor cytosols the recurrence rate after five-year follow-up was almost twice that of patients having tumors in which RI was a minority component (42% versus 27%). The difference between the subgroups was highly significant. Although there were other more powerful prognostic factors—for example, lymph node positivity and estrogen receptor status—RI expression still influenced outcome in terms of disease relapse and death in multivariate analysis.

PREDICTION OF RESPONSE TO ENDOCRINE THERAPY

Pioneering work by the Cho-Chung laboratory has shown the measurement of R subunits could aid the prediction of response to ovariectomy of carcinogen-induced rat mammary tumors.[22] Thus, while the presence of estrogen receptors (ER) predisposed tumors to response, only about 60% of the cancers regressed following castration. However, by expressing ER values as a ratio with R subunit, it was possible to increase discrimination to 95%. These results were the stimulus to perform a study in women with advanced breast cancer who received endocrine therapy as management of their disease.[23] The treatments were ovariectomy in premenopausal women and aminoglutethimide and/or tamoxifen in postmenopausal patients, to which the overall response rate was 45% (14 of 31 cases). Results relating response to ER alone and in combination with R subunits are shown in FIGURE 3. Thus ER levels were significantly higher in responding as compared to nonresponding tumors ($P < 10^{-4}$). However, there was still a major overlap in ER values between the groups. Expressing ER values as a ratio with those of R subunits markedly improved discrimination between responding and nonresponding tumors, such that it was almost complete and increased the statistical significance between the groups ($P < 10^{-7}$). These results have confirmed those of others[24] and are identical to the observations in rat mammary tumors.[22]

EFFECTS OF TREATMENT WITH ANTIESTROGEN

In order to determine the effects of endocrine therapy on tumor biology, we have made use of neoadjuvant protocols in which treatment is given to patients with the

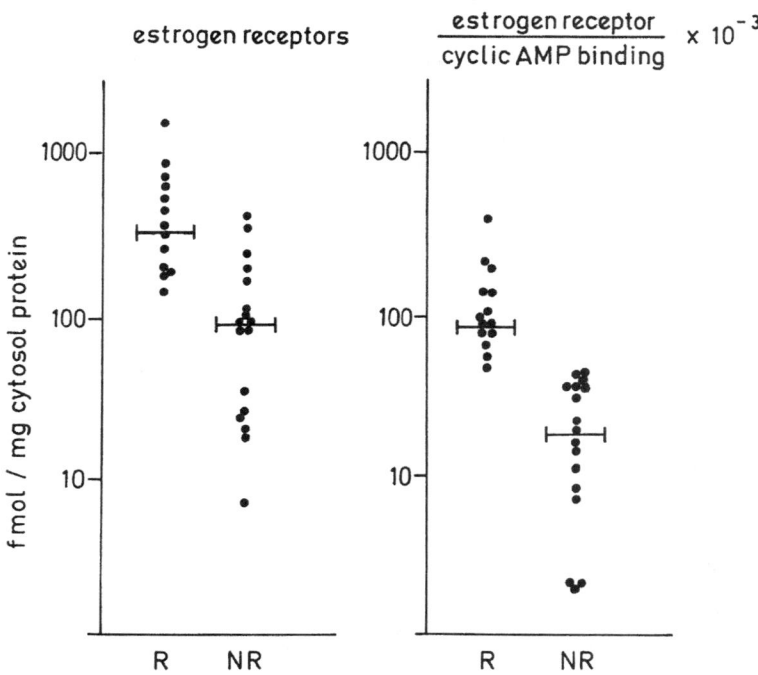

FIGURE 3. Expression of estrogen receptors (ER) in combination with that of R subunit improves the prediction of response to endocrine therapy. ER levels in responding (R) and nonresponding patients (NR) in the **left panel** and as a ratio with cyclic AMP binding proteins (R subunits) in the **right panel**. *Horizontal lines* are median P values by Wilcoxon rank test.

primary tumor *in situ* within the breast.[25,26] This may provide clinical benefits if the cancer regresses such that less extensive breast surgery is required to remove the tumor, and knowledge of the sensitivity of the cancer to a particular treatment may aid the choice of adjuvant therapy. However, there are immediate advantages to the researcher in that access to the tumor permits multiple biopsies before and during treatment so that effects of therapy on biological characteristics may be determined. The size of the primary tumor may also be monitored throughout treatment so that clinical response may be readily assessed and related to biological markers. Additionally, we developed a semiquantitative RT-PCR assay by which to measure levels of mRNA for RI subunits in small tumor samples.[27] Examples of changes in levels in two individual tumors following neoadjuvant treatment with tamoxifen are shown in FIGURE 4. The relationship between changes in RI subunit expression and response to treatment has been reported[28] and is summarized in TABLE 4. This indicates that whereas levels fell with treatment in the majority of responding tumors, they did not in any of the eight tamoxifen-resistant tumors (and, indeed, increased in

TABLE 4. Changes in RI mRNA with treatment

	Decrease	No change	Increase
Responding	15	6	2
Nonresponding	0	2	6

NOTE: Decrease represents at least a 2-fold reduction in expression. P values by Fisher's exact test. $\chi^2 = 15.5$. $P < 0.0001$.

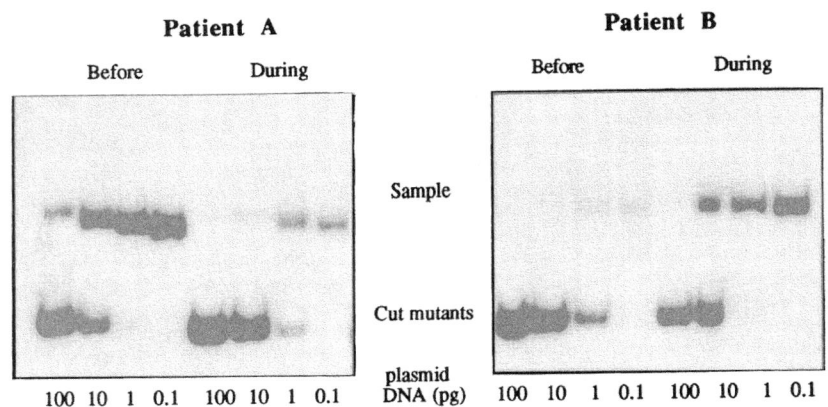

FIGURE 4. Autoradiographs from competitive RT-PCR reactions in two tumors, tissue samples being taken before and after 3 months' treatment (during) with tamoxifen. One tumor shows a decrease with treatment, and one displays an increase.

six cases). Furthermore, the degree of change in RI with treatment correlated with the magnitude of clinical response. It is impossible to discern whether the changes in responding tumors are a consequence of the response or program for the response. Thus since high RI expression is associated with dividing cells[16, 29–31] and tamoxifen has antiproliferative effects,[32] treatment might be expected to reduce RI expression. This would also be consistent with the decrease in concentration of RI subunits that accompanies castration-induced atrophy of hormone-dependent tissues.[33,34]

R SUBUNITS AND RESISTANCE TO ANTIESTROGENS

It is of interest that many of the tumors resistant to antiestrogen show an increase in RI expression because of the evidence from cell line model systems that high PKA activity through hyperphosphorylation of either ER or accessory proteins may not only induce a hormone-resistant phenotype,[35–38] but may change the pharmacology of tamoxifen from a partial estrogen agonist to an agent with full agonist activity.[39,40] Abnormal expression of RI may also be associated with other forms of resistance.[39–44] Thus strategies to reduce RI expression may reverse drug resistance.

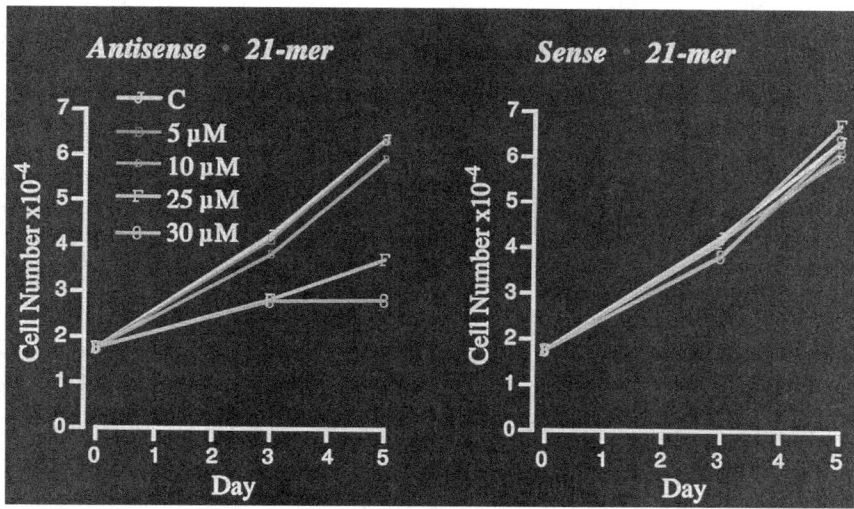

FIGURE 5. Effects of RIα oligonucleotides on the growth of MCF-7 breast cancer cells.

RI SUBUNITS AS A THERAPEUTIC TARGET

Because overexpression of R subunits, in particular RI, is associated with aggressive tumor behavior and resistance to treatment, the concept of selectively targeting RI as therapy is attractive. This has taken various forms, including the use of site-selective cyclic AMP analogues,[45] transfection of inducible vectors for RI subunits,[46] and antisense oligonucleotides for RI.[47-49] We have explored the latter strategy using phosphorothioate oligonucleotides. As is shown in FIGURE 5 the antisense 21-mer olignucleotide against codons 1–7 of RIα was capable of reducing the cell numbers of the MCF-7 breast cancer cell line in a dose-related manner (between 5 and 30 μM) as measured under standard culture conditions. A further 18-mer antisense oligonucleotide against codons 8–13 had similar effects.

In contrast, the sense 21-mer oligonucleotide had no discernible effects at these concentrations. The specificity of effect was confirmed by including random and mismatched oligonucleotides at 10 μM, which demonstrated no effect on cell number. These studies have been extended into *in vivo* studies in which nude mice bearing xenografts of MCF-7 cells have been given intraperitoneal injections of RIα of a mixed backbone antisense oligonucleotide (HYB190) daily for days 1–5 and 8–13. As is shown in FIGURE 6, xenograft growth was significantly reduced by the antisense oligonucleotide compared with animals given injection vehicle, an effect which was evident as early as day 3 and continued in the week following discontinuation of treatment. In the same experiment animals were treated with tamoxifen alone and in combination with the antisense oligonucleotide. Tamoxifen also reduced tumor growth, and the combination produced a greater (and additive) effect than either agent alone. Similar antitumor influences have been documented on other cancer xenografts.[50]

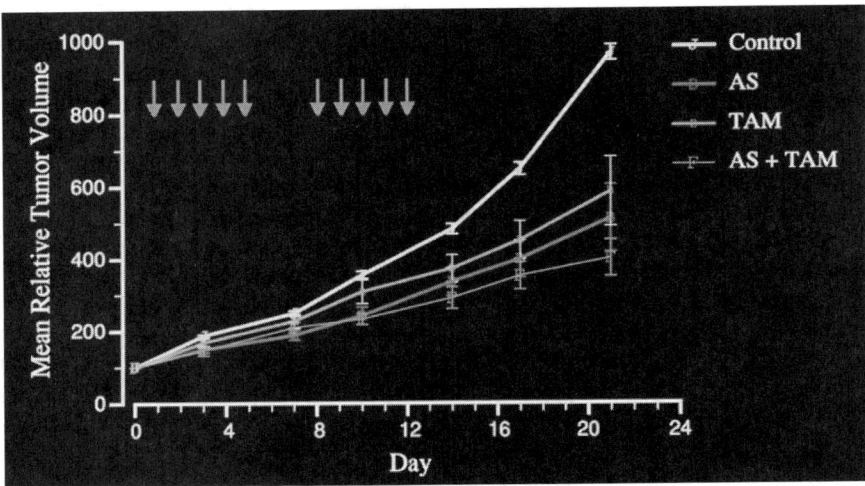

FIGURE 6. Effect of RIα antisense oligonucleotides (NYB190) on MCF-7 breast cancer xenografts. *Lines* and *bars* are mean and standard errors. *Arrows* are days of HYB190 injections. AS, HYB190 treatment; TAM, tamoxifen; AS + TAM, the agents given in combination.

REFERENCES

1. MILLER, W.R. 1996 Introduction. *In* Estrogen and Breast Cancer. W.R. Miller, Ed.: 1–13. R.G. Landes Co. Austin, TX.
2. BOYLE, D. 1998. Epidemiology of breast cancer. Bailliere's Clin. Oncol. **2:** 1–57.
3. ROBERT, M.M., F.E. ALEXANDER, R. ELTON & A. RODGER. 1990. Breast cancer stage, social class and the impact of screening on stage of presentation of breast cancer. Eur. J. Surg. Oncol. **16:** 18–21.
4. BAUM, M. 1981. Breast Cancer: the Facts. Oxford University Press. Oxford.
5. LI, M.G., C. HILL, P. REZVONI, *et al.* 1984. Long-term survival of women with breast cancer: a study of the durability of the disease. Lancet **2:** 922.
6. CARTER, C.L., C. ALLEN & D.E. HENION. 1989. Relationship of tumor size, lymph node status and survival in 24,740 breast cancer cases. Cancer **63:** 181–187.
7. HENDERSON, I.C., J.R. HARRY, D.W. KINNE, *et al.* 1989. Cancer of the breast. *In* Cancer: Principles and Practice of Oncology. 3rd edit. V. DeVita, S. Hellman & S.A. Rosenberg, Eds.: 1197–1267. Lippincot. Philadelphia.
8. ANDERSON, T.J. 1991. Genesis and source of breast cancer. Br. Med. Bull. **47:** 305–318.
9. BATTERSBY, S., T.J. ANDERSON & W.R. MILLER. 1994. Patterns of cyclic AMP binding in normal human breast. Breast Cancer Res. Treat. **30:** 153–158.
10. ANDERSON, T.J., S. BATTERSBY, R.J.B. KING, *et al.* 1989. Oral contraceptive use influences resting breast proliferation. Hum. Pathol. **20:** 1139–1144.
11. PAGE, D.L., R. VANDER ZWAAG, L.W. ROGERS, *et al.* 1978. Relation between component parts of fibrocystic disease complex and breast cancer. J. Natl. Cancer Inst. **61:** 1055–1063.
12. COHEN, S.M. & L.B. ELLWEIN. 1990. Cell proliferation in carcinogenesis. Science **249:** 1007–1011.
13. PRESTON-MARTIN, S., M.C. PIKE, R.K. ROSS, *et al.* 1990. Increased cell division as a cause of human cancer. Cancer Res. **50:** 7415–7421.
14. GORDGE, P.C., M.J. HULME, R.A. CLEGG & W.R. MILLER. 1996. Elevation of protein kinase A and protein kinase C activities in malignant as compared with normal human breast tissues. Eur. J. Cancer **32:** 2120–2126.

15. BRADBURY, A.W., D.C. CARTER, W.R. MILLER, et al. 1994. Protein kinase-A (PK-A) regulatory subunit expression in colorectal cancer and related mucosa. Br. J. Cancer **69:** 738–742.
16. CHO-CHUNG, Y.S. 1990. Role of cyclic AMP receptor proteins in growth differentiation and suppression of malignancy: new approaches to therapy. Cancer Res. **50:** 7093–7100.
17. MILLER, W.R., M.J. HULME, Y.S. CHO-CHUNG & R.A. ELTON. 1993. Types of cyclic AMP binding proteins in human breast cancers. Eur. J. Cancer **29:** 989–991.
18. HANDSCHIN, J.C., K. HANDLOSER, A. TAKAHASHI & U. EPPINBERGER. 1983. Cyclic adenosine 3':5'-monophosphate receptor proteins in dyplastic and neoplastic human breast tissue cytosol and their inverse relationship with oestrogen receptor. Cancer Res. **43:** 2947–2954.
19. MILLER, W.R., R.A. ELTON, J.M. DIXON, et al. 1990. Cyclic AMP binding proteins and prognosis in breast cancer. Br. J. Cancer **61:** 263–266.
20. MILLER, W.R., D.M.A. WATSON, W. JACK, et al. 1993. Tumour cyclic AMP binding proteins: an independent prognostic factor for disease recurrence and survival in breast cancer. Breast Cancer Res. Treat. **26:** 89–94.
21. HAWKINS, R.A., A.L. TESDALE, M.E. KILLEN, et al. 1996. Prospective evaluation of prognostic factors in operable breast cancer. Br. J. Cancer **74:** 1469–1478.
22. CHO-CHUNG, Y.S., T. CLAIR, M. SCHWIMMER, et al. 1981. Cyclic adenosine 3':5'-monophosphate receptor proteins in hormone-dependent and -independent rat mammary tumors. Cancer Res. **41:** 1840–1846.
23. WATSON, D.M.A., R.A. HAWKINS, N.J. BUNDRED, et al. 1987. Tumour cyclic AMP binding proteins and endocrine responsiveness in patients with inoperable breast cancer. Br. J. Cancer **56:** 141–142.
24. KVINNSLAND, S., R. EKANGER, S.O. DOSKELAND & T. THORSEN. 1983. Relationship of cyclic AMP binding capacity and oestrogen receptor to hormone sensitivity in human breast cancer. Breast Cancer Res. Treat. **3:** 67–72.
25. DIXON, J.M., C.D.B. LOVE, L. REMSHAW, et al. 1999. Lessons from the use of aromatase inhibitors in the neoadjuvant setting. Endocr. Relat. Cancer **6:** 227–230.
26. MILLER, W.R., T.J. ANDERSON, R.A. HAWKINS, et al. 1999. Neoadjuvant endocrine treatment; the Edinburgh experience. *In* Primary Medical Therapy for Breast Cancer. European School of Oncology Update, Vol. 4. A. Howell & M. Dowsett, Eds.: 89–99. Elsevier. Amsterdam.
27. BARTLETT, J.M.S., M.J. HULME & W.R. MILLER. 1996. Analysis of cAMP RI alpha mRNA expression in breast cancer: evaluation of quantitative polymerase chain reaction for routine use. Br. J. Cancer **73:** 1538–1544.
28. MILLER, W.R., M.J. HULME, J.M.S. BARTLETT, et al. 1997. Changes in messenger RNA expression of protein kinase A regulatory subunit Iα in breast cancer patients treated with tamoxifen. Clin. Cancer Res. **3:** 2399–2404.
29. COSTA, M., E.W. GERNER & D.H. RUSSELL. 1976. Cell cycle-specific activity of Type I and Type II cyclic adenosine 3':5'-monophosphate-dependent protein kinases in Chinese hamster ovary cells. J. Biol. Chem. **251** 3313–3319.
30. BYUS, C.V., G.R. KLUMPIL, D.O. LUCAS & D.H. RUSSELL. 1977. Type I and Type II cyclic AMP-dependent protein kinases as opposite effectors of lymphocyte mitogenesis. Nature **268:** 63–64.
31. CHO-CHUNG, Y.S., T. CLAIR, G. TORTORA & H. YOKOZAKI. 1991. Role of site-selective cAMP analogues and reversal of malignancy. Pharmacol. Ther. **50:** 1–33.
32. KEEN, J.C., J.M. DIXON, E. MILLER, et al. 1997. Expression of Ki-S1 and BCL-2 and the response to primary tamoxifen therapy in elderly patients with breast cancer. Breast Cancer Res. Treat. **44:** 123–134.
33. HOUGE, G., Y.S. CHO-CHUNG & S.O. DOSKELAND. 1992. Differential expression of cAMP-kinase subunits is correlated with growth in rat mammary carcinomas and uterus. Br. J. Cancer **66:** 1022–1029.
34. FULLER, D.J.M., C.V. BYUS & D.H. RUSSELL. 1978. Specific regulation by steroid hormones of the amount of type I cyclic AMP-dependent protein kinase holoenzyme. Proc. Natl. Acad. Sci. USA **75:** 223–227.
35. CHO, H. & B.S. KATZENELLENBOGEN. 1993. Synergistic activation of estrogen receptor mediated transcription by estradiol and protein kinase activators. Mol. Endocrinol. **7:** 441–452.

36. BUNONE, G., P.A. BRIAND, R.J. MIKSICEK & D. PICARD. 1996. Activation of the unliganded estrogen receptor by EGF involves the MAP kinase pathway and direct phosphorylation. EMBO J. **15:** 2174–2183.
37. LE GOFF, P., M.M. MONTANO, D.J. SCHODIN & B.S. KATZENELLENBOGEN. 1994. Phosphorylation of the human estrogen receptor: identification of hormone-regulated sites and examination of their influence on transcriptional activity. J. Biol. Chem. **269:** 4458–4466.
38. KATZENELLENBOGEN, B.S. 1996. Estrogen receptors: bioactivities and interactions with cell signalling pathways. Biol. Reprod. **54:** 287–293.
39. FUJIMOTO, N. & B.S. KATZENELLENBOGEN. 1994. Alteration in the agonist/antagonist balance of antiestrogens by activation of protein kinase A signalling pathways in breast cancer cells; anti-estrogen selectivity and promoter dependence. Mol. Endocrinol. **8:** 296–304.
40. ABRAHAM, I., K.V. CHIN, M.M. GOTTESMAN, *et al.* 1990. Transfection of a mutant regulatory subunit gene of cAMP-dependent protein kinase causes increased drug sensitivity and decreased expression of P glycoprotein. Exp. Cell Res. **189:** 133–141.
41. SCALA, S., A. BUDILLON, Z. ZHAN, *et al.* 1995. Downregulation of mdr-1 expression by 8 cl-cAMP in multi-drug resistant MCF-7 human breast cancer cells. J. Clin. Invest. **96:** 1026–1034.
42. ROHLFF, C., B. SAFA, A. RAHMAN, *et al.* 1993. Reversal of resistance to adriamycin by 8-chloro-cyclic AMP in adriamycin-resistant HL-60 leukemia cells is associated with reduction of type I cyclic AMP dependent protein kinase and cyclic AMP response element binding protein DNA binding activities. Mol. Pharmacol. **43:** 372–279.
43. KIM, S.H., J.J. PARK, B.S. CHUNG, *et al.* 1993. Inhibition of mdt.1 gene expression by H-87, a selective inhibitor of cAMP-dependent protein kinase. Cancer Lett. **74:** 37–41.
44. YOKOZAKI, H., A. BUDILLON, T. CLAIR, *et al.* 1993. 8-chloradenosine 3′,5′-monophosphate as a novel modulator of multi-drug resistance. Int. J. Oncol. **3:** 423–430.
45. RAMAGE, A.D., S.P. LANGDON, A.A. RITCHIE, *et al.* 1995. Growth inhibition by 8-chloro cyclic AMP of human HT29 colorectal and ZR-75-1 breast carcinoma xenografts is associated with modulation of protein kinase A isoenzymes. Eur. J. Cancer **31:** 969–973.
46. TORTORA, G., P. BUDILLAN, H. YOKOZAKI, *et al.* 1994. Retroviral vector-mediated overexpression of the RIIβ subunit of the cAMP dependent protein kinase induces differentiation in human leukaemia cells and reverts the transformed phenotype of mouse fibroblasts. Cell Growth Differ. **5:** 753–759.
47. YOKOZAKI, H., A. BUDILLON, G. TORTORA, *et al.* 1993. An antisense oligoxynucleotide that depletes RIα subunit of cyclic AMP dependent protein kinase induces growth inhibition in human cancer cells. Cancer Res. **53:** 868–872.
48. TORTORA, G., H. YOKOZAKUN, S. PEPE, *et al.* 1991. Differentiation of HL-60 leukemia by Type I regulatory subunit antisense oligodeoxynucleotide of cAMP dependent protein kinase. Proc. Natl. Acad. Sci. USA **88:** 2011–2015.
49. TORTORA, G., R. CAPUTO, V. DAMIANO, *et al.* 1997. Synergistic inhibition of human cancer cell growth by cytotoxic drugs and mixed backbone antisense oligonucleotide targeting protein kinase A. Proc. Natl. Acad. Sci. USA **94:** 12586–12591.
50. NESTEROVA, M. & Y.S. CHO-CHUNG. 1995. A single-injection to protein kinase A-directed antisense treatment to inhibit tumour growth. Nat. Med. **1:** 528–533.

Reinventing the Wheel of Cyclic AMP

Novel Mechanisms of cAMP Signaling

KHEW-VOON CHIN, WENG-LANG YANG, ROALD RAVATN,
TSUNEKAZU KITA, ELENA REITMAN, DAVID VETTORI, MARY ELLEN CVIJIC,
MICHAEL SHIN, AND LISA IACONO

Department of Medicine and Pharmacology, and The Cancer Institute of New Jersey, Robert Wood Johnson Medical School, University of Medicine and Dentistry of New Jersey, New Brunswick, New Jersey 08901, USA

ABSTRACT: Mechanisms of cAMP signal transduction have been thoroughly investigated for more than 40 years. From the binding of hormonal ligands to their receptors on the outer surface of the plasma membrane to the cytoplasmic activation of effectors, the ensuing cAMP signaling cascades and the nuclear gene regulatory functions, coupled with the structural elucidation of the cAMP-dependent protein kinase (PKA) and *in vivo* functional characterizations of each of the components of PKA by homologous recombination gene targeting, our understanding of cAMP-mediated signal transduction has reached its pinnacle. Despite this trove of knowledge, some recent findings have emerged that suggest hitherto novel and alternative mechanisms of cAMP action that could increase the signaling bandwidth of cAMP and PKA in cell growth and transcriptional regulation. This article attempts to review some of these novel and unconventional mechanisms of cAMP and PKA signaling, and to generate further enthusiasm in investigating and validating these new frontiers of the cAMP signal transduction pathway.

KEYWORDS: cyclic AMP (cAMP); cAMP-dependent protein kinase (PKA); signal transduction; regulatory subunit; function; regulation; mechanism

PROLOGUE

The paradigm of cyclic adenosine 3′, 5′-monophosphate (cAMP) signal transduction in the cell has been known for more than 40 years. Typically, the binding of hormone to its receptor leads to coupling of the receptor-hormone complex to cellular guanine nucleotide binding protein (G protein), which interacts with and activates adenylyl cyclase to produce cAMP, the cellular second messenger of hormone action. The newly minted cAMP then binds to its receptor, the cAMP-dependent protein kinase (PKA), and causes reversible phosphorylation of protein substrates that regulate a vast number of cellular processes. Three Nobel prizes have been awarded

Address for correspondence: Khew-Voon Chin, Ph.D., The Cancer Institute of New Jersey, 195 Little Albany Street, New Brunswick, NJ 08901. Voice: 732-235-6196; fax: 732-235-7493.
chinkv@rwja.umdnj.edu

for research directly linked to this signaling pathway, first to Earl Sutherland in 1971 for the discovery of cAMP, the second messenger of hormone action; then to Edmond Fischer and Edwin Krebs in 1992 for their discoveries of PKA and its role in reversible protein phosphorylation as a biological regulatory mechanism; and to Alfred Gilman and Martin Rodbell in 1994 for their discoveries of the roles of G protein in cAMP signal transduction in cells. It is now clear that reversible protein phosphorylation and G protein coupling to effector proteins are highly ubiquitous and are universal mechanisms for cellular signaling and regulation in eukaryotic cells. Despite these important advances in our understanding of the molecular mechanism of cAMP signaling, unexpected and novel insights have continued to appear in recent years, pointing to the complexity and versatility of this signaling pathway in the regulation of cellular functions.

SIGNAL TRANSDUCTION PATHWAY OF cAMP

The mechanism of cAMP signal transduction is one of the best-understood biochemical pathways. Phosphorylation mediated by the cAMP signaling pathway can be elicited by various physiological ligands in cells and is critically involved in the regulation of metabolism, cell growth and differentiation, apoptosis, and gene expression.[4–9] The effects of cAMP are mediated by its receptor PKA, which is composed of two genetically distinct subunits—regulatory (R) and catalytic (C)—, which form a tetrameric holoenzyme R_2C_2. In the presence of cAMP, the holoenzyme dissociates into an $R_2(cAMP)_4$ dimer and two free catalytically active C kinase subunits, which can phosphorylate a diverse number of target proteins both in the cytoplasmic and the nuclear compartments. In the nucleus, transcriptional regulation by cAMP is mediated by a family of cAMP-responsive nuclear factors, which bind to and regulate the expression of genes containing the cAMP-responsive element (CRE) consensus in their promoters. Phosphorylation of these CRE binding proteins (CREBs), which contain basic domain/leucine zipper motifs, by the C subunit of PKA that translocates into the nucleus following activation by cAMP, modulates their activity.

There are two major R subunit isoforms, which are further distinguished as RIα and RIβ, and RIIα and RIIβ, and three isoforms of the C subunit, Cα, Cβ, and Cγ. Two major isozymes termed type I and type II PKA were identified based on their patterns of chromatographic elution.[6,9] These isozymes may form from either homo- or heterodimers of the R subunits yielding holoenzyme complexes of PKA with a number of combinatorial configurations, including $RI\alpha_2C_2$, $RI\beta_2C_2$, $RII\alpha_2C_2$, $RII\beta_2C_2$, and $RI\alpha RI\beta C_2$.[10] The presence of multiple C subunit genes further adds to the diversity and complexity of the various holoenzyme complexes that may be found in cells.[9] In addition to differences in biochemical and functional properties, these genetically distinct isozymes have also been shown to have different patterns of expression and localization,[9,11] thus conferring the broad specificity that PKA may have on a number of physiological processes and signaling mechanisms in cells in response to the differential effects of cAMP.

NOVEL MECHANISMS OF cAMP SIGNALING

Other cAMP Receptors or Binding Proteins

Most of the effects of cAMP are mediated by PKA through phosphorylation. The R subunit has been the only known receptor for cAMP in cells, and cAMP binding to the holoenzyme has been accepted as the mechanism that regulates PKA activity. However, this dogma of cAMP signaling is being rewritten to accommodate some recent discoveries that implicate the existence of alternative mechanisms for the cAMP messenger system. It was intriguing when it was first discovered that olfactory transduction via odorant-specific olfactory receptors from ciliary plasma membrane and the pacemaker cation-permeable channels in cardiac sinoatrial node cells are regulated by direct cAMP binding without the activation of phosphorylation mediated by PKA.[12,13] Subsequently, it was found further that pacemaker ion channels in neurons are also regulated directly by cAMP binding.[14,15]

The ensuing isolation of some of these pacemaker channel genes, which encode a family of voltage-gated potassium ion channels in the brain, revealed the presence of consensus cyclic nucleotide binding site in the amino acid sequence,[16,17] suggesting their ability to bind cAMP. Furthermore, PKA phosphorylation sites are also found in different members of the ion channel family, thus explaining why, for example, the cardiac Purkinje fiber channel that contains the consensus phosphorylation site is regulated by PKA phosphorylation,[18] whereas the sinoatrial node channel is directly regulated by cAMP owing to the absence of the consensus phosphorylation site.[13] These results clearly point to dual mechanisms of cAMP signaling, one mediated directly by cAMP and the other through its binding to the PKA holoenzyme, resulting in the release and activation of the C subunit and phosphorylation.

The anomaly did not end here. Recently, another novel family of cAMP binding protein was identified. These proteins contain cAMP binding sites as well as domains homologous to the guanine nucleotide exchange factors (GEFs). These cAMP-regulated GEFs selectively activate the Ras superfamily of guanine nucleotide binding protein, Rap1, in a cAMP-dependent but PKA-independent manner.[19,20] It has already been shown that there is cross-talk between the cAMP and the Ras/mitogen activated protein (MAP) kinase signaling pathways, and cAMP can either stimulate or inhibit MAP kinase activity.[21] These findings suggest that direct coupling of cAMP to Rap1 activation by cAMP-GEFs is an important alternative cAMP messenger system.

Other studies have also implicated cAMP-induced cellular processes that seemed to be uncoupled from PKA-mediated phosphorylation. In thyroid, thyrotropin (TSH) activates function, proliferation, differentiation, and specific gene expression via cAMP. However, it was shown later that some of the effects of cAMP, especially the induction of DNA synthesis and thyroglobulin gene expression, could not be duplicated by PKA;[22] this suggests that other cAMP-dependent mechanisms could exist that are independent of PKA phosphorylation and have important roles in cAMP-signal transduction in thyroid cells. Whether other novel cAMP-receptor or binding proteins can directly regulate DNA synthesis and transcription regulation, in addition to those mediated by CREB, remains an open question.

C Subunit Partnerships

The established mechanism of C subunit kinase activation is confined to its association with the R subunits in the holoenzyme complex, which dissociates upon cAMP binding, thus liberating the C subunits that can phosphorylate their target proteins. In human, three distinct C subunit genes have been identified to date. Other than the tissue- and cell type–specific distribution of the different forms of C subunits,[9] the functional significance of the multiple forms of C subunit in cells is still not fully appreciated. Surprisingly, the C subunit protein was recently found in a ternary complex of NFκB-IκB-C subunit, which is activated in a cAMP-independent manner by cytokines and proinflammatory signals.[24] Degradation of IκB following the exposure to inducers of NFκB leads to the activation of the C subunit and subsequent phosphorylation of NFκB p65 (RelA). Therefore, this pathway represents a novel mechanism for the cAMP-independent activation of the C subunit kinase and the regulation of NFκB activity. The prevalence of this novel signaling mechanism was further bolstered by a recent study that showed that the vasoactive peptides endothelin-1 and angiotensin II similarly induced IκB degradation, thus activating the C subunit and the phosphorylation of NFκB, without stimulating the production of cAMP.[25]

Apparently, this unusual partnership of the C subunit with NFκB-IκB outside of the PKA holoenzyme complex is not uncommon. It has been shown further that the C subunit is required for transcriptional cross-talk between glucocorticoid receptor (GR) and NFκB.[26] In transcriptional cross-talk or cross-coupling, nuclear receptors are able to modulate the activity of other classes of trans-activators through DNA-binding independent mechanisms.[26] Under this circumstance, the C subunit associates with GR and cross-represses NFκB (RelA) transcriptional activity. It was further observed that a similar C subunit–dependent transrepression also occurs with NFκB and the retinoic acid receptor.[26] Other nuclear hormone receptors including the estrogen, progesterone, thyroid, and retinoid-x receptors, might also be targets for transcriptional cross-talk regulation by the C subunit kinase of PKA, independent of the PKA holoenzyme complex and cAMP activation. Therefore, the C subunit may serve a broader role as a general component for transrepression of nuclear receptors and NF-κB.

The Ins and Outs of PKA and the C Subunit

To date, all characterized biochemical mechanisms and effects of cAMP and PKA presumably occur inside cells; and PKA is known predominantly to be an intracellular enzyme.[6,9] An extracellular form of the PKA holoenzyme was first reported in the rat C-6 glioma cells in which the kinase complex was found to be associated with the outer cell surface.[27] The presence of extracellular membrane–associated PKA was also observed in rat adipocytes and spermatozoa, and human platelets and pulmonary microvascular endothelial cells.[28–32] PKA was also detected as a soluble activity in rabbit serum after thrombin stimulation of platelets.[33] The term *ectoenzyme* was thus coined for membrane-bound enzyme whose catalytic activity is localized on the extracellular cell surface; whereas the activity of exoenzyme is present in the extracellular environment without being directly associated with cells and can be isolated as soluble protein.[34,35] The above observations suggest that

PKA can be localized outside of cells either associating with the outer cell membrane as ecto-PKA or in a free soluble form of exo-PKA. The activity of ecto-PKA was demonstrated further by phosphorylation of the PKA-specific substrate Kemptide, occurring presumably at the surface of HeLa cells.[36,37] Other studies also showed that the atrial natriuretic peptide is phosphorylated in a variety of intact cells by a cAMP-dependent ectoprotein kinase.[38] More recently, it has been definitively shown that purified membrane of human LS-174T colon carcinoma cells contained PKA activity.[39] The ecto-PKA can be photoaffinity-labeled with 8-azido-[^{32}P]cAMP, and the regulatory subunits of the holoenzyme can be immunoprecipitated with specific antibodies.[39]

Since cAMP is generated in the cells by adenylyl cyclase in response to hormonal stimulation, activation of PKA that is localized in the extracellular milieu poses a logistical problem. However, both intracellular cAMP and ATP have been shown to be released into the extracellular space and are present in body fluids, thus providing the factors necessary to influence PKA activity extracellularly and therefore supporting a potential role for PKA in intercellular communication and regulation.[35,36,38] Various ecto- and exoenzymes are implicated in developmental processes, intercellular communication, feedback regulation, and events such as cell-cell interaction or receptor transduction of external stimuli. Nevertheless, the presence of PKA activity outside of cells is intriguing and not very well understood. How is the holoenzyme transported or secreted out of the cells? Are the subunits secreted independently and then assembled outside of cells? Clearly, the biogenesis and the mechanisms of activation, and, more importantly, the physiological functions and relevance of ecto- and exo-PKA need to be further investigated.

Whether ecto- and exo-PKA qualify as novel mechanisms of cAMP signaling is a subject of debate. The unusual localization of these forms of PKA holoenzyme in the extracellular milieu already raises questions about their genesis, whether the ecto- and exo-PKA activity might not have been the results of cell lysis during the course of the experiment. Further confounding this issue, it was recently found that the C subunit is also aberrantly secreted by tumor cells.[40–42] This exo-PKA is not activated by cAMP and the kinase activity is detected constitutively at various levels, in serum-free media from a variety of tissue culture cell lines. The exo-PKA specifically phosphorylates Kemptide and is inhibited by the PKA-specific peptide inhibitor (PKI) but not by the PKC-specific peptide inhibitor. Furthermore, mutation in the Cα subunit gene that disrupts its myristylation blocks the secretion of exo-PKA, suggesting that the NH$_2$-terminal myristyl group is required for export.[40] More importantly, the C subunit kinase activity is also detected in the serum sample of a large number of cancer patients compared to the low levels of kinase activity in normal volunteers.[40,41] Since the activities of the exo-PKA in conditioned media and in serum samples of cancer patients were constitutive and not activated by cAMP, it is speculated that only the C subunit is present extracellularly, independent of the PKA holoenzyme.[41] It is unclear whether the aberrant secretion of the C subunit out of the cells is associated with the pathogenesis of cancer. The presence of low levels of kinase activity detected in normal volunteers raised the possibility that the C subunit might be secreted under normal physiological conditions, either constitutively or regulated in response to hormonal stimulation. Therefore, ecto- and exo-PKA and the constitutively active free C subunit potentially provide crucial cellular regulation mediated by cAMP and phosphorylation in the extracellular milieu.

New Partnerships for R Subunit Too

Delivering the Knockout Bunch

While the C subunit kinase phosphorylates specific serine and threonine residues of target proteins, the only known function of the R subunit is that of binding cAMP and interacting with and inhibiting the activity of the C subunit. As with the C subunit, there is redundancy in the R subunit genes, as indicated above. The expression of either type I or type II PKA, or both, in cells is governed primarily by the presence of either RI or RII subunits in the holoenzyme complex. Though ubiquitously expressed, the R subunit genes are also differentially regulated in development and in specific cells and tissues.[6] Specific subcellular localization of PKA is most likely due to the interaction of the various R subunits with a complex array of A-kinase–associated proteins (AKAPs),[11,43] which serve the role of scaffold for bringing the kinase and the phosphatase, as well as the substrates, to proximity for signal processing upon hormonal stimulation. The AKAPs contain a common domain for interacting with various R subunits and a unique subcellular targeting domain.

It has also been shown that differently composed PKA holoenzymes—for example, $RII\alpha_2C_2$ and $RII\beta_2C_2$—have different biochemical and functional properties including cAMP binding and activation.[44–46] Most cells have type I and type II PKA. However, the ratio of type I to type II and the level of expression of the different subunits vary greatly between cells and tissues.[6,9] Hormones and mitogen signals have been shown to regulate the expression of the R and C subunits.[9] The redundancy and complexity of the R and C subunit genes coupled with the interactions with AKAPs are thought to confer simultaneously the broadband signaling of cAMP and the phosphorylation of specific downstream protein substrates.

Nevertheless, the functional significance of the redundancy of the R subunit genes was not fully appreciated until recently, with the advent of gene targeting or knockout transgenic technology. In mice, targeted disruption by homologous recombination of the RIβ subunit gene of PKA resulted in mice that exhibit defects in long-term depression and depotentiation at both the Schaffer collateral-CA1 synapse on pyramidal cells and the lateral perforant-path dentate synapse on granule cells, which are involved in the learning-related forms of synaptic plasticity.[47] It was also shown that RIβ-deficiency produces selective defects in mossy fiber long-term potentiation, which, however, did not affect the spatial and contextual learning in the mutant mice.[48] Furthermore, tissue injury–evoked nociceptive pain and inflammation are diminished in the RIβ-deficient mice.[49] Surprisingly, basal and cAMP-stimulated PKA activities are not affected by RIβ deletion, which is attributed to the compensatory increased in RIα expression.[47]

Ablation of the RIIβ subunit gene of PKA, in contrast, produces lean mice that resist diet-induced obesity.[50] Mutant mice also exhibit lower plasma levels of insulin and both VLDL (very low–density lipoprotein) and LDL (low-density lipoprotein) cholesterol, in comparison to wild-type mice.[51] The lean phenotype may be a result of increased lipolysis in the mutant mice.[52] Additionally, these RIIβ-deficient mice also showed an increased tendency to consume ethanol as well as decreased sensitivity to the sedative effects of ethanol, as measured by a faster recovery from ethanol-induced sleep, compared with wild-type mice.[53] Interestingly, RIβ-deficient mice did not show an increased tendency to drink ethanol, and those with targeted deletion of the $C_{\beta 1}$ subunit also showed normal voluntary consumption of ethanol,

TABLE 1. Summary of observed phenotypes and PKA activity in various R subunits-deficient mice by gene targeting

R Subunit gene	Phenotype	Total PKA activity	References
RIα	Early embryonic lethality and severe developmental defects	↑	56
RIβ	Defective hippocampal long-term depression and depotentiation	↔	47
	Defect in mossy fiber long-term potentiation		48
	Reduced inflammation and nociceptive pain		49
RIIα	Normal	↓	60,61
RIIβ	Lean phenotype, resistance to diet-induced obesity, increased lipolysis, reduced plasma insulin and VLDL and LDL cholesterol	↓	50–52
	Diminished motor learning and neuronal gene expression		54
	Loss of haloperidol induced catalepsy and gene expression		55
	Increased alcohol consumption and decreased alcohol-induced sedation		53

↑, increase; ↓, decrease; ↔, no change.

indicating that increased ethanol consumption is not a general characteristic associated with deletion of PKA subunits.[53] RIIβ deficiency is also correlated with impaired motor behavior, loss of haloperidol-induced catalepsy, and some loss of neuronal gene expression.[54,55] The RIIβ-deficient mice exhibit lower total cAMP-stimulated PKA activity in specific regions of the brain due to rapid degradation of the C subunits, resulting in significant decreases in steady state levels of both Cα and Cβ.[56] In contrast, deletion of RIβ has little impact on PKA activity.[47] Since increases in RIα and RIβ do not compensate fully for the loss of RIIβ, it was reasoned that the resulting phenotype of RIIβ deletion is attributed to the increased amount of free C subunits that are chronically unregulated in the RIIβ-deficient mice, producing a state of constitutive PKA activation in the absence of cAMP activation (ref. 56; see TABLE 1 for a summary of the PKA knockout mice and their respective phenotypes).

It has been shown that activation of skeletal muscle calcium channels is enhanced by activation of PKA and is inhibited by the Ht31 peptide, a 24 amino–acid peptide from a human thyroid AKAP, suggesting that close proximity between type IIα PKA and the L-type calcium channel, brought together by AKAP, is necessary for the phosphorylation of the channel during depolarizations.[57] It has also been shown that cAMP can influence spermatozoa motility by PKA phosphorylation of sperm proteins.[58] Coupled with the expression of RIIα in sperm,[59] type IIα PKA is thought to play a major role in sperm motility. Targeted disruption of the RIIα gene, unexpectedly, has no effects on either potentiation of the L-type calcium channel in the skeletal muscle, or fertility and sperm motility of the mutant mice.[60,61] A reduction in both C subunit protein levels and total kinase activity was observed despite a compensatory increase in the RIα protein. Even though dramatic relocalization of PKA to the cytoplasm has occurred,[61] no further discernible phenotype has been reported with the RIIα-deficient mice. These results suggest that either PKA does not play a role in potentiation of L-type calcium channel and sperm motility, or that RIIα deficiency can be compensated for by the increase in RIα subunit.

Most intriguing of all, in contrast to the deletion of the other R subunits, targeted disruption of the RIα subunit gene causes early embryonic lethality.[56,62] Since the RIα-deficient embryos can be rescued by crossing them to mice with ablation of the Cα gene, it was suggested that the excessive constitutive kinase activity resulting from RIα-deletion thus could have been attenuated by the already-reduced C subunit activity in the Cα-deficient mice. Therefore, the embryonic lethality phenotype observed with the RIα knockout mice may not be due to unknown or unique function of the RIα gene.

The Culprit: the Kinase or the R Subunit

The gene targeting experiments have yielded immense insights into the specific functions of PKA in relation to its association with various R subunit isoforms. Except for the ablation of RIβ, total and cAMP-stimulated PKA activities were affected to varying degrees in the RIα-, RIIα-, and RIIβ-subunit–deficient mice (TABLE 1), accompanied by compensatory changes in the R and the C subunits. The specific phenotypic changes associated with the knockout of each of the R subunits seemed to be accounted for in part by the alterations in the kinase activity as well as the disruption in targeting the R subunits to their specific subcellular compartments and localizations. Given the ubiquitous expression of the R subunits in various tissues and cell types, their specific subcellular localization, and the broad spectrum of physiological processes regulated by PKA in response to hormonal ligands, it is surprising that these knockout mice are not more functionally compromised than the observed phenotypes. It is possible that the severity of the functional loss of these subunits could be further revealed in the mutant mice in the presence of hormonal stimulations or physiological stresses. Where there are compensatory increases in the R and the C subunits, the knockout phenotypes are not always rescued in the mutant mice, thus suggesting that the R subunits might have other functions that cannot be replaced by merely restoring the kinase activity.

R Subunit—the Smoking Gun

The role of cAMP in cell growth has been widely studied.[63] Cyclic AMP can either positively or negatively regulate cell growth depending on cell type, genetic constitution, and cellular environment.[64] Changes in PKA activity have been well characterized during differentiation and also neoplastic transformation.[5] Differential expression of RI and RII subunits has been correlated with cell differentiation and neoplastic transformation. In fact, while RI is preferentially expressed in transformed cells, increased expression of RII is associated with terminally differentiated tissues.[5] In a large number of human cancer cell lines, RI isoform is the only R subunit of PKA detected. These findings are recapitulated in human cancer specimens as well, whereby the predominant expression of type I PKA or the RI subunit is consistently observed. It has been shown that overexpression of RIα in CHO cells rendered growth advantages in monolayer and soft agar conditions, whereas overexpression of the C subunit did not produce such consequences.[65] Similarly, overexpression of RIα, but not the C subunit, in MCF-10A cells conferred the ability to grow in serum and growth factor–free conditions.[66] It is apparent from these results that the role of cAMP in cell growth cannot be explained by changes in the kinase activity alone.

Further studies using antisense oligonucleotides directed against the RIα in human cancer cells results in inhibition of growth in monolayer culture and colony formation in soft agar, as well as tumor growth in nude mice.[67–69] In contrast, antisense targeting RIIβ stimulates cell growth[68] and overexpression of RIIβ subunit causes growth arrest and induces differentiation in the HL-60 human leukemia cells and also reverts Ki-ras-transformed NIH 3T3 cells.[70] Mutation at the autophosphorylation site of RIIβ disrupts its ability to suppress neoplastic cell growth.[71]

In fact, the speculation that the R subunit might have independent function is not new. It has been shown that R subunit could act through mechanisms other than C subunit activation. One possibility is that R subunit containing bound cAMP has functions independent of its interaction with the C subunit. For example, cAMP-bound RII subunit complex but not the C subunit nor the protein kinase holoenzyme inhibits phosphorylase phosphatase activity, leading to prolongation of the glycogen breakdown cascade.[72] The RII subunit also inhibits the activity of a purified high-molecular-weight phosphoprotein phosphatase in a cAMP-dependent manner, and the inhibited species was thought to be a RII-cAMP-phosphatase complex.[73]

The mammalian RIIβ subunit, especially the carboxyl-terminal domain containing the binding site for cAMP, has amino-acid homology with the *Escherichia coli* catabolite activator protein (CAP).[74] In addition to its ability to bind cAMP, CAP can also regulate cAMP-mediated gene expression by binding to DNA via its carboxy-terminal DNA binding domain.[75] The evolutionary conservation between RIIβ and CAP suggests that RIIβ may be able to bind DNA. It was reported recently that RIIβ could bind to and activate transcription via the CRE.[76] Binding of RIIβ to CRE is enhanced by cAMP, and a mutant lacking the autophosphorylation site exhibits reduced capacity to bind CRE and transcription from CRE.[76] Even though the role of phosphorylation induced by the kinase cannot be completely ruled out, these studies nevertheless suggest that regulation of CRE-directed transcription in eukaryotic cells may be mediated by RIIβ.

Besides RIIβ, it has also been found that RIα interacts with the ligand-activated epidermal growth factor receptor (EGFR) complex.[77] Coimmunoprecipitation with an anti-RIα antibody demonstrated the binding of RIα to the SH3 domains of the Grb2 adaptor protein, allowing the localization of the type I PKA to the activated EGFR.[77] Using affinity chromatography and immunoprecipitation, another study provided evidence for a direct interaction between RIIα and the p34cdc2 protein kinase cell cycle regulator, presenting the possibility of interdependent functioning of these two pathways in the regulation of cell division.[78]

In addition to the above biochemical and cell biological studies, genetic evidence also lends support for independent function for the R subunits. Using PKA genetic mutants derived from the CHO and the mouse Y1 adrenocortical carcinoma cells, it was found unexpectedly that PKA mutant having mutation in the RIα subunit but not in the C subunit gene exhibits increased cellular resistance to cisplatin, a DNA-damaging anticancer drug.[79,80] Moreover, wild-type CHO cells transfected with a mouse dominant mutant RIα cDNA are also more resistant to cisplatin than wild-type cells. These results raised the possibility that the RIα subunit may have functions, independent of PKA, in regulating drug resistance.

Further evaluation using various C subunit mutants in the yeast *Saccharomyces cerevisiae* (*S. cerevisiae*) also showed unambiguously that changes in PKA activity either owing to mutations in the C subunit or indirectly through alterations in other components of the cAMP signaling pathway have no effect on cisplatin sensitivity in *S. cerevisiae*.[81]

Taken together, the above studies in part suggest that the RIα subunit may have functions that are independent of PKA. The most revealing evidence came from recent reports that the RIα gene is targeted for genetic alterations in the inherited autosomal dominant Carney complex (CNC) disorder,[82,83] which is a multiple neoplasia syndrome characterized by spotty skin pigmentation, cardiac and other myxomas, endocrine tumors, and psammomatous melanotic schwannomas.[84] The germline mutations in RIα suggest its role in cell growth regulation and as a tumor suppressor, which is supported by loss of heterozygosity (LOH) studies in tumors from CNC patients.[82,83,85] The RIα mutations identified to date seemed to produce truncated protein products; thus these mutations could act by either haploinsufficiency (loss of function) or by a dominant negative effect (gain of function). The loss of RIα leads to enhanced intracellular signaling by PKA, as shown by an almost twofold greater response to cAMP in CNC tumors in comparison to non-CNC tumors.[82] These results suggest the RIα subunit may have functions independent of the PKA holoenzyme.

It has been known for a while that the cAMP/PKA signaling pathway is involved in cell growth regulation, but the mechanisms that regulate this process remain unclear.[63,64] It has also been observed that some of the effects on cell growth mediated by cAMP cannot be explained by the action of PKA.[65,66] Furthermore, there is also evidence that antisense oligonucleotide treatment of cells targeting the RIα subunit of PKA can inhibit growth.[86] Together with the recent finding that the RIα subunit is a tumor suppressor targeted for genetic alterations in the multiple neoplasia syndrome, Carney complex,[82–85] these findings suggest that the RIα subunit may have function that is independent of the C subunit kinase.

R Subunit Functions

How do R subunits exert their function outside the confines of the PKA holoenzyme? As indicated before, it is possible that R subunits may form complexes with other proteins that may be responsive to cAMP regulation, as they do with the C subunit in the formation of the PKA holoenzyme. Recent yeast two-hybrid interaction cloning experiments showed that the RIα subunit associates with the cytochrome c oxidase subunit Vb (CoxVb).[87] The mammalian cytochrome c oxidase, composed of 13 polypeptide subunits, is the terminal enzyme complex of the electron transfer chain. Interaction of CoxVb with RIα is regulated by cAMP, and elevation of cAMP levels inhibits cytochrome c oxidase activity in CHO cells.[87] RIα mutant cells show increased cytochrome c oxidase activity and decreased levels of cytochrome c in the mitochondria compared to either wild-type cells or the C subunit mutant. These results suggest a novel mechanism of cAMP signaling through the interaction of RIα with CoxVb, independent of the C subunit, thereby regulating cytochrome c oxidase activity and the release of cytochrome c in energy metabolism.[87]

Additional targets identified from the yeast two-hybrid interaction cloning includes a novel BTB-POZ domain zinc finger transcription factor, termed RIα-associated zinc finger (RIAZ).[88] Members of the family of BTB/POZ domain zinc finger transcription factors have been shown to play a variety of roles including apoptosis, transcription activation as well as repression, and oncogenesis.[89–96] Since type I PKA is predominantly localized in the cytoplasm[97] and that RIα subunit has been shown to bind tightly to the plasma membrane,[98] the interaction of RIα with RIAZ might result in sequestration of RIAZ in the cytoplasm. Overexpression of RIAZ produces a punctate pattern of nucleus distribution, and coexpression with RIα, as predicted, results in its sequestration in the cytoplasm. The cytoplasmic/nuclear translocation is inducible by cAMP. Deletion of the C-terminus of RIAZ abolishes its interaction with RIα and results in the localization of RIAZ exclusively in the nucleus.

These results, coupled with the reported mutations of RIα in neoplasia and its role as a tumor suppressor in oncogenesis and cell growth regulation[82–85] and the function of BTB-POZ domain family of zinc finger proteins in oncogenesis and growth control,[89,90] suggest a mechanism by which RIα regulates cell growth through its interaction with the BTB-POZ domain zinc finger protein. Furthermore, the discovery that other splice variants of RIAZ, including ZSG[99] and PATZ,[100] are targeted for intrachromosomal rearrangement in Ewing sarcoma[99] and the aberrant expression of RIAZ in breast cancer cell lines[88] further suggests the role of RIAZ in tumorigenesis and that the interaction of RIα with RIAZ may provide the missing link for the role of RIα as a tumor suppressor and in carcinogenesis.

EPILOGUE

The signaling transduction pathway of cAMP is an illustrious one. With the many accolades in its claims, from that of the first cellular second messenger discovered to the many surprising new facets of regulation unraveled in recent years, the signaling mechanisms of cAMP continue to provide inspiration and challenge to researchers in the field. The finding that the C subunit can function outside of the context of the PKA holoenzyme by forming a complex with NFκB-IκB points to the versatility of the components of the cAMP signaling pathway in regulating various cellular processes. Undoubtedly, the challenge of proving that the R subunits may also have functions independent of the PKA holoenzyme will be arduous, but exciting nonetheless.

REFERENCES

1. SUTHERLAND, E.W. & T.W. RALL. 1957. The properties of an adenine ribonucleotide produced with cellular particles, ATP, Mg^{++}, and epinephrine or glucagons. J. Am. Chem. Soc. **79:** 3608.
2. RALL, T.W. & E.W. SUTHERLAND. 1958. Formation of a cyclic adenine ribonucleotide by tissue particles. J. Biol. Chem. **232:** 1065–1076.

3. LIPKIN, D., W.H. COOK & R.J. MARKHAM. 1959. Adenosine-3':5'-phosphoric acid: a proof of structure. J. Am. Chem. Soc. **81:** 6198–6203.
4. SUTHERLAND, E.W., I. OYE & R.W. BUTCHER. 1965. The action of epinephrine and the role of the adenyl cyclase system in hormone action. Rec. Progr. Horm. Res. **21:** 623–646.
5. CHO-CHUNG, Y.S., S. PEEP, T. CLAIR, et al. 1995. cAMP-dependent protein kinase: role in normal and malignant growth. Crit. Rev. Oncol. Hematol. **21:** 33–61.
6. FRANCIS, S.H. & J.D. CORBIN. 1999. Cyclic nucleotide-dependent protein kinases: intracellular receptors for cAMP and cGMP action. Crit. Rev. Clin. Lab. Sci. **36:** 275–328.
7. GRAVES, J.D. & E.G. KREBS. 1999. Protein phosphorylation and signal transduction. Pharmacol. Ther. **82:** 111–121.
8. MAYR, B. & M. MONTMINY. 2001. Transcriptional regulation by the phosphorylation-dependent factor CREB. Nat. Rev. Mol. Cell Biol. **8:** 599–609.
9. SKALHEGG, B.S. & K. TASKEN. 2000. Specificity in the cAMP/PKA signaling pathway. Differential expression, regulation, and subcellular localization of subunits of PKA. Front. Biosci. **5:** D678–D693.
10. TASKEN, K., B.S. SKALHEGG, R. SOLBERG, et al. 1993. Novel isozymes of cAMP-dependent protein kinase exist in human cells due to formation of RI alpha-RI beta heterodimeric complexes. J. Biol. Chem. **268:** 21276–21283.
11. PAWSON, T., & J.D. SCOTT. 1997. Signaling through scaffold, anchoring, and adaptor proteins. Science **278:** 2075–2080.
12. NAKAMURA, T. & G.H. GOLD. 1987. A cyclic nucleotide-gated conductance in olfactory receptor cilia. Nature **325:** 442–444.
13. DIFRANCESCO, D., & P. TORTORA. 1991. Direct activation of cardiac pacemaker channels by intracellular cyclic AMP. Nature **351:** 145–147.
14. PEDARZANI, P. & J.F. STORM. 1995. Protein kinase A-independent modulation of ion channels in the brain by cyclic AMP. Proc. Natl. Acad. Sci. USA **92:** 11716–11720.
15. INGRAM, S.L. & J.T. WILLIAMS. 1996. Modulation of the hyperpolarization-activated current (Ih) by cyclic nucleotides in guinea-pig primary afferent neurons. J. Physiol. (Lond.) **492:** 97–106.
16. SANTORO, B., S.G. GRANT, D. BARTSCH & E.R. KANDEL. 1997. Interactive cloning with the SH3 domain of N-src identifies a new brain specific ion channel protein, with homology to eag and cyclic nucleotide-gated channels. Proc. Natl. Acad. Sci. USA **94:** 14815–14820.
17. SANTORO, B., D.T. LIU, H. YAO, et al. 1998. Identification of a gene encoding a hyperpolarization-activated pacemaker channel of brain. Cell **93:** 717–729.
18. CHANG, F., I.S. COHEN, D. DIFRANCESCO, et al. 1991. Effects of protein kinase inhibitors on canine Purkinje fibre pacemaker depolarization and the pacemaker current i(f). J. Physiol. **440:** 367–384.
19. DE ROOIJ, J., F.J.T. ZWARTKRUIS, M.H.G. VERHEIJEN, et al. 1998. Epac is a Rap1 guanine nucleotide-exchange factor directly activated by cyclic AMP. Nature **396:** 474–477.
20. KAWASAKI, H., G.M. SPRINGETT, N. MOCHIZUKI, et al. 1998. A family of cAMP-binding proteins that directly activates Rap1. Science **282:** 2275–2279.
21. BORNFELDT, K.E. & E.G. KREBS. 1999. Crosstalk between protein kinase A and growth factor receptor signaling pathways in arterial smooth muscle. Cell Signal. **11:** 465–477.
22. DUMONT, J.E., F. LAMY, P. ROGER & C. MAENHAUT. 1992. Physiological and pathological regulation of thyroid cell proliferation and differentiation by thyrotropin and other factors. Physiol. Rev. **72:** 667–697.
23. DREMIER, S., V. POHL, C. POTEET-SMITH, et al. 1997. Activation of cyclic AMP-dependent kinase is required but may not be sufficient to mimic cyclic AMP-dependent DNA synthesis and thyroglobulin expression in dog thyroid cells. Mol. Cell. Biol. **17:** 6717–6726.
24. ZHONG, H., H.S. YANG, H. ERDJUMENT-BROMAGE, et al. 1997. The transcriptional activity of NF-κB is regulated by the IkB-associated PKAc subunit through a cyclic AMP-independent mechanism. Cell **89:** 413–424.

25. DULIN, N.O., J. NIU, D.D. BROWNING, et al. 2001. Cyclic AMP-independent activation of protein kinase A by vasoactive peptides. J. Biol. Chem. **276:** 20827–20830.
26. DOUCAS, V., Y. SHI, S. MIYAMOTO, et al. 2000. Cytoplasmic catalytic subunit of protein kinase A mediates cross-repression by NF-kappa B and the glucocorticoid receptor. Proc. Natl. Acad. Sci. USA **97:** 11893–11898.
27. SCHLAEGER, E. & G. KOHLER. 1976. External cyclic AMP-dependent protein kinase activity in rat C-6 glioma cells. Nature **260:** 705–707.
28. MAJUMDER, G.C. 1978. Occurrence of a cyclic AMP-dependent protein kinase on the outer surface of rat epididymal spermatozoa. Biochem. Biophys. Res. Commun. **83:** 829–836.
29. KANG, E.S., R.E. GATES, T.M. CHIANG & A.H. KANG. 1979. Ectoprotein kinase activity of the isolated rat adipocyte. Biochem. Biophys. Res. Commun. **86:** 769–778.
30. NOLAND, T.D., J.D. CORBIN & D.L. GARBERS. 1986. Cyclic AMP-dependent protein kinase isozymes of bovine epididymal spermatozoa evidence against the existence of an ectokinase. Biol. Reprod. **34:** 681–689.
31. HATMI, M., J.M. GAVARET, I. ELALAMY, et al. 1996. Evidence for cAMP-dependent platelet ectoprotein kinase activity that phosphorylates platelet glycoprotein IV (CD36). J. Biol. Chem. **271:** 24776–24780.
32. ELALAMY, I., F.A. SAID, M. SINGER, et al. 2000. Inhibition by extracellular cAMP of phorbol 12-myristate 13-acetate-induced prostaglandin H synthase-2 expression in human pulmonary microvascular endothelial cells. Involvement of an ecto-protein kinase A activity. J. Biol. Chem. **275:** 13662–13667.
33. KORC-GRODZICKI, B., M. TAUBER-FINKELSTEIN & S. SHALTIEL. 1988. Platelet stimulation releases a cAMP-dependent protein kinase that specifically phosphorylates a plasma protein. Proc. Natl. Acad. Sci. USA **85:** 7541–7545.
34. EHRLICH, Y.H. 1996. Extracellular protein kinases. Science **271:** 278–279.
35. EHRLICH, Y.H., M.V. HOGAN, Z. PAWLOWSKA, et al. 1990. Ectoprotein kinase in the regulation of cellular responsiveness to extracellular ATP. Ann. N.Y. Acad. Sci. **603:** 401–416.
36. KUBLER, D., W. PYERIN, O. BILL, et al. 1983. Substrate-effected release of surface-located protein kinase from intact cells. Proc. Natl. Acad. Sci. USA **80:** 4021–4025.
37. KUBLER, D., W. PYERIN, O. BILL, et al. 1989. Evidence for ecto-protein kinase activity that phosphorylates kemptide in a cyclic-AMP dependent mode. J. Biol. Chem. **264:** 14549–14555.
38. KUBLER, D., D. REINHARDT, J. REED, et al. 1992. Atrial natriuretic peptide is phosphorylated by intact cells though cAMP-dependent ecto-protein kinase. Eur. J. Biochem. **206:** 179–186.
39. KONDRASHIN, A., M. NESTEROVA & Y.S. CHO-CHUNG. 1999. Cyclic adenosine 3′,5′-monophosphate-dependent protein kinase on the external surface of LS-174T human colon carcinoma cells. Biochemistry **38:** 172–179.
40. CHO, Y.S., Y.G. PARK, Y.N. LEE, et al. 2000. Extracellular protein kinase A as a cancer biomarker: its expression by tumor cells and reversal by a myristate-lacking Calpha and RIIbeta subunit overexpression. Proc. Natl. Acad. Sci. USA **97:** 835–840.
41. CVIJIC, M.E., T. KITA, W. SHIH, et al. 2000. Extracellular catalytic subunit activity of the cAMP-dependent protein kinase in prostate cancer. Clin. Cancer Res. **6:** 2309–2317.
42. CHO, Y.S., Y.N. LEE & Y.S. CHO-CHUNG. 2000. Biochemical characterization of extracellular cAMP-dependent protein kinase as a tumor marker. Biochem. Biophys. Res. Commun. **278:** 679–684.
43. FELICIELLO, A., M.E. GOTTESMAN & E.V. AVVEDIMENTO. 2000. The biological functions of A-kinase anchor proteins. J. Mol. Biol. **308:** 99–114.
44. OTTEN, A.D., L.A. PARENTEAU, S. DOSKELAND & G.S. MCKNIGHT. 1991. Hormonal activation of gene transcription in ras-transformed NIH3T3 cells overexpressing RII-alpha and RIIbeta subunits of the cAMP-dependent protein kinase. J. Biol. Chem. **266:** 23074–23082.
45. HUNZICKER-DUNN, M., R.E. CUTLER, JR., E.T. MAIZELS, et al. 1991. Isozymes of cAMP-dependent protein kinase present in the rat corpus luteum. J. Biol. Chem. **266:** 7166–7175.

46. CADD, G.G., M.D. UHLER & G.S. MCKNIGHT. 1990. Holoenzymes of cAMP-dependent protein kinase containing the neural form of type I regulatory subunit have an increased sensitivity to cyclic nucleotides. J. Biol. Chem. **265:** 19502–19506.
47. BRANDON, E.P., M. ZHUO, Y.Y. HUANG, et al. 1995. Hippocampal long-term depression and depotentiation are defective in mice carrying a targeted disruption of the gene encoding the RI beta subunit of cAMP-dependent protein kinase. Proc. Natl. Acad. Sci. USA **92:** 8851–8855.
48. HUANG, Y.Y., E.R. KANDEL, L. VARSHAVSKY, et al. 1995. A genetic test of the effects of mutations in PKA on mossy fiber LTP and its relation to spatial and contextual learning. Cell **83:** 1211–1222.
49. MALMBERG, A.B., E.P. BRANDON, R.L. IDZERDA, et al. 1997. Diminished inflammation and nociceptive pain with preservation of neuropathic pain in mice with a targeted mutation of the type I regulatory subunit of cAMP-dependent protein kinase. J. Neurosci. **17:** 7462–7470.
50. CUMMINGS, D.E., E.P. BRANDON, J.V. PLANAS, et al. 1996. Genetically lean mice result from targeted disruption of the RII beta subunit of protein kinase A. Nature **382:** 622–626.
51. SCHREYER, S.A., D.E. CUMMINGS, G.S. MCKNIGHT & R.C. LEBOEUF. 2001. Mutation of the RIIbeta subunit of protein kinase A prevents diet-induced insulin resistance and dyslipidemia in mice. Diabetes **50:** 2555–2562.
52. PLANAS, J.V., D.E. CUMMINGS, R.L. IDZERDA & G.S. MCKNIGHT. 1999. Mutation of the RIIbeta subunit of protein kinase A differentially affects lipolysis but not gene induction in white adipose tissue. J. Biol. Chem. **274:** 36281–36287.
53. THIELE, T.E., B. WILLIS, J. STADLER, et al. 2000. High ethanol consumption and low sensitivity to ethanol-induced sedation in protein kinase A-mutant mice. J. Neurosci. **20:** RC75.
54. BRANDON, E.P., S.F. LOGUE, M.R. ADAMS, et al. 1998. Defective motor behavior and neural gene expression in RIIbeta-protein kinase A mutant mice. J. Neurosci. **18:** 3639–3649.
55. ADAMS, M.R., E.P. BRANDON, E.H. CHARTOFF, et al. 1997. Loss of haloperidol induced gene expression and catalepsy in protein kinase A-deficient mice. Proc. Natl. Acad. Sci. USA **94:** 12157–12161.
56. AMIEUX, P.S., D.E. CUMMINGS, K. MOTAMED, et al. 1997. Compensatory regulation of RIalpha protein levels in protein kinase A mutant mice. J. Biol. Chem. **272:** 3993–3998.
57. JOHNSON, B.D., T. SCHEUER & W.A. CATTERALL. 1994. Voltage-dependent potentiation of L-type Ca^{2+} channels in skeletal muscle cells requires anchored cAMP-dependent protein kinase. Proc. Natl. Acad. Sci. USA **91:** 11492–11496.
58. TASH, J.S., & A.R. MEANS. 1982. Regulation of protein phosphorylation and motility of sperm by cyclic adenosine monophosphate and calcium. Biol. Reprod. **26:** 745–763.
59. LANDMARK, B.F., O. OYEN, B.S. SKALHEGG, et al. 1993. Cellular location and age-dependent changes of the regulatory subunits of cAMP-dependent protein kinase in rat testis. J. Reprod. Fertil. **99:** 323–334.
60. BURTON, K.A., B.D. JOHNSON, Z.E. HAUSKEN, et al. 1997. Type II regulatory subunits are not required for the anchoring-dependent modulation of Ca^{2+} channel activity by cAMP-dependent protein kinase. Proc. Natl. Acad. Sci. USA **94:** 11067–11072.
61. BURTON, K.A., B. TREASH-OSIO, C.H. MULLER, et al. 1999. Deletion of type IIalpha regulatory subunit delocalizes protein kinase A in mouse sperm without affecting motility or fertilization. J. Biol. Chem. **274:** 24131–24136.
62. AMIEUX, P.S. & G.S. MCKNIGHT. 2002. The essential role of RIα in the maintenance of regulated PKA activity. Ann. N.Y. Acad. Sci. **968:** This volume.
63. BOYNTON, A.L. & J.F. WHITFIELD. 1983. The role of cAMP in cell proliferation: a critical assessment of the evidence. Adv. Cyclic Nucleotide Res. **15:** 193–294.
64. GOTTESMAN, M.M. & R.D. FLEISHMAN. 1986. The role of cAMP in regulating tumor cell growth. Cancer Surv. **5:** 291–308.
65. TORTORA, G., S. PEPE, C. BIANCO, et al. 1994. Differential effects of protein kinase A subunits on Chinese hamster ovary cell cycle and proliferation. Int. J. Cancer **59:** 712–716.
66. TORTORA, G., S. PEPE, C. BIANCO, et al. 1994. The RI alpha subunit of protein kinase A controls serum dependency and entry into cell cycle of human mammary epithelial cells. Oncogene **9:** 3233–3240.

67. CHO-CHUNG, Y.S., M. NESTEROVA, S. PEPE, et al. 1999. Antisense DNA-targeting protein kinase A-RIA subunit: a novel approach to cancer treatment. Front. Biosci. **4:** D898–D907.
68. YOKOZAKI, H., A. BUDILLON, G. TORTORA, et al. 1993. An antisense oligodeoxynucleotide that depletes RI alpha subunit of cyclic AMP-dependent protein kinase induces growth inhibition in human cancer cells. Cancer Res. **53:** 868–872.
69. NESTEROVA, M., & Y.S. CHO-CHUNG. 2000. Oligonucleotide sequence-specific inhibition of gene expression, tumor growth inhibition, and modulation of cAMP signaling by an RNA-DNA hybrid antisense targeted to protein kinase A RIalpha subunit. Antisense Nucleic Acid Drug Dev. **10:** 423–433.
70. TORTORA, G., A. BUDILLON, H. YOKOZAKI, et al. 1994. Retroviral vector-mediated overexpression of the RII beta subunit of the cAMP-dependent protein kinase induces differentiation in human leukemia cells and reverts the transformed phenotype of mouse fibroblasts. Cell Growth Differ. **5:** 753–759.
71. NESTEROVA, M., H. YOKOZAKI, E. MCDUFFIE & Y.S. CHO-CHUNG. 1996. Overexpression of RII beta regulatory subunit of protein kinase A in human colon carcinoma cell induces growth arrest and phenotypic changes that are abolished by site-directed mutation of RII beta. Eur. J. Biochem. **235:** 486–494.
72. GERGLEY, P. & G. BOT. 1977. The control of phosphorylase phosphatase by cAMP-dependent protein kinase. FEBS Lett. **82:** 269–272.
73. KHATRA, B.S., R. PRINTZ, C.E. COBB & J.D. CORBIN. 1985. Regulatory subunit of cAMP-dependent protein kinase inhibits phosphoprotein phosphatase. Biochem. Biophy. Res. Commun. **130:** 567–572.
74. WEBER, I.T., K. TAKIO, K. TITANI & T.A. STEITZ. 1982. The cAMP-binding domains of the regulatory subunit of cAMP-dependent protein kinase and the catabolite gene activator protein are homologous. Proc. Natl. Acad. Sci. USA **79:** 7679–7683.
75. STEITZ, T.A., I.T. WEBER & J.B. MATTHEW. 1983. Catabolite gene activator protein: structure, homology with other proteins, and cyclic AMP and DNA binding. Cold Spring Harb. Symp. Quant. Biol. **47** (Pt. 1): 419–426.
76. SRIVASTAVA, R.K., Y.N. LEE, K. NOGUCHI, et al. 1998. The RIIbeta regulatory subunit of protein kinase A binds to cAMP response element: an alternative cAMP signaling pathway. Proc. Natl. Acad. Sci. USA **95:** 6687–6692.
77. TORTORA, G., V. DAMIANO, C. BIANCO, et al. 1997. The RIalpha subunit of protein kinase A (PKA) binds to Grb2 and allows PKA interaction with the activated EGF-receptor. Oncogene **14:** 923–928.
78. TOURNIER, S., F. RAYNAUD, P. GERBAUD, et al. 1991. Association of type II cAMP-dependent protein kinase with p34cdc2 protein kinase in human fibroblasts. J. Biol. Chem. **266:** 19018–19022.
79. LIU, B.-R., M.E. CVIJIC, A.E. JETZT & K.-V. CHIN. 1996. Cisplatin resistance and regulation of DNA repair in cAMP-dependent protein kinase mutants. Cell Growth Differ. **7:** 1105–1112.
80. CVIJIC, M.E. & K.-V. CHIN. 1997. Regulation of P-glycoprotein expression in cAMP dependent protein kinase mutants. Cell Growth Differ. **8:** 1243–1247.
81. CVIJIC, M.E., W.L. YANG & K.-V. CHIN. 1998. Cisplatin sensitivity in cAMP dependent protein kinase mutants of *Saccharomyces cerevisiae*. Anticancer Res. **18:** 3187–3192.
82. KIRSCHNER, L.S., J.A. CARNEY, S.D. PACK, et al. 2000. Mutations of the gene encoding the protein kinase A type I-alpha regulatory subunit in patients with the carney complex. Nat. Genet. **26:** 89–92.
83. CASEY, M., C.J. VAUGHAN, J. HE, et al. 2000. Mutations in the protein kinase A R1alpha regulatory subunit cause familial cardiac myxomas and Carney complex. J. Clin. Invest. **106:** R31–R38.
84. STRATAKIS, C.A., L.S. KIRSCHNER & J.A. CARNEY. 2001. Clinical and molecular features of the Carney complex: diagnostic criteria and recommendations for patient evaluation. J. Clin. Endocrinol. Metab. **86:** 4041–4046.
85. KIRSCHNER, L.S., F. SANDRINI, J. MONBO, et al. 2000. Genetic heterogeneity and spectrum of mutations of the PRKAR1A gene in patients with the Carney complex. Hum. Mol. Genet. **9:** 3037–3046.

86. CHO-CHUNG, Y.S. 1999. Antisense oligonucleotide inhibition of serine/threonine kinases: an innovative approach to cancer treatment. Pharmacol. Ther. **82:** 437–449.
87. YANG, W.-L., L. IACONO, W.-M. TANG & K.-V. CHIN. 1998. A novel mechanism of cAMP signaling through the interaction of the regulatory subunit of protein kinase A with cytochrome c oxidase subunit Vb. Biochemistry **37:** 14175–14180.
88. YANG, W.-L., R. RAVATN, K. KUDOH, et al. Novel cAMP signalling via the regulatory subunit of the cAMP-dependent protein kinase. Nucleic Acid Res. In press.
89. YE, B.H., F. LISTA, F. LO COCO, et al. 1993. Alterations of a zinc finger-encoding gene, BCL-6, in diffuse large-cell lymphoma. Science **262:** 747–750.
90. CHEN, Z., N.J. BRAND, A. CHEN, et al. 1993. Fusion between a novel Kruppel-like zinc finger gene and the retinoic acid receptor-alpha locus due to a variant t(11;17) translocation associated with acute promyelocytic leukaemia. EMBO J. **12:** 1161–1167.
91. ALBAGLI, O., P. DHORDAIN, C. DEWEINDT, et al. 1995. The BTB/POZ domain: a new protein-protein interaction motif common to DNA- and actin-binding proteins. Cell Growth Differ. **6:** 1193–1198.
92. REUTER, S., M. BARTELMANN, M. VOGT, et al. 1998. APM-1, a novel human gene, identified by aberrant co-transcription with papillomavirus oncogenes in a cervical carcinoma cell line, encodes a BTB/POZ-zinc finger protein with growth inhibitory activity. EMBO J. **17:** 215–222.
93. DE LA LUNA, S., K.E. ALLEN, S.L. MASON & N.B. LA THANGUE. 1999. Integration of a growth-suppressing BTB/POZ domain protein with the DP component of the E2F transcription factor. EMBO J. **18:** 212–228.
94. OKABE, S., T. FUKUDA, K. ISHIBASHI, et al. 1998. BAZF, a novel Bcl6 homologue, functions as a transcriptional repressor. Mol. Cell. Biol. **18:** 4235–4244.
95. LIN, R.J., L. NAGY, S. INOUE, et al. 1998. Role of the histone deacetylase complex in acute promyelocytic leukaemia. Nature **391:** 811–814.
96. YAMOCHI, T., Y. KANEITA, T. AKIYAMA, et al. 1999. Adenovirus-mediated high expression of BCL-6 in CV-1 cells induces apoptotic cell death accompanied by down-regulation of BCL-2 and BCL-X(L). Oncogene **18:** 487–494.
97. DEVILLER, P., P. VALLIER, J. BATA & J.M. SAEZ. 1984. Distribution and characterization of cAMP-dependent protein kinase isoenzymes in bovine adrenal cells. Mol. Cell. Endocrinol. **38:** 21–30.
98. RUBIN, C.S., J. ERLICHMAN & O.M. ROSEN. 1972. Cyclic adenosine 3′,5′-monophosphate-dependent protein kinase of human erythrocyte membranes. J. Biol. Chem. **247:** 6135–6139.
99. FEDELE, M., G. BENVENUTO, R. PERO, et al. 2000. A novel member of the BTB/POZ family, PATZ, associates with the RNF4 RING finger protein and acts as a transcriptional repressor. J. Biol. Chem. **275:** 7894–7901.
100. MASTRANGELO, T., P. MODENA, S. TORNIELLI, et al. 2000. A novel zinc finger gene is fused to EWS in small round cell tumor. Oncogene **19:** 3799–3804.

Role of the PKA-Regulated Transcription Factor CREB in Development and Tumorigenesis of Endocrine Tissues

D. ROSENBERG,[a,c] L. GROUSSIN,[a] E. JULLIAN,[a] K. PERLEMOINE,[a]
X. BERTAGNA,[a,b] AND J. BERTHERAT[a,b]

[a]*CNRS UPR1524, Institut Cochin de Génétique Moléculaire and* [b]*Endocrine Department, Hôpital Cochin, Université Paris V, Paris, France*

[c]*Ecole Nationale Vétérinaire d'Alfort, Maisons-Alfort, France*

ABSTRACT: The cAMP pathway plays a major role in the development of endocrine tissues and various molecular defects of key components of this pathway (G protein, receptors, PKA, etc.) have been observed in endocrine tumors. The ubiquitous transcription factor CREB (cAMP-response element binding protein) binds to the cAMP response element (CRE) and stimulates transcription after phosphorylation on Ser^{133} by PKA. The CREB family of transcription factors contains three members: CREB, CREM, and ATF-1. Targeted expression of dominant-negative mutants of CREB in transgenic mice leads to somatotrophs or thyroid hypoplasia. GH-secreting adenomas are benign secreting tumors expressing an activated mutant Gαs protein (Gsp) in about 40% of cases. In GH-secreting adenomas CREB is always expressed and often highly phosphorylated. The CREM isoform ICER is stimulated by cAMP, and its expression is increased in Gsp-harboring tumors. After transfection in pituitary somatotroph cells, activating mutations of Gs protein (Gsp) and overexpression of wild-type GαS stimulate transcription of various CRE-containing promoters via CREB in a Ser^{133}-specific–dependent manner. Activation of the cAMP pathway by ACTH is required for adrenal cortex (AdCx) maintenance and steroidogenesis. CREB is expressed in normal AdCx. Alterations of CRE binding proteins with loss of CREB expression and compensatory overexpression of CREMτ is observed in the human adrenocortical cancer cell line H295R. Similar alterations are found at the protein level in human malignant adrenocortical tumors. In conclusion, the CREB family of transcription factors plays an important role in the development, differentiation, and proliferation of endocrine tissues. Various alterations of the CREB family of transcription factors can be observed in endocrine tumors.

KEYWORDS: cAMP; protein kinase A (PKA); endocrine tumors; cAMP response-element binding protein (CREB)

Address for correspondence: Jérôme Bertherat, M.D., D.Sc., CNRS UPR1524, ICGM, CHU Cochin, 24, rue du Fg-St-Jacques, 75014, Paris, France. Voice: 33-1-58-41-18-95; fax: 33-1-46-33-80-60.
jerome.bertherat@cch.ap-hop-paris.fr

INTRODUCTION

Numerous peptide hormones act at the cellular level by stimulation of the cAMP signaling pathway, after binding to their specific G protein–coupled transmembrane receptors. Alterations of the cAMP pathway have been implicated in several endocrine diseases. One major cellular effect of the cAMP cascade activation is the stimulation of transcription after phosphorylation of nuclear factors by the cAMP-dependent protein kinase, PKA. The transcription factor CREB (cAMP response element binding protein) is the best characterized nuclear protein that mediates stimulation of transcription by cAMP. The important role of CREB in the development and differentiation of various endocrine tissues has been demonstrated *in vitro* and *in vivo*. In endocrine tumors alterations of CREB and related transcription factors have been reported. This paper reviews these observations underlying the physiological role of the transcription factors of the CREB family in endocrine tissues and their alterations in endocrine tumors.

REGULATION OF TRANSCRIPTION BY cAMP

Activation of the cAMP pathway by seven domains transmembrane receptors coupled to the Gαs protein leads to increased cAMP intracellular levels. The main target of cAMP is the cAMP-dependent protein kinase (PKA) (FIG. 1). In the inac-

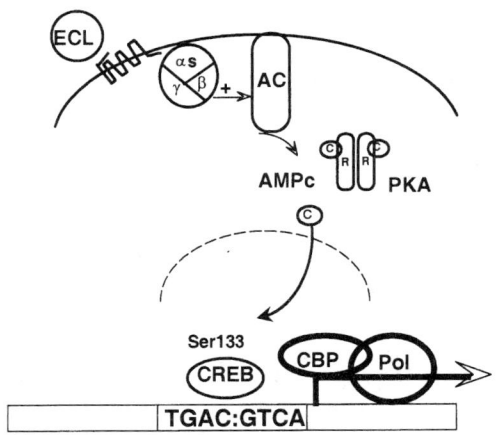

FIGURE 1. cAMP pathway from the cell surface to the nucleus. The extracellular ligand (ECL) binds to its specific seven domains transmembrane receptors, leading to activation of an heterotrimeric (α,β and γ subunit) Gs protein. Activation of the Gs protein leads to dissociation of its α subunit from the β,γ complex, stimulating adenylyl cyclase (AC) activity and therefore cAMP production. The PKA is activated after binding of cAMP. The free catalytic subunit (C) then goes into the nucleus and phosphorylates CREB at Ser[133]. Phospho-CREB binds CBP (CREB binding protein) that interacts with the polymerase (Pol), leading to transactivation.

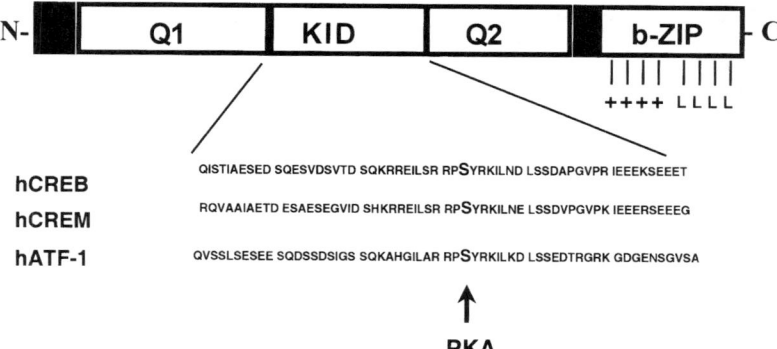

FIGURE 2. Modular structure of the transcription factors of the CREB family. Q1 and Q2 are the glutamine-rich domains of CREB. The KID is the kinase-inducible domain that is highly homologous between CREB, CREM, and ATF-1 as shown in the figure and contains the serine residue phosphorylated by PKA. The b-ZIP domain at the C terminus binds DNA and the partner of the homo- or heterodimer of transcription factors.

tive state PKA is an heterotetramer of paired regulatory and catalytic subunit (C) and is located mainly in the cytoplasm. After activation by cAMP the catalytic subunit dissociates from the regulatory subunit and diffuses into the nucleus. Subsequently, the free nuclear C subunit of PKA phosphorylates the transcription factor CREB at serine residue 133.[1,2] Expression of the *CREB* gene is ubiquitous.

The transcription factor CREB binds to the conserved consensus cAMP response element (CRE), TGACGTCA, as a dimer. A conserved palindromic CRE has been identified in the promoter of various genes regulated by cAMP (i.e., somatostatin, VIP, alphaCG, enkephalin, fos, PTH, etc.). Three proteins have been identified that bind the CRE and are phosphorylated by PKA: CREB, CREM (CRE modulator), and ATF-1.[1] While CREB and ATF-1 stimulate transcription, some isoforms of CREM can negatively regulate the CRE activity. CREB, ATF-1, and CREM are members of the bZIP superfamily of transcription factors and bind DNA as homo- or heterodimers (FIG. 2). Multiple isoforms of the CREM gene products have been described that can positively or negatively regulate gene transcription in response to cAMP. The expression pattern of CREM isoforms is modulated by cAMP and varies between tissues.[3,4]

It appears that CREB can bind the CRE even in the unphosphorylated form, but it does not stimulate transcription inasmuch as it is not phosphorylated at serine 133. CREB phosphorylation correlates well with the stimulation of transcription of CRE-containing genes. After serine 133 phosphorylation by PKA, CREB binds to CBP (CREB binding protein). Recruitment of the transcriptional coactivator CBP ultimately leads to stimulation of the transcription of cAMP responsive genes. CBP binds specifically to CREB after its phosphorylation by PKA in response to cAMP stimulation. CBP appears to be a general transcriptional coactivator for various signaling pathways.[5]

THE TRANSCRIPTION FACTOR CREB IN THE DEVELOPMENT AND CELLULAR PROLIFERATION OF ENDOCRINE TISSUES

cAMP plays an important role in cellular differentiation and has a complex action on cellular proliferation. Activation of the cAMP pathway inhibits cellular proliferation in numerous tissues, but it can also stimulate cellular proliferation in a subset of endocrine cells. For instance, cAMP increases proliferation of somatotroph and thyroid cells.[6] In these tissues, cAMP is also the well known intracellular mediator of peptidic hormones (growth hormone releasing hormone, GHRH, and thyroid stimulating hormone, TSH, respectively), which play a major role in development and differentiation. These hormones bind to seven transmembranes receptors coupled to G stimulatory (Gs) protein. Inactivating mutation of several receptors coupled to Gs, such as the GHRH, ACTH, or TSH receptors, leads to hormone resistance and somatotroph, adrenal, or thyroid deficiency, respectively. Normal function of the cAMP pathway from the cell surface to the nucleus is required for normal development of both the thyroid gland and the somatotroph cells of the pituitary.

To evaluate the physiological significance of the various components regulating the CRE, a number of transgenic models have been developed that point out the major role of cAMP-regulated transcription. CREB null mice generated by homologous recombination die just after birth from respiratory distress.[7] These animals are smaller than the controls and have central nervous system alterations (corpus callosum and anterior commissures reduction). It has been observed previously that mice lacking two of the three main splice product isoforms of CREB have a deficient long-term memory.[8] In both models of CREB knockout mice (i.e., the partial and the complete one) an overexpression of CREM and ATF-1 is observed. This suggests a compensatory mechanism between the member of the CREB/CREM/ATF-1 family. The CREB –/– mice also have impaired T cell development. Interestingly, a T cell proliferative defect and decreased interleukin 2 production are also observed in transgenic mice expressing a dominant-negative mutant of CREB under the control of the thymocyte-specific CD2 promoter.[9] In this model, the PKA phosphorylation site is mutated, and the cAMP activation of CREB is thereby abolished. The same approach demonstrated the major role of CREB in the development of the pituitary somatotroph. Transgenic mice expressing a dominant-negative mutant of CREB, which cannot be phosphorylated by PKA (CREB M1 mutated on the Ser^{133}) under the control of the GH promoter, are dwarf and present with pituitary hypoplasia due to a dramatic reduction in the number of GH cells.[10] These mice therefore present with GH deficiency. Interestingly, in these transgenic mice the dominant-negative mutant is also expressed in the lactotrophs because the GH promoter is known to be active in this pituitary cell type. Nevertheless, the development of the lactotroph lineage appears to be normal, suggesting that CREB inactivation is deleterious only for the somatotroph cells in this model.

The role of CREB appears also important in cellular proliferation of the thyroid. Transfection of a CREB mutant (KCREB) that dimerizes with endogenous wild-type CREB have been performed in the rat thyroid cell line FRTL-5. This mutant behaves in a dominant-negative fashion and inactivates the endogenous CREB, leading to a reduction in TSH-stimulated cell proliferation.[11] More recently, thyroid-targeted expression of a dominant-negative mutant of CREB have been performed in transgenic mice using the thyroglobulin promoter.[12] These mice exhibit primary hypothyroidism due to a reduced development of follicles and diminished colloid. The expression of a variety of cAMP-dependent thyroid genes was reduced in these mice, suggesting an important role for CREB and CRE binding proteins in thyroid growth and differentiation. In primary culture of dog thyroid cells, TSH induces phosphorylation of the transcription factors CREB and CREM and the expression of the CREM short isoform ICER (inducible cAMP early repressor). ICER is transcribed from an intronic promoter of the CREM gene and contains mainly a DNA binding domain without a transactivation domain. In keeping with its structure, ICER inhibits the CRE activity and its expression is stimulated by cAMP. Therefore ICER acts at the CRE level as an antagonist of CREB or the activating isoforms of CREM (i.e., CREMτ). The ectopic expression of ICER or CREM repressor isoforms (CREMα) inhibits DNA replication in dog thyroid cells.[13]

Adenovirus-mediated expression of a similar CREB-negative mutant reduces the viability of rat ovary granulosa cells, suggesting a role for CREB as a cell survival factor in the ovary.[14] Disruption of the CREM gene by homologous recombination leads to infertility in male mice due to deficient spermiogenesis without hormonal alterations of the gonadotroph axis.[15,16]

THE TRANSCRIPTION FACTORS OF THE CREB FAMILY IN ENDOCRINE TUMORS

Several diseases with endocrine gland hyperplasia or tumor and hormone hypersecretion can be secondary—that is, caused by activation of the extracellular signal that stimulates the cAMP pathway in a given tissue (FIG. 3). Eutopic or ectopic GHRH or ACTH hypersecretions cause acromegaly with somatotroph hyperplasia or Cushing's syndrome with bilateral adrenal hyperplasia. Pituitary tumors secreting TSH lead to the development of goiter and hyperthyroidism. Several molecular alterations mimicking overstimulation of the cAMP pathway by extracellular stimuli have been observed in various endocrine tumors. Somatic activating mutations of the TSH receptor causing ligand-independent activation of cAMP production by adenylyl cyclase are observed in most thyroid toxic adenoma.[17] In these tumors, activation of the cAMP pathway could be linked to both thyroid hormone hypersecretion and cellular proliferation leading to adenoma development. Germline inheritance of similar activating TSH receptor mutations have been observed in a dominant autosomic form of toxic nodular goiter. In the adrenal cortex, ectopic expression of the gastric inhibitory peptide receptor, also coupled to Gs, is associated with ACTH-independent macronodular adrenal hyperplasia or adenoma and food-dependent cortisol hypersecretion.[18–20] Alterations of the Gs protein itself have been reported more than 10 years ago in GH-secreting and thyroid tumors. These somatic activating Gs mutations (oncogene Gsp) lead to constitutive cAMP production.[21] Similar mutations have also been reported in the McCune-Albright syndrome, presenting with café au lait spot, bone fibrous dysplasia, and various endocrine tumors (GH-secreting adenomas, precocious puberty with ovarian cyst, toxic thyroid goiter, and Cushing's syndrome with ACTH-independent macronodular adrenal hyperplasia). Patients suffering from the McCune-Albright syndrome present with a mosaic distribution of the Gsp mutation.[22] Carney complex (CC) is an autosomic dominant disorder initially described as the association of myxomas, spotty skin pigmentation, and endocrine

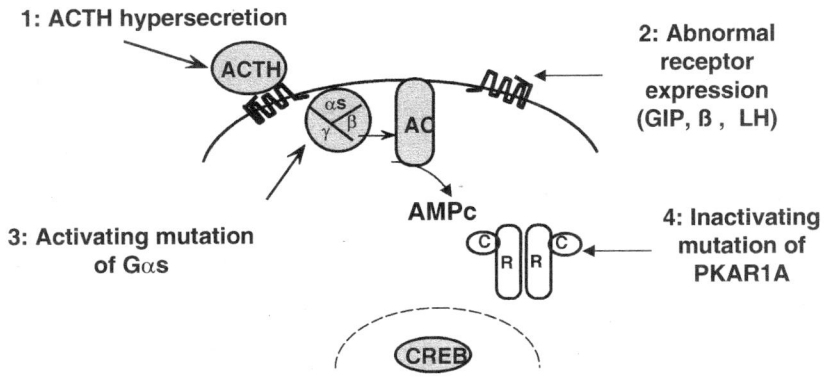

FIGURE 3. cAMP pathway activation in various causes of Cushing's syndrome. The figure shows alterations of the cAMP pathway at the extracellular (**1,2**), as well as the intracellar level (**2,3,4**), in various forms of Cushing's syndrome associated with adrenal hyperplasia, adrenal tumors, macro- or micronodular adrenocortical hyperplasia. It is important to note that **2, 3** and **4** are associated with benign tumors. **1.** Eutopic (Cushing's disease) or ectopic ACTH secretion. **2.** Abnormal expression of membrane receptor: GIP (gastric-inhibitory peptide) receptor, β-adrenergic receptor, LH receptor. **3.** Activating mutation of the α subunit of the Gs protein, termed oncogene Gsp in McCune-Albright syndrome. **4.** Inactivating mutation of the PRKAR1A gene in Carney complex.

overactivity and considered as a multiple neoplasia and lentiginosis syndrome. Various endocrine tumors are associated with the CC: GH-secreting adenomas, thyroid toxic adenomas, Sertoli cell testicular tumors, and primary pigmented nodular adrenal dysplasia with ACTH-independent Cushing's syndrome.[23] The CC shares some features of the McCune-Albright syndrome. Because of these similarities, genes involved in the cAMP pathway have been considered as putative candidates for CC. Heterozygous germline-inactivating mutations of the regulatory subunit R1A of PKA (PRKAR1A) have recently been reported in CC.[24–26]

Transfection with the constitutively activated mutant of Gαs (Gsp) permitted study of the nuclear effects of cAMP pathway alterations in endocrine tumors. Functional expression of Gsp in somatotroph cells in culture induces CREB phosphorylation at Ser^{133}. Gsp also stimulates CRE activity and CREB-dependent transcription.[27] Mutation of the Ser^{133} of CREB abolishes the transcriptional activation by Gsp. Furthermore, expression of a dominant inhibitory mutant of PKA also blocks the stimulation of gene expression by Gsp.[28–30] This suggests that the nuclear response to Gsp is mediated by PKA phosphorylation of the Ser^{133} of CREB. It is also worth noting that overexpression of the wild-type Gαs is able to stimulate CREB-dependent transcription and could therefore be another mechanism of tumorigenesis.[27,31]

Gsp stimulates the transcription of several pituitary-specific genes. For instance, the transcription from the GH, prolactin, and GHF-1/Pit-1 promoter is stimulated by Gsp.[29,30,32] The activity of the α-subunit promoter is also stimulated in transfection experiments performed in somatotroph cells. This is interesting since expression of the α-subunit is frequently observed in somatotroph tumors while it is not in normal mature GH cells. Gsp also stimulates the expression of some ubiquitous genes as the protooncogene c-fos. The transcription factors Pit-1 and c-fos are known to stimulate cell proliferation in a, respectively, cell-specific and ubiquitous manner. In keeping with the pivotal role of CREB in Gsp nuclear action, an increased Ser^{133} phosphorylation of CREB is observed in GH-secreting adenomas compared to nonfunctioning adenomas.[27] The phosphorylation status of CREB is similar in tumors harboring the Gsp mutant and those with the wild-type Gαs.[27,33] Nevertheless, it has recently been reported that ICER expression might be higher in GH adenomas with the Gsp mutation.[34] Since CREB stimulates ICER expression, this could reflect to some extent higher CREB activity in such tumors.

In the adrenal cortex, the pituitary peptide hormone adrenocorticotropin (ACTH) is the major activator of the cAMP pathway. ACTH is required for adrenal cortex activity and steroid synthesis. In humans, chronic ACTH stimulation, as observed, for instance, in Cushing's disease, leads to adrenal cortex hyperplasia and cortisol hypersecretion. Furthermore, activation of the cAMP pathway by mutation of the α subunit of G proteins (oncogene Gsp, which leads to constitutive activation of adenylyl cyclase) or seven domains transmembrane receptors coupled to Gs (i.e., gastric inhibitory peptide, LH, or β-adrenergic receptors) abnormal or ectopic expression, has been observed in cases of Cushing's syndrome with adrenal macronodular hyperplasia or adrenal tumors.[19] However, regulation of the proliferation of adrenocortical cells by ACTH and cAMP is complex, since an inhibitory role of ACTH has also been observed in some species. The H295R cell line was established from an invasive human adrenocortical carcinoma responsible for Cushing's syndrome.[35] Western blot and RT-PCR studies demonstrated that CREB is not expressed in the human adrenocortical cancer cell line H295R, while it is expressed in normal adrenal cortex.[36,37] Nevertheless, during transient transfection experiments, cAMP stimulation of two reporter genes containing canonical CRE was maintained. Cotransfection of the dominant-negative inhibitor A-CREB, which prevents transcription factors containing a CREB-like leucine zipper domain to bind DNA, completely inhibited cAMP-induced stimulation of CRE activity.[37] Western blot and RT-PCR studies showed that ATF-1, CREMα/γ, and CREMτ2α are expressed in H295R cells. High amounts of CREM proteins were present in H295R, demonstrating an overexpression of this transcription factor in the absence of CREB. Furthermore, expression of the activator isoform CREMτ was very high in H295R compared to normal adrenal cortex. Transfection assays demonstrated that CREMτ2α is a potent stimulator of CRE activity in H295R. Finally, gel retardation assays showed that CREM and ATF-1 are the nuclear proteins that specifically bind the CRE in H295R cells, while CREM binding to CRE is not observed in a CREB-expressing cell line. This loss of CREB expression and the overexpression of CREM could be linked to cellular transformation since the normal adrenal cortex expresses high levels of CREB and no or low levels of CREMτ. Interestingly, we recently observed by Western blot studies that CREB protein levels are reduced in malignant adrenocortical carcinomas by comparison with adrenal benign adenoma or normal adrenal cortex. By contrast, the level of CREB protein appears normal in adrenal cortex from patients with ACTH-dependent Cushing's syndrome or Carney complex, two diseases in which alterations of the cAMP pathway play an important role. The observation of reduced

CREB levels in adrenocortical cancer correlates with the lack of molecular alterations leading to cAMP pathway stimulation in such tumors. By contrast, a reduced expression of the ACTH receptor has been observed in adrenocortical cancer.[38] This suggests that inactivation of the cAMP pathway from the cell surface to the nucleus might be associated with a less differentiated phenotype allowing malignant growth of these very aggressive tumors.

CONCLUSION

In conclusion, the transcription factor CREB plays an important role in the development, differentiation, and proliferation of endocrine tissues. Various alterations of the CREB family of transcription factors can be observed in endocrine tumors. These alterations might be related to the various molecular defects leading to activation of the cAMP pathway in some benign and well-differentiated, slowly growing endocrine tumors. In contrast, molecular alterations stimulating the cAMP pathway are not observed in malignant endocrine adrenocortical tumors. In such endocrine tumors the loss of CREB expression might be associated with a less differentiated phenotype occurring during malignant transformation.

ACKNOWLEDGMENTS

This work was supported in part by the Plan Hospitalier de Recherche Clinique (AOM 95201 to Comete Network coordinated by Professor P. F. Plouin and dedicated to the study of adrenal tumors) and the Association pour la Recherche sur le Cancer (ARC 4225). L.G. was the recipient of a fellowship from the Association pour la Recherche sur le Cancer.

REFERENCES

1. MEYER, T.E. & J.F. HABENER. 1993. Cyclic adenosine 3',5'-monophosphate response element binding protein (CREB) and related transcription-activating deoxyribonucleic acid-binding proteins. Endocr. Rev. **14:** 269–290.
2. MAYR, B. & M. MONTMINY. 2001. Transcriptional regulation by the phosphorylation-dependent factor CREB. Nat. Rev. **2:** 599–609.
3. FOULKES, N.S., B. MELLSTROM, E. BENUSIGLIO, et al. 1992. Developmental switch of CREM function during spermatogenesis: from antagonist to activator. Nature **355:** 80–84.
4. SASSONE-CORSI, P. 1998. Coupling gene expression to cAMP signalling: role of CREB and CREM. Int. J. Biochem. Cell Biol. **30:** 27–38.
5. CHRIVIA, J.C., R.P. KWOK, N. LAMB, et al. 1993. Phosphorylated CREB binds specifically to the nuclear protein CBP. Nature **365:** 855–859.
6. ROGER, P.P., S. REUSE, C. MAENHAUT, et al. 1995. Multiple facets of the modulation of growth by cAMP. Vitam. Horm. **51:** 59–191.
7. RUDOLPH, D., A. TAFURI, P. GASS, et al. 1998. Impaired fetal T cell development and perinatal lethality in mice lacking the cAMP response element binding protein. Proc. Natl. Acad. Sci. USA **95:** 4481–4486.
8. BOURTCHULADZE, R., B. FRENGUELLI, J. BLENDY, et al. 1994. Deficient long-term memory in mice with a targeted mutation of the cAMP-responsive element-binding protein. Cell **79:** 59–68.

9. BARTON, K., N. MUTHUSAMY & M. CHANYANGAM. 1996. Defective thymocyte proliferation and IL-2 production in transgenic mice expressing a dominant-negative form of CREB. Nature **379:** 81–85.
10. STRUTHERS, R.S., W.W. VALE, C. ARIAS, et al. 1991. Somatotroph hypoplasia and dwarfism in transgenic mice expressing a non-phosphorylatable CREB mutant. Nature **350:** 622–624.
11. WOLONSHIN, P., K. WALTON, R. REHFUSS, et al. 1992. 3′,5′-cyclic adenosine monophosphate-regulated enhancer binding (CREB) activity is required for normal growth and differentiated phenotype in the FRTL5 thyroid follicular cell line. Mol. Endocrinol. **6:** 1725–1733.
12. NGUYEN, L., P. KOPP, F. MARTINSON, et al. 2000. A dominant negative CREB (cAMP response element-binding protein) isoform inhibits thyrocyte growth, thyroid-specific gene expression, differentiation, and function. Mol. Endocrinol. **14:** 1448–1461.
13. UYTTERSPROT, N., S. COSTAGLIOLA, J.E. DUMONT, et al. 1999. Requirement for cAMP-response element (CRE) binding protein CRE modulator transcription factors in thyrotropin-induced proliferation of dog thyroid cells in primary culture. Eur. J. Biochem. **259:** 370–378.
14. SOMERS, J.P., J.A. DELOIA & A.J. ZELEZNIK. 1999. Adenovirus-directed expression of a nonphosphorylatable mutant of CREB (cAMP response element-binding protein) adversely affects the survival, but not the differentiation, of rat granulosa cells. Mol. Endocrinol. **13:** 1364–1372.
15. NANTEL, F., L. MONACO, N.S. FOULKES, et al. 1996. Spermiogenesis deficiency and germ-cell apoptosis in CREM-mutant mice. Nature **380:** 159–162.
16. BERTHERAT, J. 1996. Spermiogenesis requires the transcription factor CREM (cyclic AMP-response element modulator). Eur. J. Endocrinol. **135:** 171.
17. PARMA, J., L. DUPREZ, J. VAN SANDE, et al. 1993. Somatic mutations in the thyrotropin receptor gene cause hyperfunctioning thyroid adenomas. Nature **365:** 649–651.
18. LACROIX, A., E. BOLTE, J. TREMBLAY, et al. 1992. Gastric inhibitory polypeptide-dependent cortisol hypersecretion—a new cause of Cushing's syndrome. N. Engl. J. Med. **327:** 974–980.
19. LACROIX, A., N. NDIAYE, J. TREMBLAY, et al. 2001. Ectopic and abnormal hormone receptors in adrenal Cushing's syndrome. Endocr. Rev. **22:** 75-110.
20. REZNIK, Y., V. ALLALI-ZERAH, J.A. CHAYVIALLE, et al. 1992. Food-dependent Cushing's syndrome mediated by aberrant adrenal sensitivity to gastric inhibitory polypeptide. N. Engl. J. Med. **327:** 981–986.
21. LANDIS, C.A., S.B. MASTERS, A. SPADA, et al. 1989. GTPase inhibiting mutations activate the alpha chain of Gs and stimulate adenylyl cyclase in human pituitary tumours. Nature **340:** 692–696.
22. WEINSTEIN, L.S., A. SHENKER, P.V. GEJMAN, et al. 1991. Activating mutations of the stimulatory G protein in the McCune-Albright syndrome. N. Engl. J. Med. **325:** 1688–1695.
23. CARNEY, J.A., H. GORDON, P.C. CARPENTER, et al. 1985. The complex of myxomas, spotty pigmentation, and endocrine overactivity. Medicine (Baltimore) **64:** 270–283.
24. KIRSCHNER, L.S., F. SANDRINI, J. MONBO, et al. 2000. Genetic heterogeneity and spectrum of mutations of the PRKAR1A gene in patients with the Carney complex. Hum. Mol. Genet. **9:** 3037–3046.
25. KIRSCHNER, L.S., J.A. CARNEY, S.D. PACK, et al. 2000. Mutations of the gene encoding the protein kinase A type I-alpha regulatory subunit in patients with the Carney complex. Nat. Genet. **26:** 89–92.
26. CASEY, M., C.J. VAUGHAN, J. HE, et al. 2000. Mutations in the protein kinase A R1alpha regulatory subunit cause familial cardiac myxomas and Carney complex. J. Clin. Invest. **106:** R31–R38.
27. BERTHERAT, J., P. CHANSON & M. MONTMINY. 1995. The cyclic adenosine 3′,5′-monophosphate-responsive factor CREB is constitutively activated in human somatotroph adenomas. Mol. Endocrinol. **9:** 777–783.
28. GAIDDON, C., A.L. BOUTILLIER, D. MONNIER, et al. 1994. Genomic effects of the putative oncogene G alpha s. Chronic transcriptional activation of the c-fos proto-oncogene in endocrine cells. J. Biol. Chem. **269:** 22663–22671.

29. GAIDDON, C., L. MERCKEN, C. BANCROFT, *et al.* 1995. Transcriptional effects in GH3 cells of Gs alpha mutants associated with human pituitary tumors: stimulation of adenosine 3′,5′-monophosphate response element-binding protein-mediated transcription and of prolactin and growth hormone promoter activity via protein kinase A. Endocrinology **136:** 4331–4338.
30. GAIDDON, C., J. TIAN, J.P. LOEFFLER, *et al.* 1996. Constitutively active G(S) alpha-subunits stimulate Pit-1 promoter activity via a protein kinase A-mediated pathway acting through deoxyribonucleic acid binding sites both for Pit-1 and for adenosine 3′,5′-monophosphate response element-binding protein. Endocrinology **137:** 1286–1291.
31. BERTHERAT, J. 1997. Nuclear effects of the cAMP pathway activation in somatotrophs. Horm. Res. **47:** 245–250.
32. BERTHERAT, J., M.T. BLUET-PAJOT & J. EPELBAUM. 1995. Neuroendocrine regulation of growth hormone. Eur. J. Endocrinol. **132:** 12–24.
33. PERSANI, L., S. BORGATO, A. LANIA, *et al.* 2001. Relevant cAMP-specific phosphodiesterase isoforms in human pituitary: effect of Gs(alpha) mutations. J. Clin. Endocrinol. Metab. **86:** 3795–3800.
34. PERI, A., B. CONFORTI, S. BAGLIONI-PERI, *et al.* 2001. Expression of cyclic adenosine 3′,5′-monophosphate (cAMP)-responsive element binding protein and inducible-cAMP early repressor genes in growth hormone-secreting pituitary adenomas with or without mutations of the Gsalpha gene. J. Clin. Endocrinol. Metab. **86:** 2111–2117.
35. RAINEY, W.E., I.M. BIRD & J.I. MASON. 1994. The NCI-H295 cell line: a pluripotent model for human adrenocortical studies. Mol. Cell Endocrinol. **100:** 45–50.
36. WANG, X.L., M. BASSETT, Y. ZHANG, *et al.* 2000. Transcriptional regulation of human 11beta-hydroxylase (hCYP11B1). Endocrinology **141:** 3587–3594.
37. GROUSSIN, L., J. MASSIAS, X. BERTAGNA, *et al.* 2000. Loss of expression of the ubiquitous transcription factor cAMP response element-binding protein (CREB) and compensatory overexpression of the activator CREMτ in the human adrenocortical cancer cell line H295R. J. Clin. Endocrinol. Metab. **85:** 345–354.
38. BEUSCHLEIN, F., M. FASSNACHT, A. KLINK, *et al.* 2001. ACTH-receptor expression, regulation and role in adrenocortial tumor formation. Eur. J. Endocrinol. **144:** 199–206.

The Essential Role of RIα in the Maintenance of Regulated PKA Activity

PAUL S. AMIEUX AND G. STANLEY McKNIGHT

Department of Pharmacology, University of Washington, Seattle, Washington, 98195, USA

ABSTRACT: Cloning of the individual regulatory (R) and catalytic (C) subunits of the cAMP-dependent protein kinase (PKA) and expression of these subunits in cell culture have provided mechanistic answers about the rules for PKA holoenzyme assembly. One of the central findings of these studies is the essential role of the RIα regulatory subunit in maintaining the catalytic subunit under cAMP control. The role of RIα as the key compensatory regulatory subunit in this enzyme family was confirmed by gene knockouts of the three other regulatory subunits in mice. In each case, RIα has demonstrated the capacity for significant compensatory regulation of PKA activity in tissues where the other regulatory subunits are expressed, including brain, brown and white adipose tissue, skeletal muscle, and sperm. The essential requirement of the RIα regulatory subunit in maintaining cAMP control of PKA activity was further corroborated by the knockout of RIα in mice, which results in early embryonic lethality due to failed cardiac morphogenesis. Closer examination of RIα knockout embryos at even earlier stages of development revealed profound deficits in the morphogenesis of the mesodermal embryonic germ layer, which gives rise to essential structures including the embryonic heart tube. Failure of the mesodermal germ layer in RIα knockout embryos can be rescued by crossing RIα knockout mice to Cα knockout mice, supporting the conclusion that inappropriately regulated PKA catalytic subunit activity is responsible for the phenotype. Isolation of primary embryonic fibroblasts from RIα knockout embryos reveals profound alterations in the actin-based cytoskeleton, which may account for the failure in mesoderm morphogenesis at gastrulation.

KEYWORDS: PKA holoenzyme assembly; mesoderm formation; gastrulation; cardiac morphogenesis

RIα'S ROLE AS A UNIVERSAL BUFFER AGAINST UNREGULATED PKA ACTIVITY

The cAMP-dependent protein kinase (PKA) was first described over thirty years ago, and this led to extensive characterization of the levels of regulatory (R) and catalytic (C) isoforms and the relative activation state of the kinase under a host of physiological conditions. Genetic approaches became available following the cloning of

---

Address for correspondence: Paul S. Amieux, Ph.D., Department of Pharmacology Box 357750, University of Washington, Seattle, WA 98195. Voice: 206-543-0144; fax: 206-616-4230.
pamieux@u.washington.edu

the R and C subunit genes, and this helped identify the complete family of R (RIα, RIβ, RIα, RIIβ) and C (Cα and Cβ) genes expressed in mammals. One of the first and most important phenomena observed in these studies was the substantial capacity of the PKA system to self-regulate. Overexpression of Cα or Cβ in cell culture results in significant compensation by an increase in RIα subunit protein.[1] No compensation by RIIα is observed when Cα or Cβ are overexpressed, suggesting that this phenomenon is specific to RI subunits. Although increased endogenous RIα protein did maintain the overexpressed C subunit under cAMP control, compensation could be overcome by further increasing the levels of exogenous C subunit, resulting in a pool of unregulated C subunit activity. RIβ is also capable of similar compensatory regulation, but is limited in this regard owing to its neural-specific expression in mice.[2] Intriguingly, RIα mRNA levels do not change when Cα or Cβ are overexpressed in various cell lines, suggesting a posttranscriptional mechanism of compensation.[1] Translation-rate analysis combined with pulse-chase experiments reveals that the mechanism of RIα compensation upon overexpression of C subunit is protein stabilization resulting from holoenzyme formation.[3] RIα protein has a 4–5-fold longer half-life when incorporated into holoenzyme compared with its stability as free subunit. Stabilization of RIα protein through holoenzyme formation is also supported by the observation that S49 kin⁻ mouse lymphoma cells that lack C subunit altogether have a significant reduction in steady state levels of RIα.[4] This is further corroborated by pulse-chase experiments demonstrating a 10-fold reduction in the half-life of RIα protein in these same cells.[5]

The assembly of holoenzymes *in vivo* favors the preferential formation of type II holoenzyme. Overexpression of RIIα or RIIβ in ras-transformed NIH 3T3 mouse fibroblasts completely eliminates the type I holoenzyme, resulting in a significant pool of RIα, which is rapidly turned over.[6,7] In contrast, overexpression of RIα or RIβ has no effect on type II holoenzyme formation, and a free pool of RIα or RIβ is again observed.[6,8] However, when C subunit is overexpressed in wild-type NIH 3T3 cells, which have only type II holoenzyme, type I holoenzyme now forms, suggesting that it is only when C subunit levels exceed RII subunit levels that type I holoenzyme forms.[9] The fact that RII subunits are always preferentially bound to the existing C subunit pool in the cell may explain the apparent lack of change in RII subunit levels when C subunits are overexpressed. It has been demonstrated in cell culture that when RIIβ is displaced from the C subunit by RIIα overexpression, it is rapidly turned over.[7] Furthermore, compensation by RIIα has been observed in the brain of RIIβ knockout mice, suggesting that when RII levels exceed the existing pool of free C subunit, stabilization of RII subunits may occur.[7,10]

COMPENSATION BY RIα IS RECAPITULATED *IN VIVO*

Confirmation *in vivo* of the observations made in cell culture concerning holoenzyme assembly had to await the arrival of gene knockout technology. RIβ was the first PKA subunit gene to be targeted in the mouse. Examination of cortex and hippocampus from these animals revealed significant compensation by the RIα subunit.[2] Although RIα compensation maintained all of the available C subunit under cAMP control, deficits in developmental and adult neural plasticity were still observed in various brain regions, pointing to a nonredundant biological role for the

RIβ subunit.[2,11] Previous studies looking at biochemical differences between RIα and RIβ had also strongly suggested that the RIβ subunit might form a holoenzyme that would be preferentially activated by cAMP at low concentrations.[12] Targeting of RIIβ revealed an even more striking example of compensation by RIα in brown and white adipose tissue (BAT and WAT, respectively).[13] The loss of RIIβ in BAT and WAT results in a lean phenotype in mice and complete holoenzyme reversal from RIIβ- to RIα-containing holoenzyme. Importantly, replacement of RIIβ by RIα causes a significant increase in basal PKA activity, resulting in increased UCP-1 (uncoupling protein 1) expression in BAT and altered lipolysis rates in WAT. The lower K_a of activation observed in the RIα holoenzyme versus the RIIβ holoenzyme in BAT and WAT is the likely cause of the increased basal PKA activity. Once again, fundamental differences are observed in holoenzymes formed from different R subunits in the same tissue, and these holoenzyme differences most likely involve biochemical differences in R subunits and differences in R subunit localization within the cell. In BAT and WAT, a model similar to that developed in cell culture is supported, wherein a free pool of RIα exists *in vivo*, which turns over rapidly while all available C subunit is preferentially bound to RIIβ.[3] Elimination of the major RII subunit expressed in BAT and WAT allows for the free pool of RIα to compete for available C subunit and thus be stabilized as RIα holoenzyme. Targeted disruption of RIIα, a ubiquitous R subunit like RIα, would be predicted to have profound consequences *in vivo* due to its widespread expression in all tissues. Surprisingly, RIIα mice are viable and fertile and show no obvious phenotype. Once again, however, significant compensation by RIα is observed in tissues known to have RIIα-specific functions.[14,15]

SPECIFIC LOCALIZATION OF C SUBUNIT BY RIα SUGGESTS ADDITIONAL ROLES FOR RIα BEYOND SERVING AS A UNIVERSAL BUFFER AGAINST UNREGULATED PKA ACTIVITY

For years the general notion concerning RI- versus RII-containing holoenzymes was that RII holoenzymes were localized and thus found in the particulate fraction of cellular homogenates, and RI subunits were delocalized and thus found in the cytoplasmic fraction of cellular homogenates. The literature supporting localization of RII subunits to particular organelles via the A kinase anchoring proteins (AKAPs) is extensive;[16] however, evidence has accumulated over the past several years that RIα can also be localized. AKAPs capable of binding RIα or RIIα have been described,[17,18] and AKAP-mediated localization of RIα is supported by *in vivo* studies looking at localization of RIα-containing PKA to the neuromuscular junction (NMJ) in skeletal muscle.[19,20] The physiological relevance of these findings is supported by the observation in RIIα knockout mice that RIα can functionally replace RIIα in localizing PKA to L-type calcium channels in skeletal muscle.[14] RIα has also been localized to interphase microtubules and specific regions of the mitotic spindle, and this interaction is also believed to be AKAP mediated.[21] The discovery in *C. elegans* of a specific RI-binding AKAP that does not interact with RII subunits suggests the potential for nonredundant type I PKA localization and function in cells.[22,23] Indeed, a specific role for localized RIα has been demonstrated in human T lymphocytes, where type I PKA localizes to the activated TCR complex and is required for attenuation of signals propagated through this complex.[24,25] The importance of type I PKA–mediated effects in attenuation of T cell replication has led to its consideration as a therapeutic target in combined variable immunodeficiency (CVI) and acquired immune deficiency syndrome (AIDS).[26] Furthermore, type I PKA in T cells may also serve as a potential therapeutic target in systemic lupus erythematosis (SLE).[27,28]

ROLES FOR PKA IN VERTEBRATE DEVELOPMENT

Given PKA's widespread expression in vertebrates,[29] and its multifunctional properties,[30] it would not be surprising if PKA were found to play a role during development. The vast majority of research done on the PKA system has been in adult tissues looking at the PKA-dependent regulation of metabolism in skeletal muscle, liver, adipose, and other tissues.[31] This is understandable considering its original description as a regulator of glycogen metabolism. Some interesting early reports do exist, though, on changes in PKA isozymes during the development and differentiation of tissues.[32]

PKA Is Required at Multiple Points in Oocyte Maturation

The observation that progesterone potently induced oocyte maturation *in vivo* and *in vitro* along with increasing protein phosphorylation led to the hypothesis that a classical hormone–second messenger system might be operating during oocyte maturation. It was later shown that this progesterone-induced increase in protein phosphorylation correlated with the time of GVBD (germinal vesicle breakdown) and the actions of MPF (mitosis promoting factor).[33,34] In an elegant series of experiments, Maller and Krebs demonstrated that injection of purified rabbit skeletal muscle catalytic subunit potently inhibited germinal vesicle breakdown and subsequent oocyte maturation.[33] In the converse experiment, injection of purified bovine heart type II regulatory subunit potently activated germinal vesicle breakdown and oocyte maturation. These results indicated that elevated PKA activity was necessary to maintain the normal prophase block of the oocyte. It was hypothesized that progesterone acted at the level of adenylyl cyclase and a phosphodiesterase to lower cAMP levels and thus promote increased association of R and C subunits. In a subsequent study, Maller and Krebs demonstrated that cAMP levels dropped 40 to 60 percent from basal within one minute of progesterone treatment.[35] They also showed that cyclohexamide added to the oocytes prior to injection of a heat-stable cAMP-dependent protein kinase inhibitor protein (PKI) totally blocked the ability of this inhibitor to induce maturation, indicating that the cAMP-dependent protein kinase inhibition was acting on an early step before completion of the protein synthesis necessary for maturation. From these observations, Maller and Krebs constructed a model wherein prophase arrest of the oocyte is maintained directly or indirectly by a phosphoprotein that is a substrate for the cAMP-dependent protein kinase, and this phosphoprotein was necessary and sufficient to control release of prophase arrest. They suggested that the decrease in the amount of this key phosphoprotein could result from either increased dephosphorylation or decreased phosphorylation, but the effect of removing the inhibitor was to stimulate the synthesis of an initiator protein needed for the indirect activation of stored MPF.[36]

Subsequent studies have demonstrated that the model developed by Maller and Krebs in 1980 was essentially correct, although at that time cyclins and the *cdc* kinases were still unknown. PKA activity acts at multiple points to inhibit oocyte maturation.[37] PKA activity inhibits the synthesis of the *mos* protooncogene product (a MAPKKK) and the accumulation of cyclin B1. These proteins are required during meiosis I for the activation of mitogen-activated protein kinase (MAPK) and maturation promoting factor (MPF), respectively.[38] PKA activity also inhibits the dephosphorylation of MPF ($p34^{cdc2}$) on threonine 14 and tyrosine 15 by the *cdc25* protein phosphatase, which is required for MPF activation.[37] PKA inhibition of oocyte maturation is apparently conserved in mammals, as modulation of PKA activity in mouse oocytes recapitulates what is observed in *Xenopus* and *Rana* oocytes.[39] Furthermore, it appears that in the mouse oocyte type I PKA activity is the most effective at inhibiting germinal vesicle breakdown and oocyte maturation.[40] Intriguingly, more recent studies have demonstrated that increased PKA activity is also required for inhibition of cdk2, which results in ubiquitin-mediated degradation of the cyclin B component of the cdk1-cyclin B complex and exit from mitosis into interphase.[41,42]

PKA Activity Is Required for Activation of the Zygotic Genome in the Two-Cell Mouse Embryo

Unlike *Xenopus* and *Drosophila*, developing mouse embryos activate the zygotic genome at an extremely early stage in their development—the one- to two-cell transition. Intriguingly, the timing of transcriptional activation is independent of DNA synthesis, cell division, or the nucleo/cytoplasmic ratio.[43,44] Activation of the zygotic genome is marked by the appearance of a set of proteins called the *transcription-requiring proteins* (TRPs), which are sensitive to the RNA polymerase II inhibitor, amanitin. Transplantation of two-cell amanitin-treated nuclei to one-cell mouse embryos reveals that the cytoplasm becomes transcriptionally permissive at the late one-cell stage before the first cell division.[45] N^6-monobutyryl cAMP treatment of early one-cell stage embryos that received H89-treated two-cell nuclei resulted in the premature activation of TRP synthesis.[46] Mouse one- and two-cell embryos contain significant amounts of PKA activity, which can be inhibited by PKI; and pharmacological modulation of this PKA activity causes significant changes in the phosphorylation pattern of proteins during the one- to two-cell transition in which the zygotic genome is activated.[44,47] Using three mechanistically distinct PKA inhibitors (H89, PKI, and (R_p)-cAMPs), Poueymirou and Schultz demonstrated that although cleavage of one-cell embryos to the two-cell stage occurred normally, synthesis of TRPs was blocked.[47] Experimental evidence supports the concept that this inhibition is at the level of transcription, as capping, splicing, and polyadenylation of mRNA were unaffected.[48] Furthermore, phosphoprotein analysis of two-cell embryos revealed that H89 selectively inhibited the synthesis of a specific 14-kDa protein. From experiments performed on one- and two-cell–stage mouse embryos, the following model has been suggested.[46] The capacity for gene transcription in the early embryo involves a two-step process, and protein phosphorylation may be required at either or both of these steps. The first step corresponds to the cAMP-mediated transition from a nonpermissive state to a permissive state during the one-cell stage before zygotic gene activation. The second step involves the PKA-dependent activation and/ or appearance of one or more transcription factors leading to the onset of transcription at the two-cell stage. The acquisition of a transcriptionally permissive state prior to genome activation at the two-cell stage combined with the changes in the pattern of protein phosphorylation between the one- and two-cell stage strongly suggests that PKA-dependent phosphorylation events critically regulate activation of the zygotic genome.

PKA Negatively Regulates Sonic Hedgehog *Signaling in Flies and Vertebrates*

Aside from the work being done on oocytes and one- and two-cell mouse embryos, virtually no literature existed on the role of PKA in vertebrate development until the mid-1990s. The key to unlocking a very important role for PKA in embryogenesis was the cloning and isolation of *Drosophila* mutants of the *DCO* gene, which encodes the catalytic subunit of the cAMP-dependent protein kinase.[49] Isolation of a series of mutations in the *DCO* gene, including various hypomorphic alleles, revealed a variety of developmental defects.[50] Adult females heterozygous for a strong and a weak *DCO* allele failed to lay eggs and showed a striking novel defect in oogenesis, which included formation of egg chambers containing multinucleate nurse cells. Females heterozygous for two weak *DCO* alleles produced offspring showing a variety of defects in embryogenesis, including preblastoderm arrest and alterations in cuticular patterning. Embryos zygotically null for *DCO* died as morphologically normal first-instar larvae, implying that maternally encoded protein was sufficient for embryogenesis. Finally, animals hemizygous for weak *DCO* alleles survived for several days as larvae but grew very slowly.

Using mosaic analysis with *DCO* mutant clones, researchers demonstrated that induction of PKA-deficient clones in the anterior compartment of the *Drosophila* wing, limb, and eye imaginal discs resulted in dramatic pattern duplications, while PKA-deficient clones induced within the posterior compartment or at the anterior-posterior border showed no effects on imaginal disc patterning.[51–55] These pattern alterations extended beyond the boundaries of the PKA-deficient clones, affected the growth and fate of neighboring cells, and were similar to pattern alterations associated with ectopic expression of the *Drosophila* segment-polarity gene *hedgehog*. Reduced PKA activity in these anterior imaginal disc cells led to cell-autonomous induction of *decapentaplegic*, *wingless*, and *patched* transcription, which was independent of *hedgehog*. Furthermore, expression of a mouse constitutively active catalytic subunit in the *DCO* null mutant clones compensated for the loss of endogenous PKA activity but did not block normal *hedgehog* signaling, nor did it rescue *patched* mutant clones in the anterior compartment.[51,53] In a separate experiment it was demonstrated that localized inhibition of PKA activity using a mouse dominant-negative regulatory subunit in the anterior compartment of the imaginal discs also results in pattern duplications resulting in ectopic wings and cuticular outgrowths.[53] Furthermore, localized PKA inhibition using the mouse dominant-negative regulatory subunit in a *hedgehog* null mutant background could substitute for *hedgehog* to promote disc outgrowth, suggesting a role not only in pattern specification but also in growth and proliferation.[53] From this large body of experimental evidence, a model was proposed in which PKA activity, either directly or indirectly, acts cell autonomously to suppress *hedgehog* signaling. In the normal scenario, *engrailed* expression in the posterior imaginal disc compartment induces *hedgehog* expression only in the posterior compartment. *Hedgehog* expressed in the posterior compartment acts at the anterior-posterior compartment boundary to induce expression of *wingless* and *decapentaplegic*, which organize cell proliferation and patterning in both compartments. *Patched* and PKA activity act cell autonomously to inhibit expression of *hedgehog* target genes like *wingless* and *decapentaplegic* and thus prevent ectopic organizing centers.

Given the importance of the *hedgehog* signaling pathway in organizing the development of body segments, limbs, eyes, and other organ systems in *Drosophila*, the demonstration that the *Drosophila* segment-polarity gene *hedgehog* was also conserved in vertebrates led to an explosion of interest in *hedgehog* as a central regulator of embryonic development. The cloning of multiple *hedgehog* homologues in *Xenopus*, zebrafish, chick, and mouse combined with the observation that these genes are expressed in key embryonic organizing centers such as the node, notochord, and ZPA (zone of polarizing activity) further supported the hypothesis that *hedgehog* homologues are of paramount significance in patterning the vertebrate embryo.[56–59] *Hedgehog* homologues have been demonstrated to play a variety of roles in ventral midline signaling, including specification of ventral fates in the neural tube, sclerotomal and myotomal fates in the somite, proximal fates in the eye and ventral midbrain, and anterior-posterior patterning in the limb bud.[60] Using a dominant-negative regulatory subunit and a constitutively active catalytic subunit of the cAMP-dependent protein kinase,[61,62] two groups have demonstrated in zebrafish that inhibition of PKA activity results in phenotypes that mimic ectopic *hedgehog* expression.[63,64] Expression of a mouse dominant-negative regulatory subunit of the cAMP-dependent protein kinase in zebrafish phenocopies the effect of ectopic expression of *Sonic* and *Indian hedgehog*, including expansion of proximal fates in the eye, ventral fates in the brain, and adaxial (close to the midline) fates in the somite and head mesenchyme.[64] Ectopic expression of the mouse dominant-negative regulatory subunit of PKA partially rescued optic stalk defects in *no tail* and *cyclops* mutants that lack midline structures, which normally synthesize *Sonic hedgehog*. In the converse experiment, expression of a mouse constitutively active catalytic subunit in zebrafish blocks all of the fates induced by ectopic expression of *hedgehogs* or the dominant-negative regulatory subunit. Furthermore, expression of the dominant-negative regulatory subunit activated ectopic expression of genes known to be regulated by *hedgehog*, including zebrafish *patched 1*, *myoD*, *nkx2.2*, and *pax(b)*.[63,65] In contrast, injection of zebrafish embryos with a mouse constitutively active catalytic subunit blocked the expression of all of these *hedgehog* target genes. The findings in zebrafish support the hypothesis that in flies and vertebrates PKA acts as a common negative regulator of *hedgehog* signaling. The *hedgehog* gene family provides the first clear example of a conserved signaling factor that regulates analogous processes in species of different phyla, in this case a dipteran (*Drosophila*) and vertebrates.[60]

Support for the role of *hedgehog* in midline signaling in mammals comes from several lines of evidence. First of all, when *Sonic hedgehog* from mouse and *Vhh-1* from rat were first cloned, their expression patterns were in several key vertebrate organizing centers, including the node, notochord, floorplate, and the posterior limb bud mesenchyme (ZPA).[56,59] Pharmacological manipulation of PKA activity using forskolin and cAMP analogues completely blocked the ability of *Sonic hedgehog* from the floorplate to induce dopaminergic neurons in rats.[66] Furthermore, pharmacological manipulations of PKA activity using forskolin, IBMX, and dibutyryl-cAMP in mouse presomitic explants blocked the induction of the sclerotomal markers *Pax1* and *Mtwist*, which were normally induced by the purified *Sonic hedgehog* N terminal cleavage product.[67] It was also demonstrated that agents that increase PKA activity also block the mitogenic effects of *Sonic hedgehog* in these mouse presomitic mesoderm explants. This inhibition of *hedgehog*-mediated cellular proliferation was also seen in the *Drosophila* wing imaginal disc, but in that case it was through inhibition of PKA activity with the dominant-negative regulatory subunit.[53] Increased PKA activity not only blocked the induction by purified *Sonic hedgehog* of the sclerotomal markers *Pax1* and *Mtwist*, but also restored expression of the dermomyotomal marker *Pax3*, which is normally repressed by addition of purified *Sonic hedgehog* to the presomitic mesoderm explants.[67] Using transgenic expression of the mouse dominant-negative regulatory subunit driven by the *Wnt1* enhancer, others have demonstrated ectopic activation of the *hedgehog* signaling pathway in the dorsal CNS of the mouse.[68] Transgenic expression of the dominant-negative regulatory subunit mimics the effects of ectopic expression of *Sonic hedgehog*.[56,69] This includes the dorsal activation of floorplate markers *HNF3b* and *Sonic hedgehog*, motor neuron induction, and induction of *Ptc* and *Gli*, two vertebrate homologues of *Drosophila* genes involved in the regulation and transduction of the *hedgehog* signaling pathway.[68] Finally, a targeted disruption of the *Sonic hedgehog* gene in the mouse completely recapitulates the experimental observations made in other vertebrates like zebrafish. This includes cyclopia, holoprosencephaly, axial skeleton defects, and a loss of ventral fates in the brain and spinal column.[70] Surprisingly, targeted disruption of the Cα catalytic subunit of PKA in mice does not result in ectopic *Sonic hedgehog* expression and the resultant phenotypes, including the small eye phenotype, expanded axial skeleton and muscle pioneers, and expansion of ventral midline fates in the CNS (B.S. Skalhegg *et al.*, manuscript submitted). The absence of ectopic *Sonic hedgehog* phenotypes is due to compensation by the Cβ1 subunit during embryogenesis. Indeed, it is only when successive alleles of Cα and Cβ are deleted in mice that significant patterning defects are observed (Y.S. Huang *et al.*, manuscript submitted). Mice with targeted disruptions of Cα and Cβ that retain only one Cα or Cβ1 allele (Cα$^{+/-}$Cβ1$^{-/-}$ or Cα$^{-/-}$Cβ1$^{-/+}$) show a dramatic expansion of the neural tube, which is highly regionalized and posterior to the forelimb (FIG. 1). This expansion of the neural tube in the thoracic and lumbar region is 100% penetrant in Cα$^{+/-}$Cβ1$^{-/-}$ and Cα$^{-/-}$Cβ1$^{-/+}$ mice and results in animals born with spina bifida. Twenty-five percent of Cα$^{-/-}$Cβ1$^{-/+}$ also exhibit a failure in anterior neural tube closure (exencephaly). Consistent with the existing body of evidence concerning the interplay between *Sonic hedgehog* and PKA signaling pathways, Cα$^{+/-}$Cβ1$^{-/-}$ and Cα$^{-/-}$Cβ1$^{-/+}$ mice show an expansion of ventral neural tube markers including *Sonic hedgehog*, *HNF3β*, *Islet 1/2*, and *Nkx 2.2* (FIG. 1). Conversely, dorsal neural tube markers like *Pax 6* and *Pax 7* are displaced more dorsally or not present at all. Surprisingly, sclerotomal derivatives of the somite including the vertebral arches of the axial skeleton appear perfectly normal in the regions where Cα$^{+/-}$Cβ1$^{-/-}$ and Cα$^{-/-}$Cβ1$^{-/+}$ mice manifest neural tube defects, suggesting important differences in the contribution of PKA to developmental signaling in the mesoderm-derived somite versus the ectoderm-derived neural tube. The neural tube defects ob-

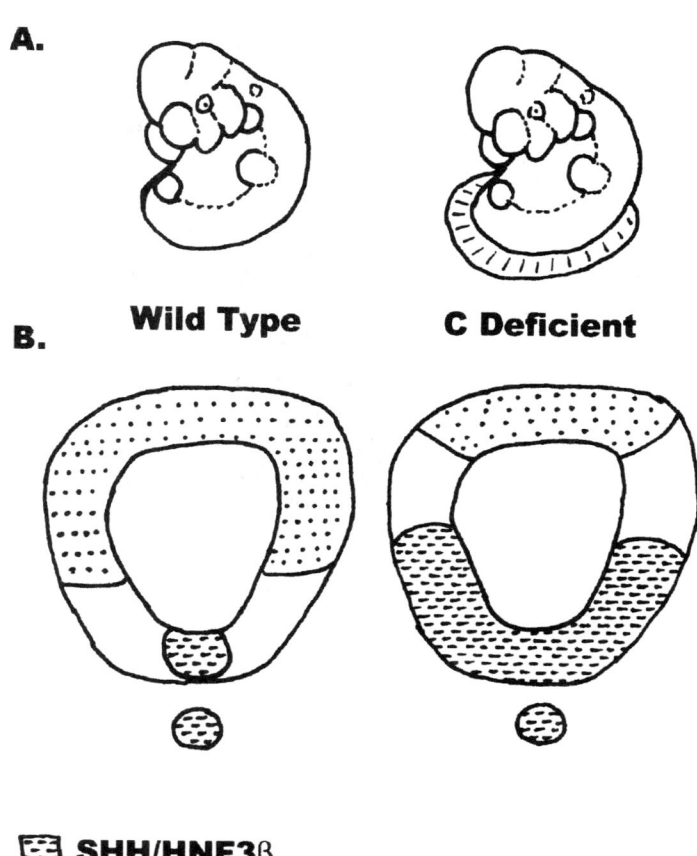

FIGURE 1. (**A**) Cartoon showing the expansion of the neural tube posterior to the forelimb in an E10.5 C-deficient embryo. (**B**) Cartoon showing expansion of ventral neural tube markers posterior to the forelimb in an E10.5 C-deficient embryo.

served in $C\alpha^{+/-}C\beta1^{-/-}$ and $C\alpha^{-/-}C\beta1^{-/+}$ mice are cell autonomous and not dependent on the *Sonic hedgehog* ligand. Generation of $C\alpha^{+/-}C\beta1^{-/-}$ and $C\alpha^{-/-}C\beta1^{-/+}$ mice on the *Sonic hedgehog* null mutant background still results in neural tube expansion and ventralization but also partially rescues the holoprosencephaly and axial midline defects observed in *Sonic hedgehog* mutant mice. These observations support the conclusion that PKA activity in vertebrates acts intracellularly to antagonize *Sonic hedgehog* signaling. The most recent studies clearly demonstrate that PKA activity directly regulates the proteolytic processing of the *Cubitus interruptus/Gli* family of transcription factors responsible for inducing *Sonic hedgehog* target genes.[71–77] PKA-mediated phosphorylation of specific *Gli* family transcription factors results in proteolytic processing generating a transcriptionally repressive form of *Gli*; lowering PKA activity alters the proteolytic processing of *Gli* resulting in a transcriptionally active form of *Gli* capable of inducing *Sonic hedgehog* target genes.

As one would predict, the complete loss of all PKA activity ($C\alpha^{-/-}C\beta1^{-/-}$ mice) results in early embryonic lethality (Y.S. Huang *et al.*, manuscript submitted). Given the wide-ranging and essential roles for PKA in DNA replication, cell cycle, and activation of the zygotic genome, it is no surprise that embryos that possess no PKA activity fail early in development.[41,47,78,79] Conversely, as one might also predict, too much PKA activity early in development also results in the failure of embryos at one of the critical early stages in mouse development.

PKA ACTIVITY NEGATIVELY REGULATES MESODERM FORMATION IN THE DEVELOPING MOUSE EMBRYO

The targeted disruption of RIα resulted in a surprising phenotype given the phenotypic consequences of targeting the other three regulatory subunits (RIβ, RIIβ, and RIIα).[2,13,14] Targeted disruption of RIα in mice leads to failure of embryos at one of the earliest stages of development.[80] The significance of tissue-restricted regulatory subunit expression as in the case of RIβ and RIIβ combined with the biochemical capacity for compensation by RI versus RII subunits helps to elucidate the causes for embryonic lethality of RIα knockout embryos and also explains the viability of the other regulatory subunit knockouts.

Western analysis of E8.5 mouse embryos reveals the presence of RIα and RIIα as the main regulatory subunits expressed during early embryogenesis (P.S. Amieux, unpublished results). Kinase assays on E8.5 RIα knockout embryos demonstrate a 40% loss of total PKA activity and a 3- to 4-fold increase in basal PKA activity when compared to wild-type and RIα heterozygote animals. Given the significant changes in basal and total PKA activity, the compensatory mechanisms available at these early stages of embryogenesis appear quite limited and explain the devastating consequences of the loss of RIα.

Examination of RIα knockout embryos from E7.5 to E10.5 reveals significant growth retardation and developmental delay. The anterior-posterior axis of RIα mutant embryos is well defined, including all axial, paraxial, and lateral plate derivatives;[80] however, all of these structures are reduced in size. One of the prominent features of the E9.5 RIα mutant embryo is the absence of a definitive heart tube. The cardiogenic plate is present, and bilateral aggregations of mesoderm are observed; but these bilateral aggregations of mesoderm have failed to fuse in the midline and form the definitive heart tube. The absence of the embryonic heart in RIα mutant embryos by E9.5 explains the rapid loss of these embryos by E10.5, as the heart is the first organ to become functional and is critical in establishing the embryonic yolk sac circulation.[81–83]

The profound deficiencies in mesoderm-derived structures, including the head mesenchyme, branchial arches, heart, and somites suggested an earlier deficit in gastrulation—an early developmental process where an epithelial-to-mesenchymal transition occurs, forming the third germ layer (mesoderm) and involving significant cell migration and proliferation.[81] By the end of gastrulation in the mouse, the embryo will go from 600 to 18,000 cells and progress from a layer of epiblast and primitive endoderm to possessing all three germ layers (definitive endoderm, mesoderm, and ectoderm).[81] Examination of RIα mutant embryos at E6.5 (gastrulation) reveals a significant reduction in the generation of mesodermal cells and aberrant movement of nascent mesoderm away from the embryological structure known as the primitive streak. These deficiencies are reflected two days later in development by greatly reduced *brachyury* and *tbx6* mRNA expression in the primitive streak and an accumulation of mesodermal cells that have failed to migrate out laterally and anteriorly.[80]

The process of gastrulation in vertebrates is critically dependent on growth factors, specifically the fibroblast growth factor (FGF) family of ligands and receptors.[84,85] Targeted disruptions of FGF4, FGF8, and FGFR1 all result in failed gastrulation of mouse embryos.[86–89] In the case of FGFR1, deficiencies in mesodermal derivatives, including head mesenchyme, branchial arches, heart, and somites, are also observed. Targeted disruption of the Shp2 tyrosine phosphatase or the FGFR1 signaling adaptor protein FRS2 results in similar deficiencies in mesodermal structures.[90–92]

The observation that receptor tyrosine kinases are critical for mesoderm formation combined with the substantial literature on PKA-dependent inhibition of receptor tyrosine kinase signaling led to the hypothesis that increased PKA activity in the RIα mutants might be antagonizing growth factor–dependent signaling in the primitive streak. This possibility is consistent with the greatly reduced *brachyury* and *tbx6* whole-mount *in situ* staining in the primitive streak of RIα mutants and the deficiencies in anterior mesoderm–derived structures.[80] The accumulation of nascent mesoderm at the base of the primitive streak also suggests deficits in integrin-dependent cell migration over the endoderm, which also depends upon growth factor receptor tyrosine kinases and extracellular matrix signals.[93] Targeted disruption in mice of several key proteins involved in integrin-mediated signaling and migration, including fibronectin and focal adhesion kinase (FAK), result in the failure of embryos at gastrulation.[94–96] FAK-deficient and fibronectin-deficient embryos bear a striking resemblance to each other and to RIα mutant embryos with regard to such criteria as growth retardation, developmental delay, failure of cardiac morphogenesis, and anterior mesoderm deficits. The similarities in the phenotypes of FAK, fibronectin, and RIα mutant embryos combined with the literature describing PKA-dependent regulation of the actin cytoskeleton support the hypothesis that growth factor/integrin signaling may be affected (TABLE 1).[9,97] Activation of PKA causes dephosphorylation of paxillin via increasing a tyrosine phosphatase activity.[98] More recent evidence demonstrates that PKA can phosphorylate and activate shp2, a tyrosine phosphatase that localizes to focal adhesion complexes and dephosphorylates both paxillin and FAK.[99,100] Related studies reveal that the p21-activated kinases (PAKs) are also critical for the integration of growth factor and integrin–extracellular matrix signals with the actin cytoskeleton.[101,102] Significantly, PKA activity also negatively regulates PAK and thus interferes with its ability to mediate anchorage-dependent growth factor responses.[103] Isolation of primary embryonic fibroblasts from E8.5 RIα mutant embryos reveals a profound disruption of the actin-based cytoskeleton. These fibroblasts bear a striking resemblance to FAK-deficient and shp2-deficient primary embryonic fibroblasts, including a dramatic increase in focal ad-

TABLE 1. Targeted mutations in mice resulting in embryonic phenotypes similiar to the RIα knockout

Targeted gene	Phenotype	Reference
FGF4/FGF8	epithelial-to-mesenchymal transition occurs in primitive streak, but mesoderm and endoderm cells do not migrate away from streak	88
FGFR1	cells in primitive streak fail to make epithelial-to-mesenchymal transition, cells accumulate in primitive streak, anterior mesoderm deficits	86,87
FRS-2α	postgastrulation lethality, severe developmental retardation migration deficits	90
SHP-2[a]	anterior mesoderm deficits, phenotype similar to FAK and fibronectin-deficient embryos, failure to make epithelial-to-mesenchymal transition, cells accumulate in the primitive streak	91,92
Fibronectin	severe defects in mesoderamally derived tissues, mesodermal tissues like heart form and express appropriate markers but do not coalesce properly	136,137,138
Focal adhesion kinase[b]	decreased mobility of mesodermal cells, increased focal adhesions, general deficiency in mesoderm	94, 96, 139–141
NIK[c]	failure of mesodermal and endodermal cells from anterior primitive streak to migrate to correct location, accumulation of cells at the primitive streak	142
SRKs(Src,Fyn,Yes)[d]	severe growth retardation, impaired cell migration	143,144

[a]Phosphorylated and activated by PKA.[100]
[b]FRNK (C terminal separately expressed noncatalytic domain of FAK[125]) phosphorylated by PKA.[145]
[c]Ste20 family member PAK is phosphorylated and inhibited by PKA.[103]
[d]Csk (COOH-terminal Src kinase) phosphorylated and activated by PKA.[126]

hesions and condensed F actin aggregation at the cell periphery.[96,104] Expression of a dominant-negative shp2 tyrosine phosphatase prevents paxillin and FAK dephosphorylation and results again in increased focal adhesions and impaired migration,[99] suggesting that successive cycles of phosphorylation/dephosphorylation are required for directed cell migration and that PKA may also be involved in this process.

In order to confirm the hypothesis that increased/inappropriately regulated PKA activity inhibits growth factor–mediated mesoderm formation in RIα mutant embryos, RIα heterozygous mice were crossed to Cα heterozygous mice to generate RIα knockouts on the Cα heterozygous and knockout background. RIα$^{-/-}$Cα$^{+/-}$ embryos displayed an intermediate phenotype and demonstrated significant rescue of mesoderm-derived structures including head mesenchyme, branchial arches, trunk and somite mesoderm, and formation of the embryonic heart tube. Surprisingly, RIα$^{-/-}$

$C\alpha^{-/-}$ embryos at E9.5 showed a near-complete rescue of all the mesoderm-derived structures and were essentially indistinguishable from wild-type embryos despite a drastically reduced level of PKA activity.[80] $RI\alpha^{-/-}C\alpha^{-/-}$ animals are still not viable and display later deficits in development that have not been well characterized. It is of interest to note that crossing $RI\alpha$ mutants to $C\alpha$ mutants confirms the hypothesis that it is increased/inappropriately regulated PKA activity that is inhibiting growth factor–mediated mesoderm formation and strongly supports the suggestion that PKA activity negatively regulates this process during gastrulation in the mouse embryo. These observations also suggest that endogenous PKA-coupled ligands and receptors might exist during these very early stages of development that could modulate growth factor–dependent signals.[105,106]

AN INTEGRATED PICTURE OF RIα AND PKA ACTIVITY IN DEVELOPMENT AND DISEASE

A series of recently published articles has uncovered the first human disease mapping to a PKA subunit–Carney complex.[107,108] Carney complex (CNC) is a multiple neoplasia syndrome characterized by spotty skin pigmentation, cardiac and skin myxomas, endocrine tumors, and psammomatous melanotic schwannomas. CNC maps to two genomic loci, 17q24 and 2p16. Familial cases mapping to the 17q24 locus reveal deletions/mutations in the RIα coding exons leading to frameshifts and premature stop codons—no mRNA and protein from the mutant alleles has been observed. In some cases it appears that haploinsufficiency due to a single mutant RIα allele is sufficient for the development of Carney complex neoplasias;[107] however, in other cases loss of heterozygosity due to somatic mutation of the remaining wild-type allele is clearly observed.[108,109]

How do we integrate what has been learned from targeted disruption of RIα in mice with Carney complex in humans? Targeted disruption of RIα in mice leads to increased basal PKA activity and inhibition of growth factor–dependent proliferation and migration of mesoderm at gastrulation. Loss of RIα in adult humans leads to increased proliferation of cells in specific tissues. The answer to these opposite effects of increased/inappropriately regulated PKA activity in nascent mesoderm versus adult endocrine and other tissues clearly lies in how this PKA activity interfaces with other signaling systems in particular cell types.[110–112] Past research has clearly demonstrated fundamental differences in the way different cell types respond to increased PKA activity with respect to mitogenesis.[112–114] In Sertoli cells, adrenocortical cells, thyrocytes, somatotrophs, Schwann cells, and melanocytes, endogenous cAMP/PKA-coupled ligands (FSH, ACTH, TSH, GHRH, neuregulin, and MSH) can potently stimulate mitogenesis.[115–120] It is certainly not coincidental nor surprising that cell types with a well-characterized mitogenic response to endogenous cAMP/PKA-coupled ligands are precisely the ones manifesting neoplastic growth in Carney complex patients heterozygous or lacking RIα altogether.[109] This observation also strongly supports the conclusion that loss of RIα in Carney complex patients leads to increased rather than decreased PKA activity, even though basal PKA activity appears unchanged in tumors from these patients.[107,108] Increased PKA activity is also supported by the increased skin pigmentation seen in Carney complex patients, as the genes required for melanin synthesis are directly regulated by PKA.[121] The unregulated PKA activity observed in endocrine and other tissues from Carney complex patients lacking RIα is consistent with the unregulated PKA activity observed in mouse embryos lacking RIα. These observations support the general conclusion that RIα is essential in the PKA system for maintaining appropriate control of PKA activity and is unique in this regard when compared to the other three regulatory subunits.[3]

The development of cardiac and skin myxomas in Carney complex patients suggests an as-yet-uncharacterized mechanism for PKA-dependent proliferation in these pluripotent primitive mesenchymal cells that can give rise to epithelial, hematopoietic, and muscle cell types.[122] A substantial body of evidence supports the hypothesis that in cell types of mesodermal origin, including vascular aortic smooth muscle cells, adipocytes, lymphocytes, and primary embryonic fibroblasts, PKA activity can potently inhibit growth factor–/receptor tyrosine kinase–mediated cell proliferation.[123–126] How do we explain the increased mitogenesis in mesoderm-derived cells from Carney complex patients? The answer to this question may lie in differences between acute versus chronic increases in PKA activity in these mesenchymal cells. The literature describing PKA-dependent antagonism of growth factor signaling and mitogenesis is invariably on an acute timescale; however, it is clear from a variety of studies that PKA activity can elicit differential effects depending on the developmental stage of a tissue and the duration of PKA activation in that tissue.[116,120,127,128] Changes in basal PKA activity can elicit profound biological consequences, including promoting or inhibiting synthesis and assembly of cell cycle components like cyclin B1 in oocytes, cyclin D1 in Schwann cells, cyclin D1/D3/p27^{kip1} in thyroid follicular cells, and cyclin D3/p27^{kip1} in T lymphocytes.[38,118,129,130] Similarly, changes in the subcellular localization of PKA can also cause alterations in the effective PKA activity in a particular cellular compartment (the nucleus, for example) and thus modify the expression of genes that signal the differentiated state of the cell.[131–134] Alterations in basal PKA activity combined with relocalization of the holoenzyme due to loss of RIα and compensation by RIIβ may both contribute to the changes observed in the differentiated state of these mesenchymal cells.

RIα HETEROZYGOTE MICE SERVE AS A MODEL FOR CARNEY COMPLEX IN HUMANS

Although no obvious Carney-like phenotypes have been observed in RIα heterozygous mice so far, we have found a strong interaction with genetic backgrounds that produces male sterility. On the mixed 129svJ:C57BL/6 genetic background, male mice are fertile, although careful analysis reveals an increase in abnormal sperm. However, male sterility has been observed when RIα heterozygous mice are mated onto the C57BL/6 genetic background. Mature sperm from these animals display highly abnormal head morphology resulting in increased susceptibility to breakage. Examination of earlier stages of spermatogenesis reveals abnormal vacuole formation in the nuclei of stage I postmitotic spermatocytes. It is intriguing to note that there is one published report of coincident male infertility and Carney complex and that in this case there were similar morphological abnormalities reported in sperm.[135] Similarly, the observation that familial transmission of Carney complex is heavily biased toward females (43 cases from a female affected parent and only 9 cases from a male affected parent) supports the possibility of a fertility defect in males that is in addition to large-cell calcifying Sertoli cell tumors.[109]

REFERENCES

1. UHLER, M.D. & G.S. MCKNIGHT. 1987. Expression of cDNAs for two isoforms of the catalytic subunit of cAMP-dependent protein kinase. J. Biol. Chem. **262:** 15202–15207.
2. BRANDON, E.P., M. ZHUO, Y.Y. HUANG, et al. 1995. Hippocampal long-term depression and depotentiation are defective in mice carrying a targeted disruption of the gene encoding the RI beta subunit of cAMP-dependent protein kinase. Proc. Natl. Acad. Sci. USA **92:** 8851–8855.
3. AMIEUX, P.S., D.E. CUMMINGS, K. MOTAMED, et al. 1997. Compensatory regulation of RIalpha? Protein levels in protein kinase A mutant mice. J. Biol. Chem. **272:** 3993–3998.
4. ORELLANA, S.A. & G.S. MCKNIGHT. 1990. The S49 Kin- cell line transcribes and translates a functional mRNA coding for the catalytic subunit of cAMP-dependent protein kinase. J. Biol. Chem. **265:** 3048–3053.
5. STEINBERG, R.A. & D.A. AGARD. 1981. Turnover of regulatory subunit of cyclic AMP-dependent protein kinase in S49 mouse lymphoma cells. Regulation by catalytic subunit and analogs of cyclic AMP. J. Biol. Chem. **256:** 10731–10734.
6. OTTEN, A.D. & G.S MCKNIGHT. 1989. Overexpression of the type II regulatory subunit of the cAMP-dependent protein kinase eliminates the type I holoenzyme in mouse cells. J. Biol. Chem. **264:** 20255–20260.
7. OTTEN, A.D., L.A. PARENTEAU, S. DOSKELAND & G.S MCKNIGHT. 1991. Hormonal activation of gene transcription in ras-transformed NIH3T3 cells overexpressing RII alpha and RII beta subunits of the cAMP-dependent protein kinase. J. Biol. Chem. **266:** 23074–23082.
8. CLEGG, C.H., G.G. CADD & G.S. MCKNIGHT. 1988. Genetic characterization of a brain-specific form of the type I regulatory subunit of cAMP-dependent protein kinase. Proc. Natl. Acad. Sci. USA **85:** 3703–3707.
9. CLEGG, C.H., W. RAN, M.D. UHLER & G.S. MCKNIGHT. 1989. A mutation in the catalytic subunit of protein kinase A prevents myristylation but does not inhibit biological activity. J. Biol. Chem. **264:** 20140–20146.
10. BRANDON, E.P., S.F. LOGUE, M.R. ADAMS, et al. 1998. Defective motor behavior and neural gene expression in RIIbeta-protein kinase A mutant mice. J. Neurosci. **18:** 3639–3649.
11. HENSCH, T.K., J.A. GORDON, E.P. BRANDON, et al. 1998. Comparison of plasticity in vivo and in vitro in the developing visual cortex of normal and protein kinase A RIbeta-deficient mice. J. Neurosci. **18:** 2108–2117.
12. CADD, G.G., M.D. UHLER & G.S. MCKNIGHT. 1990. Holoenzymes of cAMP-dependent protein kinase containing the neural form of type I regulatory subunit have an increased sensitivity to cyclic nucleotides. J. Biol. Chem. **265:** 19502–19506.
13. CUMMINGS, D.E., E.P. BRANDON, J.V. PLANAS, et al. 1996. Genetically lean mice result from targeted disruption of the RII beta subunit of protein kinase A [see comments]. Nature **382:** 622–626.
14. BURTON, K.A., B.D. JOHNSON, Z.E. HAUSKEN, et al. 1997. Type II regulatory subunits are not required for the anchoring-dependent modulation of Ca2+ channel activity by cAMP-dependent protein kinase. Proc. Natl. Acad. Sci. USA **94:** 11067–11072.
15. BURTON, K.A., B. TREASH-OSIO, C.H. MULLER, et al. 1999. Deletion of type IIalpha regulatory subunit delocalizes protein kinase A in mouse sperm without affecting motility or fertilization. J. Biol. Chem. **274:** 24131–24136.
16. FELICIELLO, A., M.E. GOTTESMAN & E.V. AVVEDIMENTO. 2001. The biological functions of A-kinase anchor proteins. J. Mol. Biol. **308:** 99–114.
17. HUANG, L.J., K. DURICK, J.A. WEINER, et al.1997. D-AKAP2, a novel protein kinase A anchoring protein with a putative RGS domain. Proc. Natl. Acad. Sci. USA **94:** 11184–11189.

18. HUANG, L.J., DURICK, K., J.A. WEINER, et al. 1997. Identification of a novel protein kinase A anchoring protein that binds both type I and type II regulatory subunits. J. Biol. Chem. **272:** 8057–8064.
19. BARRADEAU, S., T. IMAIZUMI-SCHERRER, M.C. WEISS & D.M. FAUST. 2000. Alternative 5'-exons of the mouse cAMP-dependent protein kinase subunit RIalpha gene are conserved and expressed in both a ubiquitous and tissue-restricted fashion. FEBS Lett. **476:** 272–276.
20. IMAIZUMI-SCHERRER, T., D.M. FAUST, J.C. BENICHOU, et al. 1996. Accumulation in fetal muscle and localization to the neuromuscular junction of cAMP-dependent protein kinase A regulatory and catalytic subunits RI alpha and C alpha. J. Cell. Biol. **134:** 1241–1254.
21. IMAIZUMI-SCHERRER, T., D.M. FAUST, S. BARRADEAU, et al. 2001. Type I protein kinase A is localized to interphase microtubules and strongly associated with the mitotic spindle. Exp. Cell. Res. **264:** 250–265.
22. ANGELO, R. & C.S. RUBIN. 1998. Molecular characterization of an anchor protein (AKAPCE) that binds the RI subunit (RCE) of type I protein kinase A from *Caenorhabditis elegans*. J. Biol. Chem. **273:** 14633–14643.
23. ANGELO, R.G. & C.S. RUBIN. 2000. Characterization of structural features that mediate the tethering of *Caenorhabditis elegans* protein kinase A to a novel A kinase anchor protein. Insights into the anchoring of PKAI isoforms. J. Biol. Chem. **275:** 4351–4362.
24. SKALHEGG, B.S., B.F. LANDMARK, S.O. DOSKELAND, et al. 1992. Cyclic AMP-dependent protein kinase type I mediates the inhibitory effects of 3',5'-cyclic adenosine monophosphate on cell replication in human T lymphocytes. J. Biol. Chem. **267:** 15707–15714.
25. SKALHEGG, B.S., K. TASKEN, V. HANSSON, et al. 1994. Location of cAMP-dependent protein kinase type I with the TCR-CD3 complex. Science **263:** 84–87.
26. TASKEN, K., V. HANSSON, P. AUKRUST, et al. 2000. PKAI as a potential target for therapeutic intervention. Drug News Perspect. **13:** 12–18.
27. KHAN, I.U., D. LAXMINARAYANA & G.M. KAMMER. 2001. Protein kinase A RI beta subunit deficiency in lupus T lymphocytes: bypassing a block in RI beta translation reconstitutes protein kinase A activity and augments IL-2 production. J. Immunol. **166:** 7600–7605.
28. LAXMINARAYANA, D. & G.M. KAMMER. 2000. mRNA mutations of type I protein kinase A regulatory subunit alpha in T lymphocytes of a subject with systemic lupus erythematosus. Int. Immunol. **12:** 1521–1529.
29. KUO, J.F. & P. GREENGARD. 1969. Cyclic nucleotide-dependent protein kinases. IV. Widespread occurrence of adenosine 3',5'-monophosphate-dependent protein kinase in various tissues and phyla of the animal kingdom. Proc. Natl. Acad. Sci. USA **64:** 1349–1355.
30. KREBS, E.G. & J.A. BEAVO. 1979. Phosphorylation-dephosphorylation of enzymes. Annu. Rev. Biochem. **48:** 923–959.
31. KREBS, E.G. 1972. Protein kinases. Curr. Top. Cell. Regul. **5:** 99–133.
32. KNIGHT, B.L. & J.P. SKALA. 1977. Protein kinases in brown adipose tissue of developing rats. Electrophoretic separation and assay of soluble protein kinases on polyacrylamide gels and a study of their properties and changes during development. J. Biol. Chem. **252:** 5356–5362.
33. MALLER, J.L. & E.G. KREBS. 1977. Progesterone-stimulated meiotic cell division in Xenopus oocytes. Induction by regulatory subunit and inhibition by catalytic subunit of adenosine 3':5'-monophosphate-dependent protein kinase. J. Biol. Chem. **252:** 1712–1718.
34. WALLACE, R.A. 1974. Letter: Protein phosphorylation during oocyte maturation. Nature **252:** 510–511.
35. MALLER, J.L., F.R. BUTCHER & E.G. KREBS. 1979. Early effect of progesterone on levels of cyclic adenosine 3':5'-monophosphate in Xenopus oocytes. J. Biol. Chem. **254:** 579–582.
36. MALLER, J.L., & E.G KREBS. 1980. Regulation of oocyte maturation. Curr. Top. Cell. Regul. **16:** 271–311.
37. MATTEN, W., I. DAAR & W.G.F. VANDE. 1994. Protein kinase A acts at multiple points to inhibit Xenopus oocyte maturation. Mol. Cell. Biol. **14:** 4419–4426.

38. FRANK-VAILLANT, M., C. JESSUS, R. OZON, et al. 1999. Two distinct mechanisms control the accumulation of cyclin B1 and Mos in Xenopus oocytes in response to progesterone. Mol. Biol. Cell. **10:** 3279–3288.
39. BORNSLAEGER, E.A., P. MATTEI & R.M. SCHULTZ. 1986. Involvement of cAMP-dependent protein kinase and protein phosphorylation in regulation of mouse oocyte maturation. Dev. Biol. **114:** 453–462.
40. DOWNS, S.M. & D.M. HUNZICKER. 1995. Differential regulation of oocyte maturation and cumulus expansion in the mouse oocyte-cumulus cell complex by site-selective analogs of cyclic adenosine monophosphate. Dev. Biol. **172:** 72–85.
41. D'ANGIOLELLA, V., V. COSTANZO, M.E. GOTTESMAN, et al. 2001. Role for cyclin-dependent kinase 2 in mitosis exit. Curr. Biol. **11:** 1221–1226.
42. GRIECO, D., A. PORCELLINI, E.V. AVVEDIMENTO & M.E. GOTTESMAN. 1996. Requirement for cAMP-PKA pathway activation by M phase-promoting factor in the transition from mitosis to interphase. Science **271:** 1718–1723.
43. PETZOLDT, U. & H.A. MUGGLETON. 1987. The effect of the nucleocytoplasmic ratio on protein synthesis and expression of a stage-specific antigen in early cleaving mouse embryos. Development **99:** 481–491.
44. POUEYMIROU, W.T. & R.M. SCHULTZ. 1987. Differential effects of activators of cAMP-dependent protein kinase and protein kinase C on cleavage of one-cell mouse embryos and protein synthesis and phosphorylation in one- and two-cell embryos. Dev. Biol. **121:** 489–498.
45. LATHAM, K.E., D. SOLTER & R.M. SCHULTZ. 1991. Activation of a two-cell stage-specific gene following transfer of heterologous nuclei into enucleated mouse embryos. Mol. Reprod. Dev. **30:** 182–186.
46. LATHAM, K.E., D. SOLTER & R.M. SCHULTZ. 1992. Acquisition of a transcriptionally permissive state during the 1-cell stage of mouse embryogenesis. Dev. Biol. **149:** 457–462.
47. POUEYMIROU, W.T. & R.M. SCHULTZ. 1989. Regulation of mouse preimplantation development: inhibition of synthesis of proteins in the two-cell embryo that require transcription by inhibitors of cAMP-dependent protein kinase. Dev. Biol. **133:** 588–599.
48. POUEYMIROU, W.T., J.C. CONOVER & R.M. SCHULTZ. 1989. Regulation of mouse preimplantation development: differential effects of CZB medium and Whitten's medium on rates and patterns of protein synthesis in 2-cell embryos. Biol. Reprod. **41:** 317–322.
49. KALDERON, D. & G.M. RUBIN. 1988. Isolation and characterization of Drosophila cAMP-dependent protein kinase genes. Genes Dev. **2:** 1539–1556.
50. LANE, M.E. & D. KALDERON. 1993. Genetic investigation of cAMP-dependent protein kinase function in Drosophila development. Genes Dev. **7:** 1229–1243.
51. JIANG, J. & G. STRUHL. 1995. Protein kinase A and hedgehog signaling in Drosophila limb development. Cell **80:** 563–572.
52. LEPAGE, T., S.M. COHEN, B.F.J. DIAZ & S.M. PARKHURST. 1995. Signal transduction by cAMP-dependent protein kinase A in Drosophila limb patterning [see comments]. Nature **373:** 711–715.
53. LI, W., J.T. OHLMEYER, M.E LANE & D. KALDERON. 1995. Function of protein kinase A in hedgehog signal transduction and Drosophila imaginal disc development. Cell **80:** 553–562.
54. PAN, D. & G.M. RUBIN. 1995. cAMP-dependent protein kinase and hedgehog act antagonistically in regulating decapentaplegic transcription in Drosophila imaginal discs. Cell **80:** 543–552.
55. STRUTT, D.L., V. WIERSDORF & M. MLODZIK. 1995. Regulation of furrow progression in the Drosophila eye by cAMP-dependent protein kinase A. Nature **373:** 705–709.
56. ECHELARD, Y., D.J. EPSTEIN, B. ST-JACQUES, et al. 1993. Sonic hedgehog, a member of a family of putative signaling molecules, is implicated in the regulation of CNS polarity. Cell **75:** 1417–1430.
57. KRAUSS, S., J.P. CONCORDET & P.W. INGHAM. 1993. A functionally conserved homolog of the Drosophila segment polarity gene hh is expressed in tissues with polarizing activity in zebrafish embryos. Cell **75:** 1431–1444.
58. RIDDLE, R.D., R.L. JOHNSON, E. LAUFER & C. TABIN. 1993. Sonic hedgehog mediates the polarizing activity of the ZPA. Cell **75:** 1401–1416.

59. ROELINK, H., A. AUGSBURGER, J. HEEMSKERK, et al. 1994. Floor plate and motor neuron induction by vhh-1, a vertebrate homolog of hedgehog expressed by the notochord. Cell **76:** 761–775.
60. FIETZ, M.J., J.P. CONCORDET, R. BARBOSA, et al. 1994. The hedgehog gene family in Drosophila and vertebrate development. Dev. Suppl. **1994:** 43–51.
61. CLEGG, C.H., L.A. CORRELL, G.G. CADD & G.S. MCKNIGHT. 1987. Inhibition of intracellular cAMP-dependent protein kinase using mutant genes of the regulatory type I subunit. J. Biol. Chem. **262:** 13111–13119.
62. ORELLANA, S.A. & G.S. MCKNIGHT. 1992. Mutations in the catalytic subunit of cAMP-dependent protein kinase result in unregulated biological activity. Proc. Natl. Acad. Sci. USA **89:** 4726–4730.
63. CONCORDET, J.P., K.E. LEWIS, J.W. MOORE, et al. 1996. Spatial regulation of a zebrafish patched homologue reflects the roles of sonic hedgehog and protein kinase A in neural tube and somite patterning. Development **122:** 2835–2846.
64. HAMMERSCHMIDT, M., M.J. BITGOOD & A.P. MCMAHON. 1996. Protein kinase A is a common negative regulator of Hedgehog signaling in the vertebrate embryo. Genes Dev. **10:** 647–658.
65. MUNSTERBERG, A.E., J. KITAJEWSKI, D.A. BUMCROT, et al. 1995. Combinatorial signaling by Sonic hedgehog and Wnt family members induces myogenic bHLH gene expression in the somite. Genes Dev. **9:** 2911–2922.
66. HYNES, M., J.A. PORTER, C. CHIANG, et al. 1995. Induction of midbrain dopaminergic neurons by Sonic hedgehog. Neuron **15:** 35–44.
67. FAN, C.M., J.A. PORTER, C. CHIANG, et al. 1995. Long-range sclerotome induction by sonic hedgehog: direct role of the amino-terminal cleavage product and modulation by the cyclic AMP signaling pathway. Cell **81:** 457–465.
68. EPSTEIN, D.J., E. MARTI, M.P. SCOTT, & A.P. MCMAHON. 1996. Antagonizing cAMP-dependent protein kinase A in the dorsal CNS activates a conserved Sonic hedgehog signaling pathway. Development **122:** 2885–2894.
69. GOODRICH, L.V., R.L. JOHNSON, L. MILENKOVIC, et al. 1996. Conservation of the hedgehog/patched signaling pathway from flies to mice: induction of a mouse patched gene by Hedgehog. Genes Dev. **10:** 301–312.
70. CHIANG, C., Y. LITINGTUNG, E. LEE, et al. 1996. Cyclopia and defective axial patterning in mice lacking Sonic hedgehog gene function. Nature **383:** 407–413.
71. AZA-BLANC, P., H.Y. LIN, A. RUIZ I ALTABA & T.B. KORNBERG. 2000. Expression of the vertebrate Gli proteins in Drosophila reveals a distribution of activator and repressor activities. Development **127:** 4293–4301.
72. CHEN, Y., J.R. CARDINAUX, R.H. GOODMAN & S.M. SMOLIK. 1999. Mutants of cubitus interruptus that are independent of PKA regulation are independent of hedgehog signaling. Development **126:** 3607–3616.
73. CHEN, Y., N. GALLAHER, R.H. GOODMAN & S.M. SMOLIK. 1998. Protein kinase A directly regulates the activity and proteolysis of cubitus interruptus. Proc. Natl. Acad. Sci. USA **95:** 2349–2354.
74. DAI, P., H. AKIMARU, Y. TANAKA, et al. 1999. Sonic hedgehog-induced activation of the Gli1 promoter is mediated by GLI3. J. Biol. Chem. **274:** 8143–8152.
75. RUIZ I ALTABA, A. 1999. Gli proteins encode context-dependent positive and negative functions: implications for development and disease. Development **126:** 3205–3216.
76. SHIN, S.H., P. KOGERMAN, E. LINDSTROM, et al. 1999. GLI3 mutations in human disorders mimic Drosophila cubitus interruptus protein functions and localization. Proc. Natl. Acad. Sci. USA **96:** 2880–2884.
77. WANG, B., J.F. FALLON & P.A. BEACHY. 2000. Hedgehog-regulated processing of Gli3 produces an anterior/posterior repressor gradient in the developing vertebrate limb. Cell **100:** 423–434.
78. COSTANZO, V., E.V. AVVEDIMENTO, M.E. GOTTESMAN, et al. 1999. Protein kinase A is required for chromosomal DNA replication. Curr. Biol. **9:** 903–906.
79. GRIECO, D., E.V. AVVEDIMENTO & M.E. GOTTESMAN. 1994. A role for cAMP-dependent protein kinase in early embryonic divisions. Proc. Natl. Acad. Sci. USA **91:** 9896–9900.
80. AMIEUX, P.S. 1997. RIalpha is an essential regulator of protein kinase A in the adult and developing mouse. Ph.D. thesis, University of Washington. Seattle, WA.

81. HOGAN, B., R. BEDDINGTON, F. COSTANTINI & E. LACY. 1994. Manipulating the Mouse Embryo. 2nd edit. Cold Spring Harbor Laboratory Press. New York.
82. KAUFMAN, M.H. 1992. The Atlas of Mouse Development. Academic Press. San Diego, CA.
83. THEILER, K. 1989. The House Mouse: Atlas of Embryonic Development. Springer-Verlag. New York.
84. ROSSANT, J., B. CIRUNA & J. PARTANEN. 1997. FGF signaling in mouse gastrulation and anteroposterior patterning. Cold Spring Harb. Symp. Quant. Biol. **62:** 127–133.
85. YAMAGUCHI, T.P. & J. ROSSANT. 1995. Fibroblast growth factors in mammalian development. Curr. Opin. Genet. Dev. **5:** 485–491.
86. CIRUNA, B. & J. ROSSANT. 2001. FGF signaling regulates mesoderm cell fate specification and morphogenetic movement at the primitive streak. Dev. Cell **1:** 37–49.
87. CIRUNA, B.G., L. SCHWARTZ, K. HARPAL, et al. 1997. Chimeric analysis of fibroblast growth factor receptor-1 (Fgfr1) function: a role for FGFR1 in morphogenetic movement through the primitive streak. Development **124:** 2829–2841.
88. SUN, X., E.N. MEYERS, M. LEWANDOSKI & G.R. MARTIN. 1999. Targeted disruption of Fgf8 causes failure of cell migration in the gastrulating mouse embryo. Genes Dev. **13:** 1834–1846.
89. YAMAGUCHI, T.P., K. HARPAL, M. HENKEMEYER & J. ROSSANT. 1994. Fgfr-1 is required for embryonic growth and mesodermal patterning during mouse gastrulation. Genes Dev. **8:** 3032–3044.
90. HADARI, Y.R., N. GOTOH, H. KOUHARA, et al. 2001. Critical role for the docking-protein FRS2 alpha in FGF receptor-mediated signal transduction pathways. Proc. Natl. Acad. Sci. USA **98:** 8578–8583.
91. SAXTON, T.M., M. HENKEMEYER, S. GASCA, et al. 1997. Abnormal mesodermal patterning in mouse embryos mutant for the SH2 tyrosine phosphatase Shp-2. EMBO J. **16:** 2352–2364.
92. SAXTON, T.M. & T. PAWSON. 1999. Morphogenetic movements at gastrulation require the SH2 tyrosine phosphatase Shp2. Proc. Natl. Acad. Sci. USA **96:** 3790–3795.
93. BURDSAL, C.A., C.H. DAMSKY & R.A. PEDERSEN. 1993. The role of E-cadherin and integrins in mesoderm differentiation and migration at the mammalian primitive streak. Development **118:** 829–844.
94. FURUTA, Y., D. ILIC, S. KANAZAWA, et al. 1995. Mesodermal defect in late phase of gastrulation by a targeted mutation of focal adhesion kinase, FAK. Oncogene **11:** 1989–1995.
95. GEORGES-LABOUESSE, E.N., E.L. GEORGE, H. RAYBURN & R.O. HYNES. 1996. Mesodermal development in mouse embryos mutant for fibronectin. Dev. Dyn. **207:** 145–156.
96. ILIC, D., Y. FURUTA, S. KANAZAWA, et al.1995. Reduced cell motility and enhanced focal adhesion contact formation in cells from FAK-deficient mice. Nature **377:** 539–544.
97. LAMB, N.J., A. FERNANDEZ, M.A. CONTI, et al. 1988. Regulation of actin microfilament integrity in living nonmuscle cells by the cAMP-dependent protein kinase and the myosin light chain kinase. J. Cell. Biol. **106:** 1955–1971.
98. HAN, J.D. & C.S. RUBIN. 1996. Regulation of cytoskeleton organization and paxillin dephosphorylation by cAMP. Studies on murine Y1 adrenal cells. J. Biol. Chem. **271:** 29211–29215.
99. MANES, S., E. MIRA, C. GOMEZ-MOUTON, et al. 1999. Concerted activity of tyrosine phosphatase SHP-2 and focal adhesion kinase in regulation of cell motility. Mol. Cell. Biol. **19:** 3125–3135.
100. ROCCHI, S., I. GAILLARD, E. VAN OBBERGHEN, et al. 2000. Adrenocorticotrophic hormone stimulates phosphotyrosine phosphatase SHP2 in bovine adrenocortical cells: phosphorylation and activation by cAMP-dependent protein kinase. Biochem. J. **352** Pt. 2: 483–490.
101. HOWE, A.K. 2001. Cell adhesion regulates the interaction between Nck and p21-activated kinase. J. Biol. Chem. **276:** 14541–14544.
102. LIN, T.H., Q. CHEN, A. HOWE & R.L. JULIANO. 1997. Cell anchorage permits efficient signal transduction between ras and its downstream kinases. J. Biol. Chem. **272:** 8849–8852.

103. HOWE, A.K. & R.L. JULIANO. 2000. Regulation of anchorage-dependent signal transduction by protein kinase A and p21-activated kinase. Nat. Cell. Biol. **2:** 593–600.
104. YU, D.H., C.K. QU, O. HENEGARIU, *et al.*1998. Protein-tyrosinephosphatase Shp-2 regulates cell spreading, migration, and focal adhesion. J. Biol. Chem. **273:** 21125–21131.
105. LUTTRELL, L.M., Y. DAAKA & R.J. LEFKOWITZ. 1999. Regulation of tyrosine kinase cascades by G-protein-coupled receptors. Curr. Opin. Cell. Biol. **11:** 177–183.
106. SEFTON, M., M.J. BLANCO, P. PENELA, *et al.* 2000. Expression of the G protein-coupled receptor kinase 2 during early mouse embryogenesis. Mech. Dev. **98:** 127–131.
107. CASEY, M., C.J. VAUGHAN, J. HE, *et al.* 2000. Mutations in the protein kinase A R1alpha regulatory subunit cause familial cardiac myxomas and Carney complex. J. Clin. Invest. **106:** R31–38.
108. KIRSCHNER, L.S., J.A. CARNEY, S.D. PACK, *et al.* 2000. Mutations of the gene encoding the protein kinase A type I-alpha regulatory subunit in patients with the Carney complex. Nat. Genet. **26:** 89–92.
109. STRATAKIS, C.A., L.S. KIRSCHNER & J.A. CARNEY. 2001. Clinical and molecular features of the Carney complex: diagnostic criteria and recommendations for patient evaluation. J. Clin. Endocrinol. Metab. **86:** 4041–4046.
110. BOS, J.L., J. DE ROOIJ & K.A. REEDQUIST. 2001. Rap1 signalling: adhering to new models. Nat. Rev. Mol. Cell Biol. **2:** 369–377.
111. BURGERING, B.M.T. & J.L. BOS. 1995. Regulation of Ras-mediated signalling: more than one way to skin a cat. TIBS **20:** 18–22.
112. GRAVES, L.M. & J.C. LAWRENCE. 1996. Insulin, growth factors, and cAMP. TEM **7:** 43–50.
113. BOYNTON, A.L. & J.F. WHITFIELD. 1983. The role of cyclic amp in cell proliferation: a critical assessment of the evidence. Adv. Cyclic Nucleotide Res. **15:** 192295.
114. GOTTESMAN, M.M. & R.D. FLEISCHMANN. 1986. The role of cAMP in regulating tumour cell growth. Cancer Surv. **5:** 291–308.
115. AROLA, J., P. HEIKKILA & A.I. KAHRI. 1993. Biphasic effect of ACTH on growth of rat adrenocortical cells in primary culture. Cell. Tissue Res. **271:** 169–176.
116. CREPIEUX, P., S. MARION, N. MARTINAT, *et al.* 2001. The ERK-dependent signalling is stage-specifically modulated by FSH, during primary Sertoli cell maturation. Oncogene **20:** 4696–4709.
117. KIM, H.A., J.E. DECLUE & N. RATNER. 1997. cAMP-dependent protein kinase A is required for Schwann cell growth: interactions between the cAMP and neuregulin/tyrosine kinase pathways. J. Neurosci. Res. **49:** 236–247.
118. KIMURA, T., A. VAN KEYMEULEN, J. GOLSTEIN, *et al.* 2001. Regulation of thyroid cell proliferation by tsh and other factors: a critical evaluation of in vitro models. Endocrinol. Rev. **22:** 631–656.
119. POMBO, C.M., J. ZALVIDE, B.D. GAYLINN & C. DIEGUEZ. 2000. Growth hormone-releasing hormone stimulates mitogen-activated protein kinase. Endocrinology **141:** 2113–2119.
120. SCHWAHN, D.J., W. XU, A.B. HERRIN, *et al.* 2001. Tyrosine levels regulate the melanogenic response to alpha-melanocyte-stimulating hormone in human melanocytes: implications for pigmentation and proliferation. Pigm. Cell. Res. **14:** 32–39.
121. BUSCA, R. & R. BALLOTTI. 2000. Cyclic AMP a key messenger in the regulation of skin pigmentation. Pigm. Cell. Res. **13:** 60–69.
122. COTRAN, R.S., V. KUMAR & S.L. ROBBINS, EDS. 1989. Robbins Pathologic Basis of Disease. 4th edit. W.B. Saunders Company. Philadelphia.
123. AMIEUX, P.S. 2001. Mesoderm insufficiency results in failed heart tube formation in RI alpha knockout mice. Protein Kinase A and Human Disease. National Institute of Child Health and Human Development. Washington, DC.
124. GRAVES, L.M., K.E. BORNFELDT, E.W. RAINES, *et al.* 1993. Protein kinase A antagonizes platelet-derived growth factor-induced signaling by mitogen-activated protein kinase in human arterial smooth muscle cells. Proc. Natl. Acad. Sci. USA **90:** 10300–10304.
125. SEVETSON, B.R., X. KONG & J.C.J. LAWRENCE. 1993. Increasing cAMP attenuates activation of mitogen-activated protein kinase. Proc. Natl. Acad. Sci. USA **90:** 10305–10309.

126. VANG, T., K.M. TORGERSEN, V. SUNDVOLD, et al. 2001. Activation of the COOH-terminal Src kinase (Csk) by cAMP-dependent protein kinase inhibits signaling through the T cell receptor. J. Exp. Med. **193:** 497–507.
127. AROLA, J., P. HEIKKILA, R. VOUTILAINEN & A.I. KAHRI. 1993. Role of adenylate cyclase-cyclic AMP-dependent signal transduction in the ACTH-induced biphasic growth effect of rat adrenocortical cells in primary culture. J. Endocrinol. **139:** 451–461.
128. HOWE, D.G. & K.D. MCCARTHY. 2000. Retroviral inhibition of cAMP-dependent protein kinase inhibits myelination but not Schwann cell mitosis stimulated by interaction with neurons. J. Neurosci. **20:** 3513–3521.
129. KIM, H.A., N. RATNER, T.M. ROBERTS, & C.D. STILES. 2001. Schwann cell proliferative responses to cAMP and Nf1 are mediated by cyclin D1. J. Neurosci. **21:** 1110–1116.
130. VAN OIRSCHOT, B.A., M. STAHL, S.M. LENS & R.H. MEDEMA. 2001. Protein kinase A regulates expression of p27(kip1) and cyclin D3 to suppress proliferation of leukemic T cell lines. J. Biol. Chem. **276:** 33854–33860.
131. BOSHART, M., F. WEIH, M. NICHOLS & G. SCHUTZ. 1991. The tissue-specific extinguisher locus TSE1 encodes a regulatory subunit of cAMP-dependent protein kinase. Cell **66:** 849–859.
132. FELICIELLO, A., A. GALLO, E. MELE, et al. 2000. The localization and activity of cAMP-dependent protein kinase affect cell cycle progression in thyroid cells. J. Biol. Chem. **275:** 303–311.
133. FELICIELLO, A., P. GIULIANO, A. PORCELLINI, et al. 1996. The v-Ki-Ras oncogene alters cAMP nuclear signaling by regulating the location and the expression of cAMP-dependent protein kinase IIbeta. J. Biol. Chem. **271:** 25350–25359.
134. JONES, K.W., M.H. SHAPERO, M. CHEVRETTE & R.E. FOURNIER. 1991. Subtractive hybridization cloning of a tissue-specific extinguisher: TSE1 encodes a regulatory subunit of protein kinase A. Cell **66:** 861–872.
135. LEGIUS, E., W. DAENEN, V. VANDENBERGH, et al. 1998. Syndrome of myxomas, spotty skin pigmentation, and endocrine overactivity (Carney complex). Genet. Couns. **9:** 287–290.
136. GEORGE, E.L., E.N. GEORGES-LABOUESSE, R.S. PATEL-KING, et al. 1993. Defects in mesoderm, neural tube and vascular development in mouse embryos lacking fibronectin. Development **119:** 1079–1091.
137. GEORGES-LABOUESSE, E.N., E.L. GEORGE, H. RAYBURN & R.O. HYNES. 1996. Mesodermal development in mouse embryos mutant for fibronectin. Dev. Dyn. **207:** 145–156.
138. GEORGE, E.L., H.S. BALDWIN, & R.O. HYNES. 1997. Fibronectins are essential for heart and blood vessel morphogenesis but are dispensable for initial specification of precursor cells. Blood **90:** 3073–3081.
139. SIEG, D.J., C.R. HAUCK & D.D. SCHLAEPFER. 1999. Required role of focal adhesion kinase (FAK) for integrin-stimulated cell migration. J. Cell. Sci. **112:** 2677–2691.
140. SIEG, D.J., C.R. HAUCK, D. ILIC, et al. 2000. FAK integrates growth-factor and integrin signals to promote cell migration. Nat. Cell. Biol. **2:** 249–256.
141. REN, X.D., W.B. KIOSSES, D.J. SIEG, et al. 2000. Focal adhesion kinase suppresses Rho activity to promote focal adhesion turnover. J. Cell Sci. **113:** 3673–3678.
142. XUE, Y., X. WANG, Z. LI, et al. 2001. Mesodermal patterning defect in mice lacking the Ste20 NCK interacting kinase (NIK). Development **128:** 1559–1572.
143. KLINGHOFFER, R.A., C. SACHSENMAIER, J.A. COOPER & P. SORIANO. 1999. Src family kinases are required for integrin but not PDGFR signal transduction. EMBO. J. **18:** 2459–2471.
144. LIU, J., C. HUANG & X. ZHAN. 1999. Src is required for cell migration and shape changes induced by fibroblast growth factor 1. Oncogene **18:** 6700–6706.
145. RICHARDSON, A., J.D. SHANNON, R.B. ADAMS, et al. 1997. Identification of integrin-stimulated sites of serine phosphorylation in FRNK, the separately expressed C-terminal domain of focal adhesion kinase: a potential role for protein kinase A. Biochem. J. **324:** 141–149.

Deficient Protein Kinase A in Systemic Lupus Erythematosus

A Disorder of T Lymphocyte Signal Transduction

GARY M. KAMMER

Section on Rheumatology and Clinical Immunology, Department of Internal Medicine, Wake Forest University School of Medicine, Winston-Salem, North Carolina 27157, USA

ABSTRACT: Systemic lupus erythematosus (SLE) is an idiopathic autoimmune disease characterized by impaired T lymphocyte immune effector functions. We have identified a disorder of signal transduction in SLE T cells involving the cyclic AMP/protein kinase A (cAMP/PKA) pathway. Cyclic AMP–stimulated PKA-catalyzed protein phosphorylation is markedly diminished owing to profound deficiencies of both type I (PKA-I) and type II (PKA-II) isozyme activities. Deficient PKA-I isozyme is characterized by a significant reduction in the amount of type I regulatory beta subunit (RIβ) steady state mRNA by competitive polymerase chain reaction. This is associated with a 30% decrease in RIα protein and a 65% reduction in RIβ protein. Indeed, T cells from ~25% of SLE subjects have no detectable RIβ protein. Transient transfection of T cells not expressing RIβ protein with autologous SLE RIβ cDNA bypassed the block in translation, reconstituting PKA activity and augmenting IL-2 production. Of importance was the initial identification of novel RIα mRNA mutations characterized by heterogeneous transcript mutations, including deletions, transitions, and transversions. Most mutations are clustered adjacent to GAGAG motifs and CT repeats. By contrast, deficient PKA-II activity is the result of spontaneous dissociation of the cytosolic RIIβ$_2$C$_2$ holoenzyme, aberrant RIIβ translocation to the nucleus from the cytosol, and retention of RIIβ in the nucleus. In conclusion, distinct mechanisms account for deficient PKA-I and PKA-II isozyme activities in SLE T cells.

KEYWORDS: autoimmunity; T lymphocytes; protein phosphorylation; transcription; translation

INTRODUCTION

Systemic lupus erythematosus (SLE) is an acute and chronic idiopathic autoimmune disease characterized by diverse abnormal T lymphocyte effector functions.[1] One mechanism that may contribute to abnormal CD4$^+$ T cell helper and CD8$^+$ T cell cytotoxic cell functions is aberrant signal transduction.[2–4] Our initial identification of a signaling disorder in SLE T cells revealed impaired cAMP-dependent protein

Address for correspondence: Gary M. Kammer, M.D., Section on Rheumatology & Clinical Immunology, Wake Forest University School of Medicine, Medical Center Blvd., Winston-Salem, NC 27157. Voice: 336-716-4209; fax: 336-716-9821.
gmkammer@wfubmc.edu

Ann. N.Y. Acad. Sci. 968: 96–105 (2002). © 2002 New York Academy of Sciences.

phosphorylation.[5,6] Because this defective signaling was associated with abnormal $CD8^+$ cytotoxic/suppressor activity[7] and altered mobility of transmembrane molecules,[8,9] subsequent analyses were performed to identify the molecular mechanisms underlying diminished cAMP-dependent protein phosphorylation in SLE T cells. Our results have elucidated a profound loss of total PKA activity due to deficiencies of both the type I (PKA-I) and type II (PKA-II) isozymes.

DEFICIENT PKA-I ISOZYME ACTIVITY IN SLE T CELLS

Reduced PKA-I Isozyme Activity

Recognition that total PKA activity and PKA-catalyzed protein phosphorylation[6] were significantly diminished in SLE T cells prompted us to quantify PKA-I and PKA-II isozyme activities. Compared to healthy and rheumatoid arthritis controls, PKA-I–catalyzed protein phosphorylation was significantly diminished in the T cell plasma membrane, where the isozyme is chiefly localized.[10,11] Separation of the isozymes by column chromatography and quantification of isozyme activities revealed that the ratio of PKA-I:PKA-II activity was reduced from 4.2:1 in normal T cells to ~1:1 in SLE T cells. Interestingly, we observed that a ~14-kDa plasma membrane–associated protein was heavily phosphorylated following activation of PKA-I via cAMP in normal specimens. Later work revealed that this phosphorylated protein is histone 2B (H2B), which is chaperoned in the plasma membrane by heat shock protein 60 (hsp60).[12] However, in most SLE plasma membrane specimens PKA-catalyzed phosphorylation of H2B was undetectable. In fact, cAMP-dependent phosphorylation of most plasma membrane substrates was strikingly diminished.[10] These results suggested that SLE T cells have deficient PKA-I activity in the plasma membrane fraction and that this deficiency contributes to the overall reduction of PKA activity.

Abnormal PKA-I Isozyme Kinetics

To understand the mechanism underlying deficient PKA-I activity, kinetic analyses of the isozyme were performed. As shown in TABLE 1, we confirmed a significant decrement of PKA-I activity in the T cells of a second cohort of SLE subjects. There was a statistically significant increase in the Michaelis-Menten (K_m) and apparent association constant for cAMP ($K_{a(cAMP)}$), but a decrement in the mean maximal binding of cAMP ($B_{max(cAMP)}$) to the RI subunit in SLE T cells compared to normal control cells. Reduction of the Hill coefficient from 1.2 in normal cells to 0.7 in SLE T cells indicated a loss of positive cooperativity between cAMP binding sites A and B of the RI subunits. Taken together, these findings suggested that either a RI subunit mutation(s) exists or the amount of intracellular RI protein is markedly reduced.[13]

Reduced PKA RIα and RIβ mRNA and Protein Content

The PKA-I isozyme is composed of two holoenzymes on the basis of their R subunit isoform: $RI\alpha_2C_2$ and $RI\beta_2C_2$.[14] To determine whether deficient PKA-I activity reflected altered content of either holoenzyme, we initially determined the presence or absence of each R- and C-subunit isoform transcript. The results revealed that

TABLE 1. T cell PKA-I isozyme kinetics in SLE and healthy controls[a]

	PKA-I (pmol/min/mg)	K_m^b (nM)	V_{max}^c (pmol/min/mg)	$K_{a(cAMP)}^d$ (nM)	$B_{max}(cAMP)^e$ (pmol/mg)
Controls	1,542 ± 423	79	1,391	98.2	5.9
SLE	429 ± 350	178	368	245	2.9
P value	<0.001	0.07	<0.001	<0.009	<0.03

[a] $N = 16$ SLE subjects and controls each.
[b] Michaelis-Menten constant.
[c] Maximal enzyme velocity.
[d] Association constant of cAMP for RI-subunit.
[e] Maximal binding capacity of cAMP for RI-subunit.

TABLE 2. Mechanisms of PKA-I and PKA-II isozyme deficiencies

Isozyme	Identified/proposed mechanism(s)	References
PKA-I	↓ RIα mRNA and translation[a]	15,17
	↓↓ RIβ mRNA and translation[a]	15,17
	RIα mRNA mutations	26
PKA-II	Aberrant nuclear translocation of RIIβ from cytosol and nuclear retention	29

[a] ↓ and ↓↓ refer to the reduction of RIβ mRNA or protein relative to RIα mRNA or protein.

SLE T cells expressed all seven of these isoforms: RIα, RIβ, RIIα, RIIβ, Cα, Cβ, and Cγ. We then performed competitive polymerase chain reaction (C-PCR) to quantify the steady state transcripts of each RI isoform. Compared to normal and Sjögren syndrome control T cells, SLE T cells have a 20% and 49% reduction in the amount of RIα and RIβ transcripts, respectively. The amount of RIβ transcript is significantly less than control T cells ($P = 0.008$). This result raised the possibility of disordered transcriptional regulation of RIβ and/or altered transcript stability.

We subsequently quantified the amounts of both RIα and RIβ protein. Interestingly, we found that, on average, SLE T cells have a 30% decrease in RIα protein ($P = 0.002$) and a 65% decrease in RIβ ($P < 0.001$) protein, shifting the ratio of RIα:RIβ to 6.5:1 from 3.2:1. Indeed, T cells from 25% of SLE subjects lacked any detectable RIβ protein as measured by immunoblotting of T cell lysates. When several IL-2–dependent SLE T cell lines that lacked RIβ protein were restudied after propagation in culture for 10 passages, we found a persistent absence of RIβ protein in these T cell progeny, excluding a potential effect of disease activity or the lupus microenvironment.[15]

High Prevalence of Deficient T Cell PKA-I Activity

Demonstration of a deficiency of PKA-I activity in SLE T cells prompted us to estimate its prevalence among SLE subjects. In a sample of 35 consecutive, unselected SLE subjects from our lupus clinic over a four-year interval, the prevalence of deficient PKA-I activity was 80%. During this period, PKA-I activities remained significantly reduced compared to T cells from healthy controls. By contrast, scores from the systemic lupus erythematosus disease activity index (SLEDAI) significantly improved. There was no identifiable relationship between deficient PKA-I activity and either SLEDAI scores or the proportion of T cells bearing certain activation markers, such as CD25 or HLA-DR. These results revealed that there is a very high prevalence of deficient T cell PKA-I activity among SLE subjects that persists over time and appears to be independent of SLE disease activity.[16]

Bypassing a Block in RIβ Translation Reconstitutes PKA Activity and Augments IL-2 Production

Considering the high prevalence of deficient PKA-I activity in the T cells of subjects with SLE, we explored potential mechanisms that might contribute to reduced RIα and RIβ protein content. One mechanism is impaired translation of RIβ protein. Before addressing this question, however, we investigated two other potential mechanisms for the loss of RIβ protein. Our original identification of low intracellular RIβ protein was by one-dimensional (1-D) gel electrophoresis and immunoblotting. However, we considered the possibility that a charge shift of RIβ could alter its mobility in gels and even account for its entire absence following isolation of the protein by fast-protein liquid chromatography. Using two-dimensional (2-D) gel electrophoresis, however, we found that reduced RIβ protein does not reflect a charge shift.[17] Rather, this reflects a true decrement or absence of RIβ protein. By contrast, Cα-subunit protein, which together with other isoforms of the C subunit form the $RIβ_2C_2$ holoenzyme, was present in amounts comparable to normal control T cells. However, because there are no monoclonal antibodies directed at the β or γ isoforms of C subunit, we cannot be certain at this time that these proteins are expressed in physiologic amounts. Notwithstanding this, our current data support the idea that reduced or absent RIβ subunit in the presence of physiologic amounts of Cα subunit accounts in large part for deficient PKA-I isozyme activity in SLE T cells.

A second mechanism that could lead to reduced intracellular RIβ protein is proteolysis and ubiquitination.[18] Using both proteasome and cysteine protease inhibitors, however, we found no evidence of accelerated proteolysis and proteasome degradation.[17]

Having excluded those putative mechanisms, we assessed the capacity of SLE T cells to translate RIα, RIβ, and Cα proteins. Cells were biosynthetically labeled with ^{35}S-methionine in the presence of dibutyryl cAMP and isobutylmethylxanthine and chased, and the kinetics of protein synthesis were analyzed over 48 h. Using SLE T cells in which little or no RIβ was detected by 2-D gel electrophoresis, we found no detectable RIβ produced, and the amount of RIα synthesized was significantly reduced compared to normal control T cells. By contrast, Cα protein was synthesized in amounts comparable to control cells. Thus, impaired synthesis of RIα and RIβ proteins may account for the markedly skewed ratio of these proteins and, therefore, the profound deficiency of the PKA-I isozyme in SLE vs. healthy T cells.[15] A diagram summarizing the proposed mechanism of inhibition of RIβ translation in SLE T cells is shown in FIGURE 1.

To determine whether the defective translational process could be bypassed, we transiently transfected SLE T cells with the pCR3.1/RIβ construct or an empty vector and quantified RIβ protein in immunoblots over the subsequent 24- and 48-h intervals. Five pCR3.1/RIβ constructs were made from RIβ cDNAs obtained from five individual SLE subjects whose T cells revealed minimal or no RIβ protein. RIβ cDNAs spanned the coding region and were under the control of a CMV promoter. Constructs were transiently transfected into the T cells of SLE subjects from which the RIβ cDNAs were prepared. Over the time span tested, T cells transfected with mock vector continued to express RIα, but not RIβ protein. By contrast, T cells transfected with pCR3.1/RIβ constructs revealed an 8–10-fold increase in RIβ production. When PKA activity was quantified in cells producing RIβ protein, there was a mean 73% increase in activity compared to either nontransfected or mock-transfected cells. These experiments unequivocally demonstrated that a block in RIβ translation could be bypassed by providing the SLE subject's own RIβ cDNA driven by an exogenous promoter and that this significantly elevated PKA activity.[17]

It has long been held that, in certain nonhematopoietic cells, the PKA-I isozyme promotes cell growth and proliferation.[19] However, the role of this isozyme in T cell growth still remains uncertain. It has also long been known that SLE T cells stimulated via the T cell receptor/CD3 complex *in vitro* produce significantly lower amounts of interleukin 2 (IL-2) than normal control T cells.[20] If PKA-I promotes T cell growth, then it is conceivable that one mechanism by which this might occur is via regulation of IL-2 production. To determine whether enhanced RIβ synthesis and, consequently, increased RIβ$_2$C$_2$ holoenzyme in SLE T cells would alter IL-2 production, we activated nontransfected or transfected SLE T cells with anti-CD3/anti-CD28/recombinant IL-1α over 48 h. Indeed, SLE T cells that expressed 10-fold more RIβ and had a significant increase in their PKA activities also increased IL-2

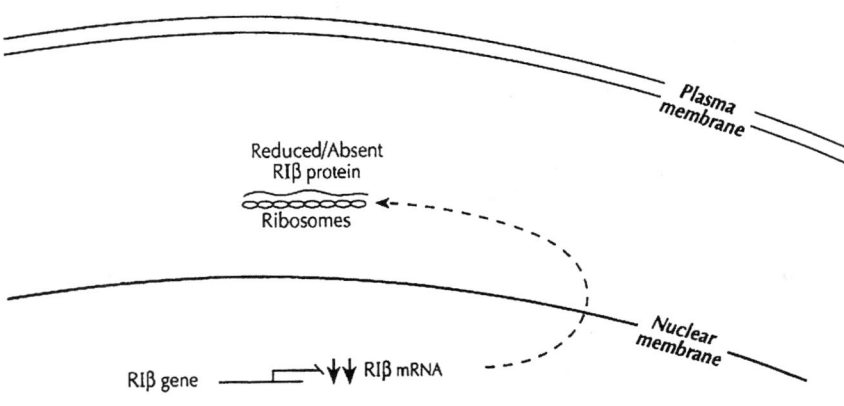

FIGURE 1. Proposed mechanism of impaired RIβ protein translation in SLE T cells.

production ~4-fold. By contrast, nontransfected or mock-transfected SLE T cells showed no change in IL-2 production over this interval. Although the amount of IL-2 produced by transfected SLE T cells is 17-fold lower than that of activated normal T cells, the capacity of SLE T cells to significantly increase IL-2 synthesis suggests that $RI\beta_2C_2$ holoenzyme may convey a signal integral to *IL-2* regulation.

It is well recognized that regulation of *IL-2* transcription is complex. Its promoter/enhancer region encodes response elements for several transcription factors, including AP-1, NF-κB, and NFAT. Recently, a CREB/CREM binding element was also identified at −180 relative to the start site.[21] Our recent efforts to understand why IL-2 production by SLE T cells is impaired revealed two particularly interesting findings. First, T cells from many SLE subjects express reduced or undetectable amounts of the p65 subunit of the NF-κB p65/p50 heterodimer.[22] Reduced or absent p65 subunit would be expected to significantly hinder the capacity of the heterodimer to effectively translocate to the nucleus from the cytosol upon its activation[23] and would, therefore, impede *IL-2* transactivation. Second, substitution of phosphorylated CREM for phosphorylated CREB binding to the −180 site of the *IL-2* enhancer/promoter would also be expected to down-regulate transcription and, therefore, to contribute to reduced IL-2 production.[21,24,25] Thus, even if the PKA-I isozyme does promote T cell growth, its repletion in SLE T cells is likely to have a limited salutary effect in the presence of these two disorders of transcription factors. Moreover, other regulatory defects of the transcriptional machinery, including specifically that of *RIβ*, are likely to be uncovered in the near future that will impact on gene expression or repression and, ultimately, T cell effector functions.

RIα Transcript Mutations

Our documentation of abnormal PKA-I isozyme kinetics, discussed above, led us to consider the possibility of a structural abnormality of the RIα and/or RIβ isoform due to a mutation(s).[13] We initially used single-strand conformation polymorphism (SSCP) analyses to detect such structural changes in the RIα cDNA. Of the first 10 SLE subjects studied, cDNAs from a single individual revealed a shifted band. Repeat SSCP analyses of several independently obtained cDNAs from this subject resulted in isolation of the same shifted band. Sequence analyses revealed that this shifted band carried heterogeneous transcript mutations, including deletions, transitions, and transversions. Of particular interest is that these putative transcript mutations are clustered adjacent to GAGAG motifs and CT repeats—hot spots susceptible to transcript editing and/or molecular misreading. However, careful, detailed analyses revealed no genomic mutations whatsoever. This is the first recognition of transcript mutations in a human autoimmune disease.[26] Such findings suggest that mRNA editing and/or defective function of RNA polymerase could be an underlying mechanism. Pathophysiologically, mutant RIα transcripts may be significant, for they can encode diverse, aberrant RIα isoforms, including truncated, dominant-negative subunits, resulting in deficient PKA-I activity. Based on experiments in the murine system, such mutated proteins would not be expected to efficiently bind to C subunits to form $RI\alpha_2C_2$ holoenzyme.[27,28] Experiments are currently underway to identify the presence of such transcript mutations of RIβ and to further investigate the genetic mechanism(s) involved.

DEFICIENT PKA-II ISOZYME ACTIVITY IN SLE T CELLS

Prevalence of Deficient PKA-II Activity and Relationship to Disease Activity

During the course of our work, it became apparent that the marked reduction of total PKA phosphotransferase activity could not be entirely accounted for by deficient PKA-I isozyme activity. For instance, we observed both reduced T cell PKA-I and PKA-II activities in some SLE subjects by DEAE-cellulose chromatography and ^{32}P-N$_3$-cAMP photoaffinity binding experiments (see fig. 4 in ref. 10). Control experiments in a later analysis confirmed these results[16] and led to a more detailed study.

T cell PKA-II activities from a cohort of 35 unselected SLE subjects with varying disease activity were compared to age-, sex,- and race-matched healthy controls. The findings revealed a statistically significant reduction of PKA-II activity in SLE T cells compared with controls. The prevalence of this deficiency was ~40%. As with PKA-I activity,[16] we found no relationship between PKA-II activity and SLE disease activity, duration of disease, or therapy. Moreover, there was no significant difference in the extent of deficient PKA-II activity and race. Thus, in 15 SLE subjects followed over a four-year interval, there were no significant differences in PKA-II activities even though the disease activity significantly improved.[29]

DEFICIENT PKA-II ACTIVITY IS A DISORDER PRIMARY TO THE SLE T CELL

It was conceivable that low PKA-II activity was a product of the T cell's lupus microenvironment and, therefore, might be reversible. To investigate this possibility, we studied IL-2– and IL-4–dependent T cell lines that were propagated over 10 passages *in vitro*. These lines were derived from the PBMCs of SLE subjects with markedly diminished PKA-II activities as well as healthy controls. The rationale for this approach was to study the progeny of SLE T cells that were originally isolated from PBMC but had not been exposed to the disease process *in vivo*. Thus, any defects cannot be attributed to extracellular stimuli in the lupus microenvironment.

Activation and proliferation of normal T cells results in ~58% reduction of cAMP-activatable PKA-II holoenzyme due to the isozyme's activation. When these cells are rested and returned to the G_0/G_1 phase of the cell cycle, cAMP-activatable PKA-II activity rises toward baseline levels. By contrast, cycling SLE T cell lines had ~29% reduction of PKA-II activity. This difference from normal cycling T cells may reflect the preexisting deficiency of PKA-II holoenzyme. When these cell lines were rested and returned to the G_0/G_1 phase of the cell cycle, however, PKA-II activity remained persistently depressed. These data suggest that the progeny of SLE T cells have a persistent deficiency of PKA-II activity that is independent of either mitogenesis or the lupus microenvironment.

Spontaneous Translocation of RIIβ to the Nucleus from the Cytosol

To determine the mechanism of deficient PKA-II activity in SLE T cells, we initially quantified the amount of RIIα and RIIβ proteins in nuclei-free cell lysates. Freshly isolated normal T cells express both R-subunit isoforms by immunoblot, and have a mean RIIβ:RIIα ratio of 3.95:1. Sjögren syndrome disease controls also possess both R-subunit isoforms, but there is an increased amount of RIIβ cytosolic protein, resulting in a mean RIIβ:RIIα ratio of 5.35:1. By contrast, lysates from SLE T cells revealed markedly reduced or no RIIβ protein by immunoblot, giving a mean RIIβ:RIIα ratio of 1.28:1. Of interest is that in 29% of subjects there was no detectable cytosolic RIIβ protein. These results demonstrate that deficient PKA-II activity in SLE T cells is associated with reduced/absent cytosolic RIIβ protein.

Using confocal immunofluorescence microscopy, we have demonstrated that activation of the PKA-II isozyme in normal T cells by either 8-Cl-cAMP or anti-CD3/anti-CD28/rIL-1α induced rapid nuclear translocation of RIIβ. In SLE T cells with low PKA-II activity, we observed spontaneous translocation of RIIβ from the cytosol to the nucleus in the absence of *in vitro* cell activation. This was reflected by both confocal microscopy and quantification of RIIβ protein in nuclear immunoprecipitates. By contrast, RIIα protein remained cytosolic, and Cα-subunit could be identified in both the cytosol and nucleus. This compartmentalization was similar to that observed in normal control T cells. Moreover, analyses of both freshly isolated and *in vitro* propagated SLE T cells suggested that RIIβ may be retained in the nucleus rather than shuttle between the cytosol and nucleus as seems to be the case under physiologic conditions.

SUMMARY

We have identified a signaling disorder intrinsic to the SLE T cell characterized by markedly impaired cAMP-activatable, PKA-catalyzed protein phosphorylation. Our current understanding of this disorder is that the signaling abnormality (a) is present in T cells from subjects with active or inactive SLE independent of therapy; (b) persists over time independent of SLE disease activity; and (c) reflects deficiencies of both PKA-I and PKA-II activities. The prevalence of T cell PKA-I and PKA-II deficiencies in SLE approximates 80% and 40%, respectively.

The pathophysiologic mechanisms underlying each PKA isozyme deficiency are different. Deficient PKA-I activity is the result of a decrease in RIβ that is greater than that of RIα protein; T cells from ~25% of subjects have no detectable RIβ protein. Because free C subunit would be expected to exist in excess of both RI isoforms, limiting amounts of these isoforms would be anticipated to lead to reduced formation of $RIα_2C_2$ and $RIβ_2C_2$ holoenzymes, resulting in low PKA-I activity. Although disorders at the level of both transcription and translation appear to exist in SLE T cells, the precise mechanisms resulting in low steady state RIα and RIβ transcripts as well as those impairing RIα and RIβ translation have yet to be identified.

By contrast, deficient PKA-II activity is a consequence of spontaneous activation of the holoenzyme, release of RIIβ, and its translocation to and retention in the nucleus. Long-term overexpression of nuclear RIIβ may modify transcriptional activation of genes such as *c-Fos*.[29]

ACKNOWLEDGMENTS

This work was supported by grants from the National Institutes of Health (RO1 AR39501, RO1 AI42269 and RO1 AI46526) and the Lupus Foundation of America;

by an Intramural Research Grant from the Wake Forest University School of Medicine; and by a grant from the General Clinical Research Center of the Wake Forest University School of Medicine (MO1 RR07122). I thank members of the Kammer laboratory for their review of this manuscript and for the many thoughtful and constructive discussions during the course of this work.

REFERENCES

1. KAMMER, G.M. & G.C. TSOKOS, EDS. 1999. Lupus: Molecular and Cellular Pathogenesis. Humana Press. Totowa, NJ.
2. DAYAL, A.K. & G.M. KAMMER. 1996. The T cell enigma in lupus. Arthritis Rheum. **39:** 23–33.
3. TSOKOS, G.C. & S.-N.C. LIOSSIS. 1999. Immune cell signaling defects in lupus: activation, anergy and death. Immunol. Today **20:** 119–124.
4. TSOKOS, G.C. & G.M. KAMMER. 2000. Molecular aberrations in human systemic lupus erythematosus. Mol. Med. Today **6:** 418–424.
5. MANDLER, R., R.E. BIRCH, S. POLMAR, et al. 1982. Abnormal adenosine-induced immunosuppression and cAMP metabolism in T lymphocytes of patients with systemic lupus erythematosus. Proc. Natl. Acad. Sci. USA **79:** 7542–7546.
6. HASLER, P., L.A. SCHULTZ & G.M. KAMMER. 1990. Defective cAMP-dependent phosphorylation of intact T lymphocytes in active systemic lupus erythematosus Proc. Natl. Acad. Sci. USA **87:** 1978–1982.
7. KAMMER, G.M., R.E. BIRCH & S. POLMAR. 1983. Impaired immunoregulation in systemic lupus erythematosus: defective adenosine-induced suppressor T lymphocyte generation. J. Immunol. **130:** 1706–1712.
8. KAMMER, G.M. 1983. Impaired T cell capping and receptor regeneration in active systemic lupus erythematosus: evidence for a disorder intrinsic to the T lymphocyte. J. Clin. Invest. **72:** 1686–1697.
9. KAMMER, G.M. & E. MITCHELL. 1988. Impaired mobility of human T lymphocyte surface molecules during inactive systemic lupus erythematosus. Relationship to a defective cAMP pathway. Arthritis Rheum. **31:** 88–98.
10. KAMMER, G.M., I.U. KHAN & C.J. MALEMUD. 1994. Deficient type I protein kinase A isozyme activity in systemic lupus erythematosus T lymphocytes. J. Clin. Invest. **94:** 422–430.
11. HASLER, P., J.J. MOORE & G.M. KAMMER. 1992. Human T lymphocyte cAMP-dependent protein kinase: subcellular distributions and activity ranges of type I and type II isozymes FASEB J. **6:** 2735–2741.
12. KHAN, I.U., R. WALLIN, R.S. GUPTA & G.M. KAMMER. 1998. Protein kinase A-catalyzed phosphorylation of heat shock protein 60 chaperone regulates its attachment to histone 2B in the T lymphocyte plasma membrane. Proc. Natl. Acad. Sci. USA **95:** 10425–10430.
13. KAMMER, G.M., I.U. KHAN, J.A. KAMMER, et al. 1996. Deficient type I protein kinase A activity in systemic lupus erythematosus T lymphocytes. II. Abnormal isozyme kinetics. J. Immunol. **157:** 2690–2698.
14. SKALHEGG, B.S. & K. TASKEN. 2000. Specificity in the cAMP/PKA signaling pathway. Differential expression, regulation, and subcellular localization of subunits of PKA. Front. Biosci. **5:** D678–D693.
15. LAXMINARAYANA, D., I.U. KHAN, N. MISHRA, et al. 1999. Diminished levels of protein kinase A RIα and RIβ transcripts and proteins in systemic lupus erythematosus T lymphocytes. J. Immunol. **162:** 5639–5648.
16. KAMMER, G.M. 1999. High prevalence of T cell type I protein kinase A deficiency in systemic lupus erythematosus. Arthritis Rheum. **42:** 1458–1465.
17. KHAN, I.U., D. LAXMINARAYANA & G.M. KAMMER. 2001. Protein kinase A RIβ subunit deficiency in lupus T lymphocytes: bypassing a block in RIβ translation reconstitutes protein kinase A activity and augments IL-2 production. J. Immunol. **166:** 7600–7605.

18. HEGDE, A.N., A.L. GOLDBERG & J.H. SCHWARTZ. 1993. Regulatory subunits of cAMP-dependent protein kinases are degraded after conjugation to ubiquitin: a molecular mechanism underlying long-term synaptic plasticity. Proc. Natl. Acad. Sci. USA **90:** 7436–7440.
19. CHO-CHUNG, Y.S., S. PEPE, T. CLAIR, et al. 1995. cAMP-dependent protein kinase: role in normal and malignant growth. Crit. Rev. Oncol. Hematol. **21:** 33–61.
20. HANDWERGER, B.S., I.G. LUZINA, L.D.A. SILVA, et al. 1999. Cytokines in the immunopathogenesis of lupus. *In* Lupus: Molecular and Cellular Pathogenesis. G.M. Kammer & G.C. Tsokos, Eds. 321–340. Humana Press. Totowa, NJ.
21. POWELL, J.D., C.G. LERNER, G.R. EWOLDT & R.H. SCHWARTZ. 1999. The −180 site of the IL-2 promoter is the target of CREB/CREM binding in T cell anergy. J. Immunol. **163:** 6631–6639.
22. WONG, H.K., G.M. KAMMER, G. DENNIS & G.C. TSOKOS. 1999. Abnormal NF-κB activity in T lymphocytes from patients with systemic lupus erythematosus is associated with decreased p65-RelA protein expression. J. Immunol. **163:** 1682–1689.
23. GHOSH, S., M.J. MAY & E.B. KOPP. 1998. NF-κB and Rel proteins: evolutionarily conserved mediators of immune responses. Ann. Rev. Immunol. **16:** 225–260.
24. SASSONE-CORSI, P. 1998. Coupling gene expression to cAMP signalling: role of CREB and CREM. Int. J. Biochem. Cell Biol. **30:** 27–38.
25. SOLOMOU, E.E., Y.-T.JUANG, M.F. GOURLEY, et al. 2001. Molecular basis of deficient IL-2 production in T cells from patients with systemic lupus erythematosus. J. Immunol. **166:** 4216–4222.
26. LAXMINARAYANA, D. & G.M. KAMMER. 2000. mRNA mutations of the RIα-subunit of type I protein kinase A in T lymphocytes of subjects with systemic lupus erythematosus. Int. Immunol. **12:** 1521–1529.
27. CORRELL, L.A., T.A. WOODFORD, J.D. CORBIN, et al. 1989. Functional characterization of cAMP-binding mutations in type I protein kinase. J. Biol. Chem. **264:** 16672–16678.
28. WOODFORD, T.A., L.A. CORRELL, G.S. MCKNIGHT & J.D. CORBIN. 1989. Expression and characterization of mutant forms of the type I regulatory subunit of cAMP-dependent protein kinase. J. Biol. Chem. **264:** 13321–13328.
29. MISHRA, N., I.U. KHAN, G.C. TSOKOS & G.M. KAMMER. 2000. Association of deficient type II protein kinase A activity with aberrant nuclear translocation of the RIIb-subunit in systemic lupus erythematosus T lymphocytes. J. Immunol. **165:** 2830–2840.

The Role of Cyclic AMP and Its Effect on Protein Kinase A in the Mitogenic Action of Thyrotropin on the Thyroid Cell

S. DREMIER, K. COULONVAL, S. PERPETE, F. VANDEPUT, N. FORTEMAISON, A. VAN KEYMEULEN, S. DELEU, C. LEDENT, S. CLÉMENT, S. SCHURMANS, J. E. DUMONT, F. LAMY, P. P. ROGER, AND C. MAENHAUT

Institute of Interdisciplinary Research (IRIBHN), Université of Brussels, School of Medicine, Campus Erasme, B 1070 Brussels, Belgium

ABSTRACT: Cyclic AMP has been shown to inhibit cell proliferation in many cell types and to activate it in some. The latter has been recognized only lately, thanks in large part to studies on the regulation of thyroid cell proliferation in dog thyroid cells. The steps that led to this conclusion are outlined. Thyrotropin activates cyclic accumulation in thyroid cells of all the studied species and also phospholipase C in human cells. It activates directly cell proliferation in rat cell lines, dog, and human thyroid cells but not in bovine or pig cells. The action of cyclic AMP is responsible for the proliferative effect of TSH. It accounts for several human diseases: congenital hyperthyroidism, autonomous adenomas, and Graves' disease; and, by default, for hypothyroidism by TSH receptor defect. Cyclic AMP proliferative action requires the activation of protein kinase A, but this effect is not sufficient to explain it. Cyclic AMP action also requires the permissive effect of IGF-1 or insulin through their receptors, mostly as a consequence of PI3 kinase activation. The mechanism of these effects at the level of cyclin and cyclin-dependent protein kinases involves an induction of cyclin D3 by IGF-1 and the cyclic AMP–elicited generation and activation of the cyclin D3-CDK4 complex.

KEYWORDS: cyclic AMP; thyroid; thyrotropin; protein kinase A; mitogenic; cell proliferation; function; differentiation; human

THE STORY OF cAMP–INDUCED CELL PROLIFERATION

Cyclic AMP as the Antiproliferative Intracellular Signal

Soon after the discovery of cyclic AMP as a signal relaying the effects of hormones and neurotransmitters on their membrane receptors in the cell, Sutherland and his colleagues generalized their early findings to as many extracellular signals and as many hormonal actions as possible.[1] For this they set up rigorous criteria to be validated before an action could be accepted as mediated by cyclic AMP. One of

---

Address for correspondence: Dr. Jacques Dumont, IRIBHN, ULB, Campus Erasme, Bldg. C (CP 602), Route de Lennik 808, B 1070 Brussels, Belgium. Voice: 32 2 555 41 33; fax: 32 2 555 46 55.

jedumont@ulb.ac.be

their early followers, I. Pastan, was the first to ask the question of the role of cyclic AMP in cell proliferation. For this, he used either fibroblasts or cell lines derived mostly from fibroblasts. In such cells, he showed that cyclic AMP levels decreased during proliferation in culture. Moreover, analogues of cyclic AMP decreased cell proliferation. Similar results were obtained in other model systems, mostly cell lines, which soon led to the dogma of cyclic AMP as the antimitogenic, antiproliferative, intracellular signal.[2] Although a few scattered findings suggested that the truth was not so simple, the dogma held sway during the 1970s and early 1980s.[3]

Cyclic AMP as the Intracellular Functional Signal in the Thyroid Cell

After initial experiments by Pastan and Gilman showing that the thyrotropic hormone TSH stimulated cyclic AMP generation by bovine thyroid membranes and its accumulation in bovine thyroid slices,[4–6] our laboratory became involved in the laborious task of validating Sutherland criteria for the various functional effects of TSH on dog thyroid cells. First, we demonstrated that TSH activated adenylate cyclase in membrane preparations and caused cyclic AMP accumulation in thyroid slices. In slices, two effects of TSH—the stimulation of iodide organification, which is the first step in thyroid hormone synthesis, and the stimulation of thyroid hormone secretion—were shown to be reproduced by an external nonspecific stimulator of adenylate cyclase (cholera toxin) and by an analogue of cyclic AMP (dibutyryl cyclic AMP). The concentration-effect curves and the kinetics of action were compatible with a cause-effect relationship. Moreover, the proved inhibition of cyclic AMP hydrolysis by phosphodiesterase inhibitors potentiated the effects of TSH on iodination and secretion.[7–9] Similar experiments with other activators of adenylate cyclase (forskolin) and other analogues of cyclic AMP, carried out on this and other experimental systems, demonstrated a host of other cyclic AMP–mediated TSH effects (e.g., on thyroglobulin secretion, specific gene expression, including iodide transport). Thus, cyclic AMP became recognized as the mediator of TSH functional effects on the thyroid.[9] This generalization later had to be limited, when it was found that in human cells TSH also activated the phospholipase C cascade and consequently thyroid hormone synthesis[10] (FIG. 1).

At this stage, the idea that TSH also stimulates thyroid growth, while well supported by various experiments on different models *in vivo*, was not easily accepted by specialists of cell cultures, who could not reproduce this effect *in vitro*.[11–14] The effect *in vivo* could well be indirect. Moreover, considering the prevalent dogma, the idea that a direct effect was mediated by cyclic AMP was a heresy.

Cyclic AMP as the Intracellular Mitogenic Signal in Thyroid Cells

In the Brussels group, we decided to test definitively whether TSH directly activated thyroid cell proliferation and whether cyclic AMP mediated or inhibited this effect. Physiology implies that when a system is functionally stimulated, high or/and prolonged levels of stimulation should cause the growth of the involved organ. It could therefore be expected that physiologically related functional and mitogenic stimulations would rely on the same mechanism. The choice of the experimental model was obvious: by far the best functional responses to TSH had been obtained on dog rather than human, pig, bovine, horse, or sheep thyroid slices. We therefore

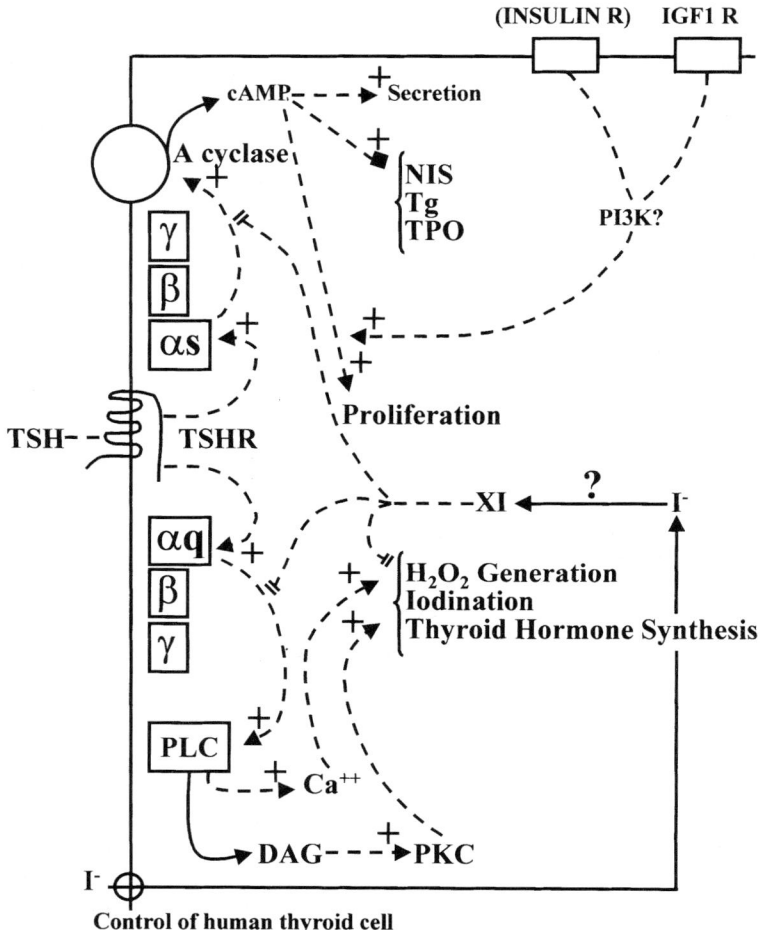

FIGURE 1. TSH regulation of the human thyroid cells. α, β, γ subunits of GTP binding proteins, $α_s$ stimulating adenylate cyclase, $α_q$ stimulating phospholipase C (PLC). DAG: diacylglycerol; PI3K: phosphatidylinositol 3 kinase; PKC: protein kinase C; NIS: Na+/I- symporter; Tg: thyroglobulin; TPO: thyroperoxidase. → Chemical transformation; ---+→ positive control; --- ‖ negative control; ---+◆ gene expression.[87]

set up a system of dog thyroid cell primary cultures[15,16] based on the previous methodologies of Kerkof and Fayet.[17–19] To measure cell proliferation, cell counting proved to be unreliable, as the stickiness of thyroid cells did not allow the counting of separate cells. Two other methods were set up: the calculation of cell numbers from measurement of the DNA of the culture[15] and the detection of the number of cells entering into DNA synthesis after a stimulus either by autoradiography after

^{3}H-thymidine labeling[16] or by immunohistochemistry after bromodeoxyuridine incorporation.

Dog thyroid cells in control medium but with insulin and a low concentration of fetal calf serum did not grow markedly. In the presence of TSH, the cells entered into DNA synthesis, and the DNA content of the culture increased. Pure preparations of TSH, when administered to quiescent cells in a serum-free medium supplemented with insulin, enhanced the proliferation of dog thyroid cells in primary culture in the presence of serum[15] and triggered DNA synthesis,[16] cell cycle progression,[20] and limited cell proliferation. These effects were reproduced with cholera toxin, dibutyryl cyclic AMP, and later by forskolin.[15,16,21] Thus, dog thyroid cell proliferation was induced by TSH through the cyclic AMP system.

When we submitted our article to various high-prestige journals, it was refused on the basis of the negative dogma. When finally submitted to *Molecular and Cellular Endocrinology*, our manuscript elicited two opposite comments by referees: One wrote that we showed that TSH stimulates thyroid cell proliferation, which everybody knew and was not original. The other referee commented that it was well known that cyclic AMP inhibited cell proliferation, and therefore our results were impossible. The editor argued that, if his referees disagreed that much, our results had to be interesting. It is quite remarkable that this paper, which started a whole new field, finally appeared in *Molecular and Cellular Endocrinology*, while the negative findings of our competitors came out in *Proceedings of the National Academy of Sciences*.[11] So much for the commercially driven emphasis on "impact factors."

The first study by a large group to test this concept on rat thyroid cell lines failed to support it. In FRTL5 cells, the most popular cell line, TSH induced cell proliferation (the cells had been selected for this property),[22] but the effects of TSH were not mimicked by cyclic AMP analogues.[23] The title of the article was quite assertive—"The relationship of growth and adenylate cyclase activity in cultured thyroid cells: separate bioeffects of thyrotropin." This is astonishing, as we now know that in these cells, TSH does not activate any other signal transduction cascade. We proceeded to develop our concept without bothering to argue. Later studies on the FRTL5 cells and on other cell lines showed that the effects of TSH were, indeed, mimicked by cyclic AMP analogues, cholera toxin, and forskolin.[24–27] Other actions of TSH might be involved. The TSH receptor activates Gi in dog thyroid cells, which could trigger other mitogenic cascades. However, pretreatment of the thyrocytes with pertussis toxin, which demonstrably blocks Gi effects in these cells, does not inhibit the TSH induction of DNA synthesis.[28]

We showed that dog thyroid cells, in primary culture, were stimulated to multiply by other growth factors such as EGF,[29] bFGF,[21] and HGF,[30] and also by the PKC-activating tumor promoter (phorbol myristate ester, TPA).[31] All the factors, except HGF, required the complementary action of IGF-1 or insulin at high concentrations acting through the IGF-1 receptor.[16,32–34] However, while TSH and the cyclic AMP cascade were inducing proliferation and specific thyroid gene expression (thyroglobulin,[35,36] thyroperoxidase,[37] the sodium iodide transporter NIS, and the H_2O_2 generating system[38])—that is, the expression of differentiation—growth factors and TPA led to proliferation and to a reversible inhibition of differentiation.[29–31,35] All these conclusions were later shown to apply to the human thyroid cell in primary culture.[39–41]

In Vivo *Demonstration of the Validity of the Concept of the Mitogenic and Differentiating Action of TSH and the Cyclic AMP Cascade in Thyroid Cells*

The mitogenic and differentiating actions of TSH through cyclic AMP on thyroid cells were demonstrated for dog and human thyroid cells in primary cultures and for rat thyroid cell lines (FRTL5, PC Cl3, WRT), but bovine or pig thyroid cells did not respond to TSH as a mitogenic stimulus. Nevertheless, based on these *in vitro* findings, we proposed in 1989 that one could extrapolate to *in vivo* human diseases and predict that, whereas constitutive activation of growth factor cascades by the classical oncogenes would generate undifferentiated thyroid cancers, constitutive activation of the TSH–cyclic AMP cascade could explain the benign thyroid hyperfunctioning autonomous adenomas.[42] Such predictions are daring, as they extrapolate from an artificial model of cells in primary culture to the organ *in vivo*, from experiments on a time scale of 8 days to the time scale of the organ, which encompasses a growth period of 20 years and at least 30 consecutive cell divisions (from 10^{-12} L to a lesion of 10^{-3} L).

The first question was: does the concept apply to the thyroid *in vivo*? To answer this question we relied on the transgenic mouse technology. In our first attempts to clone the cDNA of the TSH receptor, we had cloned four unknown serpentine receptors.[43] We identified one of these as the adenosine A2 receptor.[44] This receptor was difficult to characterize initially because it conferred on cells expressing it an apparently constitutive adenylate cyclase activation. Consequently, to show its stimulation by its agonists we had to use pharmacological activators in the presence of excess adenosine deaminase. The adenosine deaminase was necessary to constantly remove the adenosine spilling over from the cells. Though barely constitutive by itself, the receptor was activated by this normal metabolite and was barely desensitized over long-term incubations. It therefore behaved in cells not specially endowed with removal mechanisms as a "constitutive activator." This receptor is normally not expressed in thyroid cells. We therefore expressed the adenosine A2a receptor specifically in the thyroid gland of transgenic mice using a construct coding for the receptor downstream from the thyroglobulin promoter. The resulting mice became hyperthyroid and developed a goiter[45]—that is, they demonstrated the predicted phenotype of chronic activation of the cyclic AMP cascade: increased levels of cyclic AMP, hyperfunction, and increased cell proliferation. This was a first proof of the principle of our concept *in vivo*. Similar results were later obtained by other groups using the same type of construct with constitutively activated Gs_α, the G protein directly activating adenylate cyclase, or cholera toxin, a specific activator of the Gs_α protein.[46,47]

The phenotype of TgA2R mice, in fact, reproduced at the level of the whole thyroid the characteristics of human thyroid autonomous adenomas: growth and hyperfunction in the absence of TSH stimulation. This supported our previous prediction that such adenomas might result from the constitutive activation of the TSH–cyclic AMP cascade.

At this time, systematic mutagenesis of the $\alpha 1$ and the $\beta 2$ adrenergic receptors by the Lefkowitz group showed that some point mutations could confer a constitutive activation to these serpentine receptors.[48] We had just sequenced the DNA of dog and human TSH receptors.[49,50] We therefore decided to search for mutations in the human TSH receptor expressed in the autonomous adenomas. The results were clear-cut: in 6 out of 8 adenomas, a mutation was found in one half or less of the DNA of the adenoma but not in the corresponding adjacent tissue. This showed that the mutation was somatic. It also suggested that the lesion affected one allele in all the tumor cells—that is, that it was a dominant mutation affecting all the cells of the tumor and also that the tumor was monoclonal. It was shown, in transfected COS and CHO cells, that the mutations increased the constitutive activity of the receptor vs. cyclic AMP accumulation.[51,52] Moreover, when microinjected in dog thyroid cells, mRNA coding for a mutated receptor induced DNA synthesis without any agonist addition to the medium.[53]

Subsequent studies in our laboratory and in others have now demonstrated more than 30 mutations conferring constitutive activity to the TSH receptor for adenylate cyclase, but only three of them also increased phospholipase C activity.[54] Similar mutations account for cases of hereditary hyperthyroidism (Leclere's disease) and neomutations for congenital hyperthyroidism.[55,56] In these cases, the mutations affect all cells of the thyroid and therefore induce, as do the A_2R genes in transgenic mice, an autonomous adenoma involving the whole gland. Similarly, patients with McCune-Albright's disease, which is caused by mutations conferring constitutive activity to Gs_α, the G protein activating adenylate cyclase, also exhibit thyroid enlargement and hyperthyroidism.[57,58] Further argument for the mitogenic role of the cAMP cascade in human thyrocytes is provided by patients with Graves' disease who develop a goiter, caused by autoantibodies stimulating this cascade through the TSH receptor, the TSAbs.[59]

Thus, the mitogenic role of the cyclic AMP cascade in the thyroid was demonstrated in (1) *in vitro* models, the dog and human thyroid cells in primary culture and the rat thyroid cell lines (FRTL-5, PCCl3, WRT); (2) in *in vivo* models, the A_2R, constitutive Gs_α and cholera toxin transgenic mice; and (3) in human diseases, the autonomous adenomas, the thyroids of congenital hyperthyroidism, and McCune-Albright and Graves' diseases.

This concept has been extended, albeit with fewer arguments, to other cell types, such as the pituitary somatotrophs and the ovary granulosa cells.[42,60,61]

MECHANISM OF THE MITOGENIC ACTION OF CYCLIC AMP

Studies in the Thyrocytes

In the thyroid, cyclic AMP–induced mitogenesis is a differentiation characteristic. It does not operate in some species: pig and bovine thyroid cells do not proliferate in response to TSH and cyclic AMP. It is lost after treatment of the cells with EGF[62] and in thyroid cancers. Its mechanism is therefore apt to be different in different cell types. Indeed, it is different in FRTL5 and WRT cell lines and in primary cultures of dog and human cells.[63] In this review, we concentrate on the latter.

So far as they have been studied, the action and composition of the classical growth factor pathways in the dog thyrocyte correspond to previously described findings in other systems (FIG. 2). EGF and HGF stimulate, each to a different extent, both the Ras-Raf-MAP kinase[30,64,65] and the PI3K-PKB phosphorylation cascades.[66] Both trigger cell proliferation and repress, to various extents, the expression of specific genes—that is, of differentiation characteristics. While the stimulation of

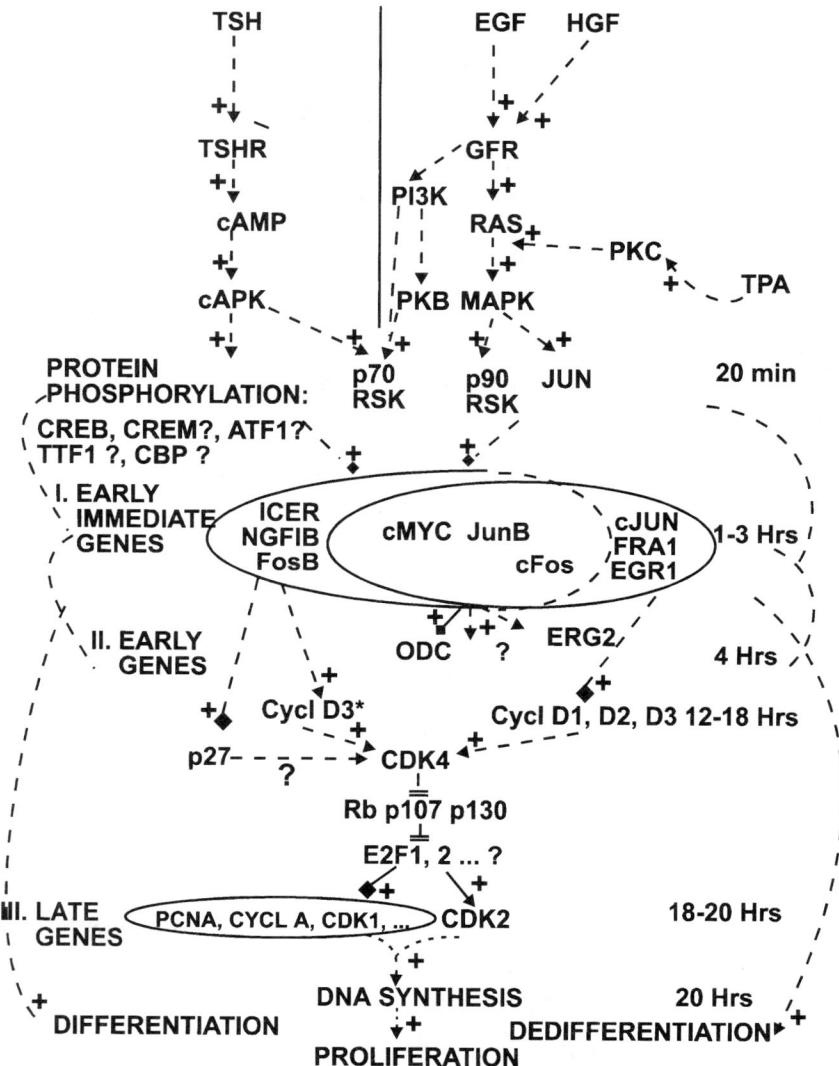

FIGURE 2. Mitogenic cascades of the dog thyroid cell. CAPK: protein kinase A; PKC, PKB: protein kinases C, B; RSK: ribosomal S6 kinase; ODC: ornithine decarboxylase; Cycl: cyclin; CDK: cyclin dependent kinase. ---+→ stimulation; --- || inhibition; ---+◆ gene expression.[87]

proliferation by EGF requires the presence of IGF-1 or insulin, HGF is fully mitogenic by itself.[34] In human thyrocytes, EGF acts as it does in dog thyrocytes. In our hands, HGF is inactive in human thyrocytes, although it represses differentiation (unpublished data); but others reported a marked stimulation in the presence of high serum concentrations.[67]

In dog and human thyrocytes TSH stimulates cell function, triggers DNA synthesis and cell proliferation, and induces specific gene expression. These effects are reproduced by cholera toxin, forskolin, and cyclic AMP analogues and thus are mediated by cyclic AMP. Similar conclusions apply to human thyroid cells. TSH and cyclic AMP do not activate either PI3kinase[66] or the Ras-Raf-MAP kinase cascade[64,65] in dog thyroid cells. In human thyrocytes, TSH and cyclic AMP also do not activate the p42/p44 MAP kinases. The previously described cyclic AMP–independent activation of MAP kinase by TSH[68] has been shown to be due to a contaminant (Vandeput *et al.*, submitted).

The mechanism of action of cyclic AMP involves protein kinase A:

- Both TSH and cyclic AMP analogues stimulate the phosphorylation of a specific set of proteins.[69]
- Studies with various pairs of cAMP analogues suggest that in dog thyrocytes both protein kinases A (I and II) are involved in the stimulation of protein iodination and thyroid hormone secretion, but that mitogenesis is more specifically induced by analogues acting on protein kinase AI.[70]
- Both TSH and forskolin induce the translocation of C subunits of protein kinase A into the nucleus.[53]
- Both TSH and forskolin induce the phosphorylation of the CREB transcription factor.[53,71]
- All the effects of TSH and forskolin observed in dog thyroid cell cultures (morphological changes, increased thyroglobulin expression, and mitogenesis) are inhibited by the microinjection of PKI, the "Walsh" PKA-specific protein inhibitor.[53]
- The effects of TSH that have been studied in this regard, especially the mitogenic effect, are inhibited by the supposedly PKA-specific inhibitor H89.

Not only is PKA necessary for the mitogenic actions of TSH, but its phosphorylating effect on CREB is also necessary. Indeed, the induction of DNA synthesis by TSH is inhibited by overexpression of the CREB natural inhibitors ICERIγ and CREMα isoforms.[71]

The next question was: if PKA appears to be necessary for all TSH and cAMP actions, is it sufficient to elicit them? This has been explored by microinjecting dog thyrocytes with an expression plasmid encoding the C subunit of PKA or the protein itself. In both cases, the continuous presence of C during the normal prereplicative phase was checked. Although microinjected C subunits reproduced the morphological effects of TSH as well as thyroperoxidase mRNA induction, as revealed by *in situ* hybridization, neither thyroglobulin expression nor DNA synthesis were stimulated in these cells. The data suggest that PKA is necessary but not sufficient for these effects.

An independent effect of the R of PKA is still possible. However $R_I \Delta_{1-95}$, which does not inhibit C, had no complementary effect to C on the triggering of DNA synthesis. Moreover, when R and C were microinjected in dog thyroid cells, the cells became sensitized to low, normally inactive concentrations of TSH for the morphologic effects of TSH but not for its induction of DNA synthesis.[53]

The conclusion that PKA was insufficient to account for some effects of cyclic AMP raised the question of other mediators of cyclic AMP action. It led to the discovery by two groups of guanyl nucleotide exchange factors (EPAC1 and EPAC2) activating the small G protein Rap.[72] The important expression of EPAC1 in the human and dog thyrocytes supports a role for this effector in the action of TSH. However, until now no conclusive evidence of such a role has been obtained. In particular, microinjection of expression plasmids coding for wild-type EPAC1 or for its constitutively active form, constitutive EPAC, did not complement C expression in inducing DNA synthesis or thyroglobulin expression. Moreover, in dog thyroid cells, Rap1 is activated by all the stimulants tested—EGF, phorbol myristate ester insulin, and others, which do not complement the action of C. Other possible cAMP-regulated proteins—for instance, the cyclic nucleotide–activated ion channels and protein kinase G—have also been considered; but cAMP analogues activating the former or cGMP analogues stimulating the latter tested in combination with the C subunit of PKA do not trigger DNA synthesis (unpublished data). A few years ago, before our results were published, most of the cAMP effects were assumed to be mediated by the activation of PKA. Now, various cAMP PKA–independent events have also been described by others.[88,89]

The Complementary Roles of TSH-Cyclic AMP and of IGF-1 in Thyrocyte Mitogenesis

In most work on cells in culture, the media used are routinely supplemented with various substrates but also with serum or high concentrations of insulin. Thus, from the beginning of thyroid cell cultures in reduced serum or serum-free conditions, insulin was present in the culture media. For FRTL5 cells, as well as for dog (see FIG. 3) and human cells, its presence was shown to be required for the growth effects of TSH and growth factors. Insulin alone did not elicit DNA synthesis. As both TSH and insulin were required for cell multiplication and as TSH is the physiological regulator of thyroid cells, we called TSH the stimulatory signal and insulin the permissive factor.[73] Their relative roles have been validated in *in vivo* transgenic mice. Mice expressing both human IGF-1 and IGF-1 receptor in their thyroids develop a certain degree of autonomy, retaining their weight and their thyroid hormone secretion in the presence of lower plasma TSH concentrations.[74] Similar findings have been made in acromegalic patients with chronically elevated levels of GH and IGF-1.[75] When stimulated by TSH, indirectly by antithyroid drug treatment, such mice develop a significant goiter.

In an analysis of insulin action, we were able to show that insulin at the high concentration used was, in fact, acting through the IGF-1 receptor and that IGF-1 was, in fact, much more potent than insulin.[33,34] However, when dog and human thyroid cells were pretreated with TSH or cyclic AMP enhancers, insulin receptors were induced; insulin at the lower concentrations active on these receptors also becomes permissive for the mitogenic stimulation by TSH.[33,76]

Insulin or IGF-1 through the IGF-1 receptor permits the mitogenic effect of TSH and cyclic AMP and stimulates by itself protein synthesis, protein accumulation, and cell growth.[34] Carbamylcholine also has a permissive action on the mitogenic effect of the cyclic AMP cascade[77] and stimulates cell growth.[34] Growth is a normal prerequisite for cell division. It was, therefore, logical to hypothesize that the permis-

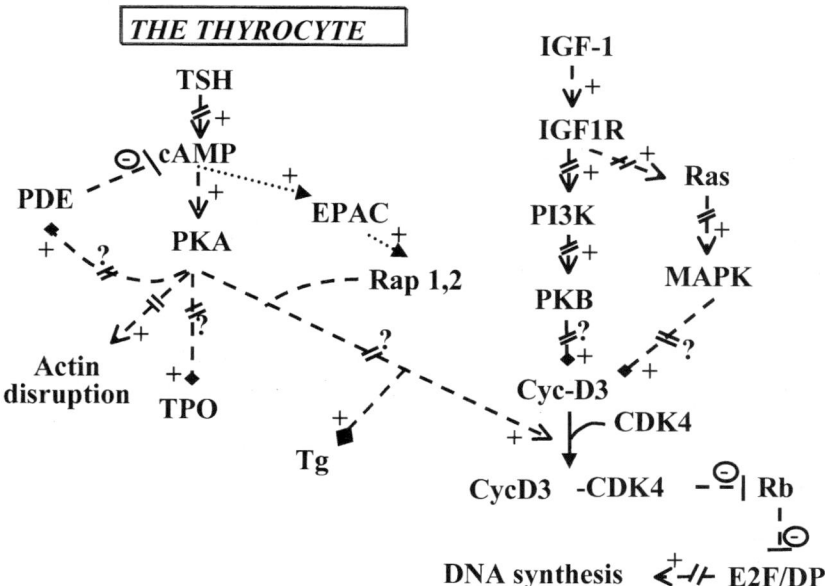

FIGURE 3. Role of IGF-1 and TSH in the triggering of DNA synthesis in the dog thyroid cell. PKA, B: protein kinases A, B; PDE: cyclic AMP phosphodiesterase; EPAC: Rap guanyl nucleotide exchange factor; Tg: thyroglobulin; TPO: thyroperoxidase; Cycl: cyclin; CDK: cyclin dependent kinase. ---+→ stimulation; ---− | inhibition; ---+◆ gene expression; --- // --- several steps omitted.[87]

sive role of the IGF-1 receptor was due to its growth promoting effect. However, short-term washing of thyroid cells largely impaired the permissive action of insulin even though the size of the cells did not diminish.[34] Similar cleaning experiments showed that atropine immediately suppressed the permissive effect of carbamylcholine in hypertrophied thyroid cells.[77] Moreover, both insulin and carbamylcholine have permissive effects when added with TSH or cyclic AMP analogues—that is, before any significant hypertrophy has taken place.[34,77] Therefore, although a general increase in protein synthesis may be necessary for TSH cyclic AMP mitogenic action, cell hypertrophy is not.

Insulin or IGF-1 activates in many cells both the Ras-Raf-MAP kinase and the PI3K-PKB pathways. In dog thyroid cells, some activity of both of these pathways is necessary for the mitogenic action of TSH–cyclic AMP, as inhibition of either pathway inhibits the initiation of DNA synthesis (ref. 66; Vandeput, unpublished). However, there are some indications that the role of PI3K and PKB activations are determinant in this process. Insulin and HGF activate the PI3K pathway strongly and for several hours.[66] By contrast, EGF activates this pathway, to a minor extent and for a short duration; while phorbol myristate ester and TSH do not.[66] On the other hand EGF, TPA, and HGF activate the Ras/MAP kinase pathway strongly and for several hours, while the effect of insulin is weak and short.[64–66] These findings suggest that the strong and sustained stimulation of the PI3K/PKB pathway may account for the permissive action of insulin/IGF-1 in the proliferation of thyrocytes and that the stimulation of this pathway by HGF may explain why this agent does not require insulin or IGF-1 for its mitogenic action.

The pleotropic and partly overlapping patterns of immediate early gene induction by TSH, insulin, EGF, HGF, and TPA do not allow one to distinguish at this level inductions that would be specific for the mitogenic and the permissive actions of these agents.[78–81] Nevertheless, further downstream the roles of these pathways can be distinguished. Classically, the triggering of DNA synthesis results from the release of E2F transcription factors from their inhibitory binding to proteins of the Rb family—Rb, p107, p130. This results from the phosphorylation of these proteins by complexes of cyclin Ds and cyclin E with the cyclin-dependent protein kinases CDK4 and CDK2. The formation of CDK4 complexes is limited by the supply of cyclin Ds. The effect of mitogenic pathways is to induce the synthesis of these cyclins. In agreement with this scheme, there is a perfect qualitative correlation between the coordinated phosphorylation of Rb, p107, and p130 and the triggering of DNA synthesis in dog thyroid cells: all the combinations of agents that trigger DNA synthesis also induce the prior phosphorylation of the three proteins, and vice versa.[82]

The major cyclin D expressed in the thyroid is cyclin D3.[83,84] Cyclin D synthesis is, indeed, caused by HGF, EGF, phorbol myristate ester, and insulin.[83,85] This common induction by agents able to trigger DNA synthesis, like HGF, or unable to do it, like EGF or TPA, shows that cyclin D synthesis by itself is insufficient to induce DNA synthesis. The positive effect of combinations of EGF + insulin and TPA + insulin on DNA synthesis might be explained by a synergy at the level of cyclin D synthesis. Unlike growth factors, TSH and cyclic AMP do not induce the synthesis of the three cyclin Ds.[83] Nevertheless, cyclin D3 is required for the proliferation of dog thyrocytes stimulated by TSH and cyclic AMP, but not in the proliferation of dog thyrocytes stimulated by EGF or HGF, which induce cyclins D1 and D2.[83] The role of TSH and cAMP is to stimulate the assembly and nuclear translocation of active cyclin D3–CDK4 complexes.[83,86] This was the first demonstration of physiological mitogenic stimulation targeting the assembly of such complexes.

The formation and nuclear translocation of essential cyclin D3–CDK4 complexes depend on the synergistic interaction of TSH and insulin.[85] These complexes are absent in cells stimulated by TSH or insulin alone. Paradoxically, in the absence of insulin, TSH strongly inhibits the basal accumulation of cyclin D3.[85] In contrast, insulin alone stimulates the required cyclin D3 accumulation, and it overcomes in large part the inhibition by TSH; but it is unable to assemble cyclin D3–CDK4 complexes in the absence of TSH.[85] Similar observations were made when carbamylcholine was used to replace insulin as a permissive factor for cyclic AMP–dependent mitogenesis.[77]

Thus the system behaves as if insulin or carbamylcholine would generate a sufficient supply of cyclin D3 but the cAMP cascade would trigger active cyclin D3–CDK4 complex formation and activation. The permissive role of insulin would be to supply cyclin D3; the mitogenic triggering of the cyclic AMP cascade would generate and activate the complex. By analogy to a car, insulin provides the fuel, but the TSH cyclic AMP pathway is the starter.

In conclusion, a clear but still incomplete scheme of the regulation of human and dog thyroid cell proliferation by TSH and cyclic AMP is now emerging. This scheme is, however, different from the schemes that are delineated in rat thyroid cell lines[63] and in granulosa cells.[61] As stated earlier,[60] the pathway is a cell-specific differentiation characteristic.

ACKNOWLEDGMENTS

The work reported in this article has been supported over the years by the Service du Premier Ministre Affaires Scientifiques, Techniques et Culturelles SSTC (PAI), EU Quality of Life and Radioprotection programs, the Fonds National de la Recherche Scientifique (FNRS), Fonds de la Recherche Scientifique Médicale (FRSM), Fonds Cancérologique Fortis, Opération Télévie, and Fédération Belge contre le Cancer.

REFERENCES

1. ROBISON, G.A., R.W. BUTCHER & E.W. SUTHERLAND. 1968. Cyclic AMP. Annu. Rev. Biochem. **37:** 149–174.
2. PASTAN, I. & G.S. JOHNSON. 1974. Cyclic AMP and the transformation of fibroblasts. Adv. Cancer Res. **19:** 303–329.
3. BOYNTON, A.L., J.F. WHITFIELD & L.P. KLEINE. 1983. Ca2+/phospholipid-dependent protein kinase activity correlates to the ability of transformed liver cells to proliferate in Ca2+-deficient medium. Biochem. Biophys. Res. Commun. **115:** 383–390.
4. PASTAN, I. & R. KATZEN. 1967. Activation of adenyl cyclase in thyroid homogenates by thyroid-stimulating hormone. Biochem. Biophys. Res. Commun. **26:** 792–798.
5. GILMAN, A.G. & T.W. RALL. 1968. The role of adenosine 3′,5′-phosphate in mediating effects of thyroid-stimulating hormone on carbohydrate metabolism of bovine thyroid slices. J. Biol. Chem. **243:** 5872–5881.
6. GILMAN, A.G. & T.W. RALL. 1968. Factors influencing adenosine 3′,5′-phosphate accumulation in bovine thyroid slices. J. Biol. Chem. **243:** 5867–5871.
7. RODESCH, F., P. NEVE, C. WILLEMS, et al. 1969. Stimulation of thyroid metabolism by thyrotropin, cyclic 3′,5′ AMP dibutyryl cyclic 3′,5′ AMP and prostaglandin E1. Eur. J. Biochem. **8:** 26–32.
8. DUMONT, J.E., C. WILLEMS, J. VAN SANDE, et al. 1971. Regulation of the release of thyroid hormones: role of cyclic AMP. Ann. N.Y. Acad. Sci. **185:** 291–316.
9. DUMONT, J.E. 1971. The action of thyrotropin on thyroid metabolism. Vitam. Horm. **29:** 287–412.
10. LAURENT, E., J. MOCKEL, J. VAN SANDE, et al. 1987. Dual activation by thyrotropin of the phospholipase C and cAMP cascades in human thyroid. Mol. Cell. Endocrinol. **52:** 273–278.
11. WESTERMARK, B., F.A. KARLSSON & O. WALINDER. 1979. Thyrotropin is not a growth factor for human thyroid cells in culture. Proc. Natl. Acad. Sci. USA **76:** 2022–2026.
12. GÄRTNER, R., W. GREIL, P. DEMHARTER, et al. 1985. Involvement of cyclic AMP, iodine and metabolites of arachidonic acid in the regulation of cell proliferation of isolated porcine thyroid follicles. Mol. Cell. Endocrinol. **42:** 145–155.
13. EGGO, M.C., L.R. BACHRACH, G. FAYET, et al. 1984. The effects of growth factors and serum on DNA synthesis and differentiation in thyroid cells in culture. Mol. Cell. Endocrinol. **38:** 141–150.
14. ERRICK, J.E., K.W.A. ING, M.C. EGGO, et al. 1986. Growth and differentiation in cultured human thyroid cells: effects of epidermal growth factor and thyrotropin. In Vitro **22:** 28–36.
15. ROGER, P.P., A. HOTIMSKY, C. MOREAU, et al. 1982. Stimulation by thyrotropin, cholera toxin and dibutyryl cyclic AMP of the multiplication of differentiated thyroid cells in vitro. Mol. Cell. Endocrinol. **26:** 165–172.

16. ROGER, P.P., P. SERVAIS & J.E. DUMONT. 1983. Stimulation by thyrotropin and cyclic AMP of the proliferation of quiescent canine thyroid cells cultured in a defined medium containing insulin. FEBS Lett. **157:** 323–329.
17. KERKOF, P.R., P.J. LONG & I.L. CHAIKOFF. 1964. In vitro effects of thyrotropin hormone. I. On the pattern of organization of monolayer cultures of isolated sheep thyroid gland. Endocrinology **74:** 170–179.
18. FAYET, G., M. MICHEL-BECHET & S. LISSITZKY. 1971. Thyrotropin-induced aggregation and reorganization into follicles of isolated porcine thyroid cells in culture. Eur. J. Biochem. **24:** 100–111.
19. RAPOPORT, B. 1976. Dog thyroid cells in monolayer tissue culture: adenosine 3'-5'-cyclic monophosphate response to thyrotropic hormone. Endocrinology **98:** 1189–1197.
20. BAPTIST, M., J.E. DUMONT & P.P. ROGER. 1993. Demonstration of cell cycle kinetics in thyroid primary culture by immunostaining of proliferating cell nuclear antigen: differences in cyclic AMP-dependent and -independent mitogenic stimulations. J. Cell Sci. **105:** 69–80.
21. ROGER, P.P. & J.E. DUMONT. 1984. Factors controlling proliferation and differentiation of canine thyroid cells cultured in reduced serum conditions: effects of thyrotropin, cyclic AMP and growth factors. Mol. Cell. Endocrinol. **36:** 79–93.
22. AMBESI-IMPIOMBATO, F.S., L.A.M. PARKS & H.G. COON. 1980. Culture of hormone-dependent functional epithelial cells froml rat thyroids. Proc. Natl. Acad. Sci. USA **77:** 3455–3459.
23. VALENTE, W.A., P. VITTI, L.D. KOHN, et al. 1983. The relationship of growth and adenylate cyclase activity in cultured thyroid cells. Separate bioeffects of thyrotropin. Endocrinology **112:** 71–79.
24. DERE, W.H. & B. RAPOPORT. 1986. Control of growth in cultured rat thyroid cells. Mol. Cell. Endocrinol. **44:** 195–199.
25. JIN, S., F.J. HORNICEK, D. NEYLAN, et al. 1986. Evidence that adenosine 3',5'-monophosphate mediates stimulation of thyroid growth in FRTL-5 cells. Endocrinology **119:** 802–810.
26. TRAMONTANO, D., A.C. MOSES & S.H. INGBAR. 1988. The role of adenosine 3',5'-monophosphate in the regulation of receptors for thyrotropin and insulin-like growth factor I in the FRTL-5 rat thyroid follicular cell. Endocrinology **122:** 133–136.
27. TRAMONTANO, D., A.C. MOSES, B.M. VENEZIANI, et al. 1988. Adenosine 3',5'-monophosphate mediates both the mitogenic effect of thyrotropin and its ability to amplify the response to insulin-like growth factor I in FRTL-5 cells. Endocrinology **122:** 127–132.
28. ALLGEIER, A., K.L. LAUGWITZ, J. VAN SANDE, et al. 1997. Multiple G-protein coupling of the dog thyrotropin receptor. Mol. Cell. Endocrinol. **127:** 81–90.
29. ROGER, P.P. & J.E. DUMONT. 1982. Epidermal growth factor controls the proliferation and the expression of differentiation in canine thyroid cells in primary culture. FEBS Lett. **144:** 209–212.
30. DREMIER, S., M. TATON, K. COULONVAL, et al. 1994. Mitogenic, dedifferentiating, and scattering effects of hepatocyte growth factor on dog thyroid cells. Endocrinology **135:** 135–140.
31. ROGER, P.P., S. REUSE, P. SERVAIS, et al. 1986. Stimulation of cell proliferation and inhibition of differentiation expression by tumor-promoting phorbol esters in dog thyroid cells in primary culture. Cancer Res. **46:** 898–906.
32. ROGER, P.P., P. SERVAIS & J.E. DUMONT. 1987. Induction of DNA synthesis in dog thyrocytes in primary culture: synergistic effects of thyrotropin and cyclic AMP with epidermal growth factor and insulin. J. Cell. Physiol. **130:** 58–67.
33. BURIKHANOV, R., K. COULONVAL, I. PIRSON, et al. 1996. Thyrotropin via cyclic AMP induces insulin receptor expression and insulin co-stimulation of growth and amplifies insulin and insulin-like growth factor signaling pathways in dog thyroid epithelial cells. J. Biol. Chem. **271:** 29400–29406.
34. DELEU, S., I. PIRSON, F. CLERMONT, et al. 1999. Immediate early gene expression in dog thyrocytes in response to growth, proliferation, and differentiation stimuli. J. Cell. Physiol. **181:** 342–354.

35. ROGER, P.P., B. VAN HEUVERSWYN, C. LAMBERT, et al. 1985. Antagonistic effects of thyrotropin and epidermal growth factor on thyroglobulin mRNA level in cultured thyroid cells. Eur. J. Biochem. **152:** 239–245.
36. POHL, V., P.P. ROGER, D. CHRISTOPHE, et al. 1990. Differentiation expression during proliferative activity induced through different pathways: in situ hybridization study of thyroglobulin gene expression in thyroid epithelial cells. J. Cell Biol. **111:** 663–672.
37. GÉRARD, C., A. LEFORT, D. CHRISTOPHE, et al. 1989. Control of thyroperoxidase and thyroglobulin transcription by cAMP: evidence for distinct regulatory mechanisms. Mol. Endocrinol. **3:** 2110–2118.
38. DE DEKEN, X., D. WANG, M.C. MANY, et al. 2000. Cloning of two human thyroid cDNAs encoding new members of the NADPH oxidase family. J. Biol. Chem. **275:** 23227–23233.
39. ROGER, P.P., M. TATON, J. VAN SANDE, et al. 1988. Mitogenic effects of thyrotropin and adenosine 3′,5′-monophosphate in differentiated normal human thyroid cells in vitro. J. Clin. Endocrinol. Metab. **66:** 1158–1165.
40. LAMY, F., M. TATON, J.E. DUMONT, et al. 1990. Control of protein synthesis by thyrotropin and epidermal growth factor in human thyrocytes: role of morphological changes. Mol. Cell Endocrinol. **73:** 195–209.
41. MAENHAUT, C., G. BRABANT, G. VASSART, et al. 1992. In vitro and in vivo regulation of thyrotropin receptor mRNA levels in dog and human thyroid cells. J. Biol. Chem. **267:** 3000–3007.
42. DUMONT, J.E., J.C. JAUNIAUX & P.P. ROGER. 1989. The cyclic AMP-mediated stimulation of cell proliferation. Trends Biochem. Sci. **14:** 67–71.
43. LIBERT, F., M. PARMENTIER, A. LEFORT, et al. 1989. Selective amplification and cloning of four new members of the G protein-coupled receptor family. Science **244:** 569–572.
44. MAENHAUT, C., J. VAN SANDE, F. LIBERT, et al. 1990. RDC8 codes for an adenosine A2 receptor with physiological constitutive activity. Biochem. Biophys. Res. Commun. **173:** 1169–1178.
45. LEDENT, C., J.E. DUMONT, G. VASSART, et al. 1991. Thyroid expression of an A2 adenosine receptor transgene induces thyroid hyperplasia and hyperthyroidism. EMBO J. **11:** 537–542.
46. MICHIELS, F.M., B. CAILLOU, M. TALBOT, et al. 1994. Oncogenic potential of guanine nucleotide stimulatory factor alpha subunit in thyroid glands of transgenic mice. Proc. Natl. Acad. Sci. USA **91:** 10488–10492.
47. ZEIGER, M.A., M. SAJI, Y. GUSEV, et al. 1997. Thyroid-specific expression of cholera toxin A1 subunit causes thyroid hyperplasia and hyperthyroidism in transgenic mice. Endocrinology **138:** 3133–3140.
48. KJELSBERG, M.A., S. COTECCHIA, J. OSTROWSKI, et al. 1992. Constitutive activation of the alpha 1B-adrenergic receptor by all amino acid substitutions at a single site. Evidence for a region which constrains receptor activation. J. Biol. Chem. **267:** 1430–1433.
49. PARMENTIER, M., F. LIBERT, C. MAENHAUT, et al. 1989. Molecular cloning of the thyrotropin (TSH) receptor. Science **296:** 1620–1622.
50. PARMENTIER, M., F. LIBERT, C. MAENHAUT, et al. 1989. Nucleotide sequence of the dog thyrotropin receptor cDNA. Nucleic Acids Res. **17:** 10493–10494.
51. PARMA, J., L. DUPREZ, J. VAN SANDE, et al. 1993. Somatic mutations in the thyrotropin receptor gene cause hyperfunctioning thyroid adenomas. Nature **365:** 649–651.
52. PARMA, J., J. VAN SANDE, S. SWILLENS, et al. 1995. Somatic mutations causing constitutive activity of the thyrotropin receptor are the major cause of hyperfunctioning thyroid adenomas: identification of additional mutations activating both the cyclic adenosine 3′,5′-monophosphate and inositol phosphate-Ca2+ cascades. Mol. Endocrinol. **9:** 725–733.
53. DREMIER, S., V. POHL, C. POTEET-SMITH, et al. 1997. Activation of cyclic AMP-dependent kinase is required but may not be sufficient to mimic cyclic AMP-dependent DNA synthesis and thyroglobulin expression in dog thyroid cells. Mol. Cell. Biol. **17:** 6717–6726.
54. DUPREZ, L., J. PARMA, S. COSTAGLIOLA, et al. 1997. Constitutive activation of the TSH receptor by spontaneous mutations affecting the N-terminal extracellular domain. FEBS Lett. **409:** 469–474.

55. DUPREZ, L., J. PARMA, J. VAN SANDE, et al. 1994. Germline mutations in the thyrotropin receptor gene cause non-autoimmune autosomal dominant hyperthyroidism. Nat. Genet. **7:** 396–401.
56. ABRAMOWICZ, M.J., L. DUPREZ, J. PARMA, et al. 1997. Familial congenital hypothyroidism due to inactivating mutation of the thyrotropin receptor causing profound hypoplasia of the thyroid gland. J. Clin. Invest. **99:** 3018–3024.
57. WEINSTEIN, L.S., A. SHENKER, P.V. GEJMAN, et al. 1991. Activating mutations of the stimulatory G protein in the McCune-Albright syndrome. N. Engl. J. Med. **325:** 1688–1695.
58. MASTORAKOS, G., N.S. MITSIADES, A.G. DOUFAS, et al. 1997. Hyperthyroidism in McCune-Albright syndrome with a review of thyroid abnormalities sixty years after the first report. Thyroid **7:** 433–439.
59. GUPTA, M.K. 2000. Thyrotropin-receptor antibodies in thyroid diseases: advances in detection techniques and clinical applications. Clin. Chim. Acta **293:** 1–29.
60. ROGER, P.P., S. REUSE, C. MAENHAUT, et al. 1995. Multiple facets of the modulation of growth by cAMP. Vitam. Horm. **51:** 59–191.
61. RICHARDS, J.S. 2001. New signaling pathways for hormones and cyclic adenosine 3′,5′-monophosphate action in endocrine cells. Mol. Endocrinol. **15:** 209–218.
62. ROGER, P.P., M. BAPTIST & J.E. DUMONT. 1992. A mechanism generating heterogeneity in thyroid epithelial cells: suppression of the thyrotropin/cAMP-dependent mitogenic pathway after cell division induced by cAMP-independent factors. J. Cell Biol. **117:** 383–393.
63. KIMURA, T., A. VAN KEYMEULEN, J. GOLSTEIN, et al. 2001. Regulation of thyroid cell proliferation by tsh and other factors: a critical evaluation of in vitro models. Endocr. Rev. **22:** 631–656.
64. LAMY, F., F. WILKIN, M. BAPTIST, et al. 1993. Phosphorylation of mitogen-activated protein kinases is involved in the epidermal growth factor and phorbol ester, but not in the thyrotropin/cAMP, thyroid mitogenic pathways. J. Biol. Chem. **268:** 8398–8401.
65. VAN KEYMEULEN, A., P.P. ROGER, J.E. DUMONT, et al. 2000. TSH and cAMP do not signal mitogenesis through Ras activation. Biochem. Biophys. Res. Commun. **273:** 154–158.
66. COULONVAL, K., F. VANDEPUT, R.C. STEIN, et al. 2000. Phosphatidylinositol 3-kinase, protein kinase B and ribosomal S6 kinases in the stimulation of thyroid epithelial cell proliferation by cAMP and growth factors in the presence of insulin. Biochem. J. **348** Pt. 2: 351–358.
67. ECCLES, N., M. IVAN & D. WYNFORD-THOMAS. 1996. Mitogenic stimulation of normal and oncogene-transformed human thyroid epithelial cells by hepatocyte growth factor. Mol. Cell. Endocrinol. **117:** 247–251.
68. SAUNIER, B., C. TOURNIER, C. JACQUEMIN, et al. 1995. Stimulation of mitogen-activated protein kinase by thyrotropin in primary cultured human thyroid follicles. J. Biol. Chem. **270:** 3693–3697.
69. CONTOR, L., F. LAMY, R. LECOCQ, et al. 1988. Differential protein phosphorylation in induction of thyroid cell proliferation by thyrotropin, epidermal growth factor, or phorbol ester. Mol. Cell. Biol. **8:** 2494–2503.
70. VAN SANDE, J., A. LEFORT, S. BEEBE, et al. 1989. Pairs of cyclic AMP analogs, that are specifically synergistic for type I and type II cAMP-dependent protein kinases, mimic thyrotropin effects on the function, differentiation expression and mitogenesis of dog thyroid cells. Eur. J. Biochem. **183:** 699–708.
71. UYTTERSPROT, N., S. COSTAGLIOLA, J.E. DUMONT, et al. 1999. Requirement for cAMP-response element (CRE) binding protein/CRE modulator transcription factors in thyrotropin-induced proliferation of dog thyroid cells in primary culture. Eur. J. Biochem. **259:** 370–378.
72. DE ROOIJ, J., F.J. ZWARTKRUIS, M.H. VERHEIJEN, et al. 1998. Epac is a Rap1 guanine-nucleotide-exchange factor directly activated by cyclic AMP. Nature **396:** 474–477.
73. DUMONT, J.E., F. LAMY, P.P. ROGER, et al. 1992. Physiological and pathological regulation of thyroid cell proliferation and differentiation by thyrotropin and other factors. Physiol. Rev. **72:** 667–697.

74. CLÉMENT, S., S. REFETOFF, B. ROBAYE, et al. 2001. Low TSH requirement and goiter in transgenic mice overexpressing IGF-I and IGF-I receptor in the thyroid gland. Endocrinology **142**: 5131–5139.
75. CHEUNG, N.W., J.C. LOU & S.C. BOYAGES. 1996. Growth hormone does not increase thyroid size in the absence of thyrotropin: a study in adults with hypopituitarism. J. Clin. Endocrinol. Metab. **81**: 1179–1183.
76. VAN KEYMEULEN, A., J.E. DUMONT & P.P. ROGER. 2000. TSH induces insulin receptors that mediate insulin costimulation of growth in normal thyroid cells. Biochem. Biophys. Res. Commun. **279**: 202–207.
77. VAN KEYMEULEN, A., S. DELEU, J. BARTEK, et al. 2001. Respective roles of carbamylcholine and cyclic adenosine monophosphate in their synergistic regulation of cell cycle in thyroid primary cultures. Endocrinology **142**: 1251–1259.
78. REUSE, S., C. MAENHAUT J.E. & DUMONT. 1990. Regulation of protooncogenes c-fos and c-myc expressions by protein tyrosine kinase, protein kinase C, and cyclic AMP mitogenic pathways in dog primary thyrocytes: a positive and negative control by cyclic AMP on c-myc expression. Exp. Cell Res. **189**: 33–40.
79. REUSE, S., I. PIRSON & J.E. DUMONT. 1991. Differential regulation of protooncogenes c-jun and jun D expressions by protein tyrosine kinase, protein kinase C, and cyclic-AMP mitogenic pathways in dog primary thyrocytes: TSH and cyclic-AMP induce proliferation but downregulate C-jun expression. Exp. Cell Res. **196**: 210–215.
80. DELEU, S., I. PIRSON, K. COULONVAL, et al. 1999. IGF-1 or insulin, and the TSH cyclic AMP cascade separately control dog and human thyroid cell growth and DNA synthesis, and complement each other in inducing mitogenesis. Mol. Cell Endocrinol. **149**: 41–51.
81. PIRSON, I., K. COULONVAL, F. LAMY, et al. 1996. C-myc expression is controlled by the mitogenic cAMP-cascade in thyrocytes. J. Cell. Physiol. **168**: 59–70.
82. COULONVAL, K., C. MAENHAUT, J.E. DUMONT, et al. 1997. Phosphorylation of the three Rb protein family members is a common step of the cAMP-, the growth factor, and the phorbol ester-mitogenic cascades but is not necessary for the hypertrophy induced by insulin. Exp. Cell Res. **233**: 395–398.
83. DEPOORTERE, F., A. VAN KEYMEULEN, J. LUKAS, et al. 1998. A requirement for cyclin D3-cyclin-dependent kinase (cdk)-4 assembly in the cyclic adenosine monophosphate-dependent proliferation of thyrocytes. J. Cell Biol. **140**: 1427–1439.
84. COPPÉE, F., F. DEPOORTERE, J. BARTEK, et al. 1998. Differential patterns of cell cycle regulatory proteins expression in transgenic models of thyroid tumours. Oncogene **17**: 631–641.
85. VAN KEYMEULEN, A., J. BARTEK, J.E. DUMONT, et al. 1999. Cyclin D3 accumulation and activity integrate and rank the comitogenic pathways of thyrotropin and insulin in thyrocytes in primary culture. Oncogene **18**: 7351–7359.
86. DEPOORTERE, F., I. PIRSON, J. BARTEK, et al. 2000. Transforming growth factor β1 selectively inhibits the cyclic AMP-dependent proliferation of primary thyroid epithelial cells by preventing the association of cyclin D3-cdk4 with nuclear p27kip1. Mol. Cell Biol. **11**: 1061–1076.
87. PIRSON, I., N. FORTEMAISON, C. JACOBS, et al. 2000. The visual display of regulatory information and networks. Trends Cell Biol. **10**: 404–408.
88. STAPLES, K.J., M. BERGMANN, K TOMITA, et al. 2001. Adenosine 3′,5′-cyclic monophosphate (cAMP)-dependent inhibition of IL-5 from human T lymphocytes is not mediated by the cAMP-dependent protein kinase A. J. Immunol. **167**(4): 2074–2080.
89. WANG, L., F. LIU & M.L. ADAMO. 2001. Cyclic AMP inhibits extracellular signal-regulated kinase and phosphatidylinositol 3-kinase/Akt pathways by inhibiting Rap 1. J. Biol. Chem. **276**(40): 37242–37249.

Cyclic AMP and the Reverse Transformation Reaction

THEODORE T. PUCK,[a,b] PATRICIA WEBB,[a] AND ROBERT JOHNSON[a]

[a]*Eleanor Roosevelt Institute, Denver, Colorado 80206, USA*

[b]*Department of Medicine and Cancer Center, University of Colorado Health Sciences Center, Denver, Colorado 80262, USA*

ABSTRACT: Traditional methods for cancer treatment have been aimed at killing the cancer cells. Unfortunately this approach all too often is accompanied by harmful killing of normal cells. The present paper describes an experimental program in our laboratory in which cancer cells are treated so as to revert to normal cell behavior. This process, which we have named *reverse transformation*, appears to offer considerable hope in the treatment of a large number of malignancies.

KEYWORDS: cyclic AMP; bromo-cAMP; dibutyryl cAMP; fibroblasts; malignancy; reverse transformation

EFFECT ON THE TRANSFORMED CHO CELL

When cAMP is added to Chinese hamster ovary (CHO) cells, the cells lose characteristics associated with malignancy and adopt characteristics of a normal fibroblast. The morphology changes from compact, randomly oriented to elongated fibroblast-like cells growing parallel to their long dimensions (FIG. 1). Cell surface knobs or blebs disappear (FIGS. 2, 3). There is an increase in the number of microtubules per unit volume of cytoplasm and a change in their distribution from a random arrangement to an orderly alignment parallel to each other and to the long axis of the cell. The cells acquire contact inhibition of growth so that growth becomes strictly monolayered instead of three dimensional (FIG. 1). The ability to grow in suspension is eliminated without altering the capacity to grow on surfaces (TABLE 1). The ability to be agglutinated and rounded by plant agglutinins and specific cell antibodies is lost. Collagen and other protein syntheses such as fibronectin are induced. Active transport of alpha ^{14}C aminobutyrate is markedly augmented. These changes are consistent with the conversion from a malignant to a normal fibroblastic state. The conversion is under genetic control as demonstrated by the production of specific mutants with altered characteristics with respect to this reaction. The effects of the change are recognizable within an hour and affect cells throughout all or most of the cell cycle. The changes in cultured Chinese hamster ovary cells produced by the ad-

Address for correspondence: Dr. Theodore T. Puck, Eleanor Roosevelt Institute, 1899 Gaylord St., Denver, CO 80206. Voice: 303-336-5654; fax: 303-333-8423.

peggy@eri.uchsc.edu

Ann. N.Y. Acad. Sci. 968: 122–138 (2002). © 2002 New York Academy of Sciences.

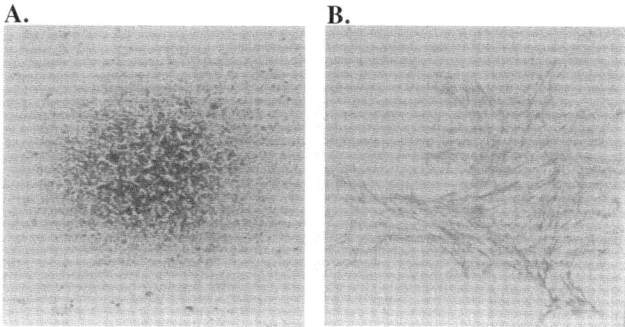

FIGURE 1. Effect of DBcAMP on cell and colonial morphology of transformed CHO cells. **A.** Colony grown in standard medium from single cell. **B.** colony grown from single cell in same medium plus 10^{-3} M DBcAMP.

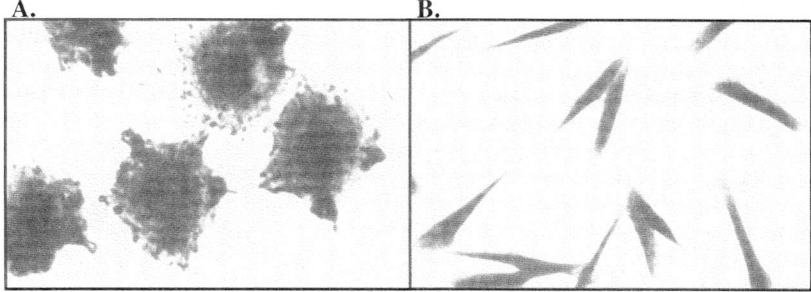

FIGURE 2. A. Demonstration of the knob-like pseudopodal structures in transformed CHO cells. **B.** The disappearance of these structures when DBcAMP plus testosterone is added to the medium.

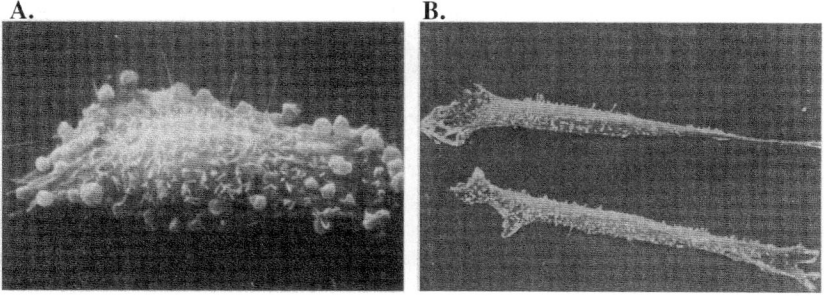

FIGURE 3. Scanning electron micrographs showing: **A.** The spherical knobs or blebs of about 2 μm in diameter. **B.** Their disappearance as the cell elongates under the influence of DBcAMP. Time-lapse photography shows these knob-like structures to be in constant oscillatory motion perpendicular to the cell surface, with a period varying from 15 to 60 seconds.

TABLE 1. Effect of DBcAMP on single cell growth of CHO cells on surfaces and in suspension

	On surface	In agar suspension
No addition	189	193
DBcAMP 5×10^{-4} testololactone 10 µg/mL	124	0

NOTE: Basal medium = F12FCM5 throughout. Number of colonies is indicated as a result of plating 200 cells.

dition and removal of dibutyryl cAMP plus testosterone are unaffected by inhibition of either RNA or protein synthesis.[1-9]

Testosterone propionate has a smaller but similar effect to cAMP when used at high concentrations with a longer incubation and promotes the action of cAMP when added at very low concentrations. The response to testosterone is specific since steroids like estradiol and hydrocortisone are inactive.

Agents like colcemid or cytochalasin B, which disorganize microtubules and microfilaments, respectively, change reverse transformed CHO cells back to the parental CHO form. They also change permanent fibroblast-like cells with smooth membranes into knobbed, epithelial-like cells similar to transformed CHO. Colcemid and cytochalasin B antagonize the cAMP action. Therefore, the conversion by cAMP of cells from the transformed to the reverse transformed or normal state is strongly dependent on the cell cytoskeleton.[1-9]

The knobs that characterize the transformed cell are in constant, violent vibratory movement, imparting to the membrane oscillatory activities in which individual knob-like structures protrude into the extracellular space to about 1/10 of the cell diameter and return within a time period of 15–30 seconds. The disappearance of these knobs and the concomitant tranquilization of the membrane is the first manifestation of the cAMP-induced reverse transformation reaction(TABLE 2). We have proposed that disintegration of the microtubular-microfibrillar system causes the actin-containing contractile elements to dissociate from the organized network and to connect at discrete points along the membrane. In these regions the contractile protein is no longer dynamically coupled to the rigid tubular structures and is free to contract with little or no load. The result, then, is a series of continuously oscillating foci at points along the membrane. Both the moderate activity that occurs in transformed cells like CHO and the more intense activity that can be elicited by the addition of sufficient cytochalasin B or colcemid to these cells or even to normal fibroblasts appear to fit this explanation.[10]

An experimental test of this model was carried out. The model demands that actin be localized in a pattern similar to that of the knobs in transformed cells and even in normal cells treated with colcemid or cytochalasin B. In a typical experiment, normal Chinese hamster fibroblasts were treated with 2 micromolar cytochalasin B for 4 hours to produce intense knob activity. Rabbit antiserum to actin was then added, followed by fluorescent goat antiserum to rabbit immunoglobulin. It was demonstrated that deposits of actin are formed over the cell in punctate regions similar to those for the distribution of knob activity. Similar deposits were also demonstrated with tubulin. These considerations also explain another cellular phenomenon. Normal fibroblasts do not display oscillatory knob or bleb activity. However, at the end

TABLE 2. The change with time after addition of cyclic AMP, of specific cellular and molecular parameters in a cell which undergoes reverse transformation

Approximate time after addition of cAMP to culture	Cellular effect observed	Molecular effect observed
15–20 min.	membrane tranquilization (disappearance of oscillating knobs on cell membrane)	phosphorylation of vimentin
0.5–6 hours	cell stretching and orientation	synthesis and phosphorylation of a cyclic AMP–induced protein (CIP)
6–48 hours	growth control instituted gene exposure reaction instituted	synthesis of several new proteins
48–96 hours	fibronectin deposition at cell membrane	synthesis of still newer proteins

of mitosis, the spindle is decomposed while the interphase microtubular-microfibrillar system is being reformed. Thus at the very end of mitosis, a state of disorganization of the microtubular microfibrillar system exists so that the temporary appearance of oscillatory knob activity even in normal cells may be expected. Time-lapse experiments revealed that in every cell examined an explosive outbreak of typical knob activity of very short duration (seconds to minutes) occurs at the end of mitosis.

cAMP also strongly affects the behavior of cell surface antigens. Thus, the addition of dibutyryl cAMP (DBcAMP) plus testololactone prevents cell killing that would otherwise occur by the addition of specific antibody and complement to transformed CHO cells. This effect, too, was shown to be completely reversible, normal killing again resulting when the cAMP derivative was removed from the medium. Dibutyryl cAMP is somewhat more effective than cAMP itself in bringing out reverse transformation. Experiments indicate that at least one role of the butyryl group is to protect cAMP from degradation by phosphodiesterase.

Attachment of mammalian cells to solid substrates requires divalent cations such as calcium and magnesium. Removal of the divalent cations forces transformed CHO cells to round and assume a spherical shape. Upon rounding, the cells lose the characteristic knob-like pseudopodal structures. The fibroblast form of CHO cells achieved with cAMP treatment is rounded little, if any, by the removal of divalent cations from the medium.

Administration of dibutyryl cAMP to transformed CHO cells induces a calcium flux reaction in which free calcium is liberated from bound reservoirs within the cell within a matter of seconds and then gradually returns to a bound state within the next 120 seconds.[11] Presumably, the phosphorylating action of cAMP triggered release of calcium to bind to phosphate groups, as is postulated further in this paper.

Cells grown in the presence of cAMP and testololactone are more tightly bound to the surfaces on which they are grown than in the absence of these agents. Since one of the properties of malignant cells is their tendency to dissociate from their tissue of origin and set up new clonal growth in other regions of the body, this would also seem an important aspect of cAMP action on the system.

We have postulated an intimate association between the cytoskeletal fibers and the nuclear envelope on one end and cell surface components on the other. Malignant cells have lost control of cell reproduction and exhibit a propensity to break away from their original anchorage and set up foci of growth elsewhere in the body. Both of these characteristics are associated with the acquisition of the transformation habitus in which the normal cytoskeletal network has suffered some degree of disorganization. The fact that cells attach to and stretch out on solid surfaces and assume a spherical shape under the action of lectins, antibodies, and proteases implies that at least an important fraction of the surface structures with which elements of the cytoskeletal system are connected are antigenic and involve protein and carbohydrate moieties.[9,12,13]

Transformed cells frequently exhibit instability in chromosome number even in clonal populations. Defects in the cytoskeletal structure could well be reflected in corresponding defects in the spindle, which is made of microtubules and which would result in errors of chromosomal distribution among the cell progeny. Thus, this formulation proposes that cytoskeletal fibers are involved not only with the regulation of cell reproduction, but also in preventing aneuploidy. Both processes are distorted when microtubules are disorganized, as in the case of cancer.

Treatment of CHO cells with cAMP restores the deposition of fibronectin around the cell membrane, a property that has been lost in the transformed cell. Colcemid, which disrupts microtubules, causes little or no disruption of the fibronectin deposit induced by cAMP derivatives; but cytochalasin D, which disrupts 5-nm microfilaments, eliminates the deposit completely. Thus, different components of the cytoskeleton appear to exert specific affects on the reverse transformation reaction.[14]

Ornithine decarboxylase (ODC) is synthesized when cells traverse the ending of the G1 phase and enter S for the initiation of DNA synthesis. Low concentrations of microtubule depolymerizing agents like colchicine suppress ODC induction almost completely in normal fibroblasts although hardly at all in the CHO cell. In fact, the transformed cell reveals a superinduction of ODC at a very low colchicine level, 10^{-7} mole. Higher concentrations of colchicine suppress ODC induction in all cells. Experiments with actinomycin-D and cycloheximide indicate that the principal colchicine action involves inhibition at the level of protein or mRNA synthesis rather than inactivation of already-synthesized enzymes. These results indicate that a microtubular system is needed to reinitiate certain steps associated with growth in G1-blocked normal cells and that a second microtubular action-terminating enzyme biosynthesis may exist. This complex microtubular control appears to be defective in transformed CHO cells.[15]

Work from a number of laboratories has demonstrated that the cytoskeleton exerts a variety of different and complex regulatory metabolic actions in the biochemical economy of the mammalian cell. Among these are the demonstration of a DBcAMP-caused increase in tyrosineaminotransferase activity in rat hepatoma cells. However, in this case colcemid and DBcAMP reinforce rather than antagonize each other.[16]

Experiments have demonstrated that bromo-cyclic AMP (BrcAMP) can induce apoptosis in transformed CHO cells. The proliferation of suspension cultures of transformed CHO cells was inhibited by 0.5 µm BrcAMP treatment and restored by

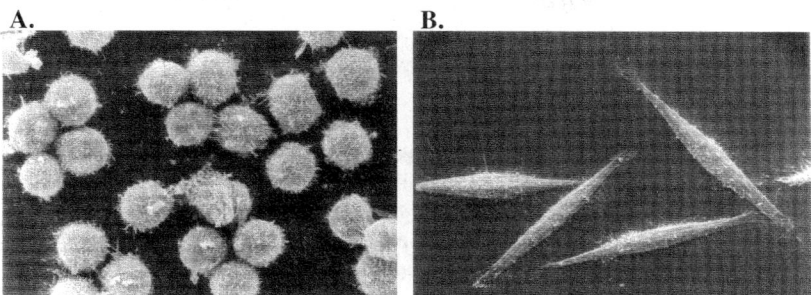

FIGURE 4. A. Vole cells transformed by avian sarcoma virus. **B.** Cells as in A treated with DBcAMP plus testololactone.

its removal. This treatment also inhibited histone H1 phosphorylation completely, reduced histones H2A and H4 phosphorylations, induced DNA degradation, and produced cells containing micronuclei. Agarose gel electrophoresis of the degraded DNA fragments produced the typical ladder pattern, confirming that the cells were undergoing apoptosis. BrcAMP did not stop cells in late G1, S, G2, or M from traversing the cell cycle and dividing, but induced apoptosis following mitosis. The restriction point of BrcAMP arrest was located in the middle of the wider band of G1 arrest induced by isoleucine deprivation. Cells synchronized in G1 before the restriction point were held in G1 arrest by BrcAMP and spared apoptotic death. The restoration of apoptosis to the transformed CHO is another indicator of the return to normal cell behavior.[17]

CHO mutants were selected for unresponsiveness to the reverse transformation reaction of cAMP derivatives. Phosphorylation analysis was carried out by 2D gel electrophoresis. Seven differences in protein phosphorylations in the parental CHO cell were identified as a result of treatment with dibutyryl cAMP. Unresponsive mutants of two kinds were found. One had lost all seven of the phosphorylation changes induced by dibutyryl cAMP in the wild-type cell. However, the other type, equally resistant to reverse transformation by cAMP derivatives, differs from the parental cell in only one (or possibly two) phosphorylation events involving a 55,000-dalton protein. Phosphorylation of this protein, which is not yet identified, may therefore be directly related to transformation and its reversal in the CHO cell.[18]

Not all cancers respond to cAMP. The HeLa cell, for example, does not display a typical reverse transformation response. On the other hand, vole cells transformed by avian sarcoma virus carrying the *src* gene display typical reverse transformation when treated with cAMP. The transformed cells, which have lost their fibroblastic morphology, reorganize the cytoskeleton with a typical fibronectin deposit around the cell membrane and no longer grow in agar (FIG. 4). Moreover, the cAMP treatment does not inactivate the *src* gene product. These data imply that cAMP exerts its effect in the same pathway at or after the point in this regulatory chain influenced by the *src* gene product, which mainly phosphorylates tyrosine moieties. The decision to adapt the transformed or the normal state may be determined by the degree to which the *src* gene on the one hand, or cAMP-mediated kinase activities on the other, predominates in the cell. However, the development of all the transformation characteristics noted as a result of introduction of the *src* gene and their coordinate reversal by cAMP support the thesis that in the normal fibroblast all these properties are part of a common cAMP-induced regulatory system involving the cytoskeleton.[9,12,13,19]

A theory devised to account for these and other facts of differentiation and its distortion in malignancy was developed as follows. There are at least two levels of regulation of gene expression in mammalian cells. The first is conversion of the genes in question from a sequestered state, in which they are unable to interact with molecules in the environment, to an exposed state in which they can be so influenced. The next step is the turning of such genes on and off by appropriate effector molecules in the nucleus. We proposed that the cytoskeleton transmits information from the cell membrane to chromosomal loci in the nucleus and is an important element in regulation of the exposure process. Thus, the cytoskeleton is a critical structure both in normal differentiation and in its distortions like that of cancer. Of course, cancer could result in damage to regulatory processes beyond the cytoskeletal stage, and such cancer cells would fail to display the phenomena of reverse transformation when treated with agents like cAMP.[9,12,13]

This picture predicts that the CHO and similar cells that undergo reverse transformation should also display increased genome exposure when treated with reverse transformation agents. Such was, indeed, found to be the case. Genome exposure was measured quantitatively by isolating cell nuclei and treating them for a measured period with DNase I, an enzyme that hydrolyzes any DNA with which it makes contact. The small–molecular weight DNA released into the supernatant by this treatment was then quantitatively measured spectrophotometrically. A typical experiment demonstrating the increase in genome exposure in a series of transformed cells under the influence of their specific reverse transformation agents is shown in TABLE 3.[20]

The general pattern of DNA hydrolysis exhibited by all cells consists of a curve that at first rises sharply within increased DNase I and then becomes almost horizontal, indicating that roughly half of the nuclear DNA is highly sequestered. When CHO cells were compared with their reverse-transformed homologues achieved by dibutyryl cAMP, the transformed form displayed less genome exposure at every

TABLE 3. Demonstration that four transformed cell strains display reduced nuclear DNA exposure as compared with their more normal or reverse-transformed counterparts

Malignant form of cell	More normal form	Range of DNase I concentrations	Mean OD_{260} of malignant form	Mean OD_{260} of normal form
Raszip6	3T3	60–120 U/mL	0.120 ± 0.065	0.395 ± 0.125
CHO	CHO + DBcAMP	25–60 U/mL	0.145 ± 0.051	0.349 ± 0.085
HL60	HL60 + retinoic acid	20–50 U/mL	0.195 ± 0.088	0.635 ± 0.333
PC12	PC12 + NGF	25–80 U/mL	0.066 ± 0.020	0.221 ± 0.119

NOTE: CHO cells were treated with 1.5 mM DBcAMP for 24 h; HL60 were treated with 3 μM retinoic acid for 72 h; PC12 were treated with 0.77 nM NGF for 19–72 h.

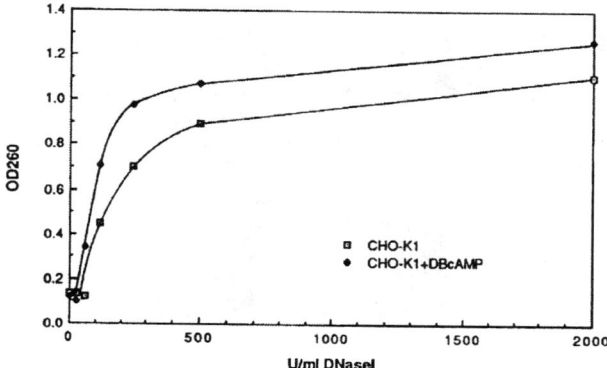

FIGURE 5. A typical experiment measuring the amount of DNA released after treatment of isolated nuclei with varying concentrations of DNase I. The *top curve* presents data from CHO cells treated with 1.5 mM DBcAMP, while the *bottom curve* is for the same cells in standard growth medium. Other pairs of transformed and normal or reverse-transformed cells demonstrate similar curves but with minor variations. While there were minor displacements of such pairs of curves from day to day, the general relationship between the two curves was highly reproducible. An OD_{260} of 1000 represents approximately 50% of the total cellular DNA.

DNase I dose tested (FIG. 5). Colcemid prevents the increase of exposure of CHO by DBcAMP, but it must be administered before or simultaneously with the latter compound. Calcium is also necessary for maintenance of DNA in the more highly exposed state characteristic of the reverse-transformed phenotype. Thus, if calcium is excluded from the final buffer in which the nuclei are suspended, no cAMP effect on genome exposure is obtained.[20]

Experiments have demonstrated that the action of cAMP in increasing genome exposure is gene specific. Thus, in the examination of 47 different genetic loci, some, like ribosomal RNA genes, were found to be sensitive to DNase I hydrolysis, in both the absence and the presence of cAMP; some were resistant under both conditions; and some were resistant in the untreated cell but became sensitive after cAMP treatment. This specificity of gene exposure due to the reverse transformation process emphasizes its role in genome regulation.[21]

In situ nick translation experiments show three principal genome exposure regions in the nucleus: around the periphery, around the nucleoli, and in punctate positions in the interior of the nucleus. Confocal microscopy illuminates the characteristic genome exposure patterns of normal, transformed, and reverse-transformed cells. In FIGURE 6 are shown confocal sections through the transformed CHO and the reverse-transformed cell. Normal cells present a pattern like those of FIGURE 6B. The normal cell and the reverse-transformed cell yield similar patterns, showing a highly exposed rim around the nuclear periphery. Both the normal cell and the reverse-transformed cell exhibit this pattern. The transformed cell reveals an absence or sometimes great attenuation of the peripheral rim but does show internal sites of exposure, particularly around nucleoli. Confocal microscopy, by permitting examination of sections throughout the nucleus, makes possible clear identification of re-

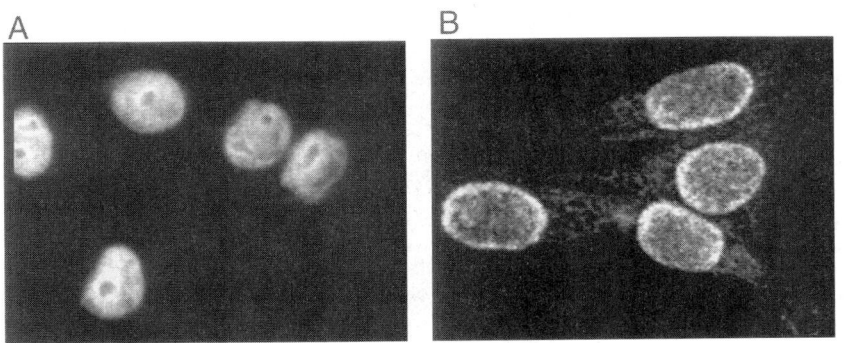

FIGURE 6. Confocal center sections of (**A**) untreated CHO-K1 with the cancerous phenotype and (**B**) reverse-transformed CHO-K1.

gions of exposed and sequestered DNA. The data support the picture that the peripheral nuclear shell of exposed DNA contains differentiation-specific genes that include the specific growth control genes that are functional in normal cells but not in cancer. The exposed genes surrounding the nucleoli may well represent housekeeping genes active in both normal and cancer cells. The DNase I–resistant DNA in the interior of the nucleus is postulated to consist, for the most part, of genes specific to alternative differentiation states and to be sequestered and inactive in the particular cell studied.[22]

This picture would predict aspects of the behavior of the X chromosomes in normal female cells, one of which is strongly active while the other is largely, but not completely, inactive. FISH experiments with X chromosome probes applied to normal human female cells indicated that both X chromosomes were in the periphery of the nucleus. However, one of these was always highly condensed, while the second one was diffused and indicated the presence of invisible regions of the chromosome interspersed with the fluorescent ones. This is exactly the behavior to be expected since one X chromosome, the Barr body, is largely inactive and therefore would be expected to be highly condensed and so to yield a large fluorescent signal. The other X chromosome, having many active DNA sites, would have these exposed as uncondensed DNA whose fluorescence was not great enough to exceed the background level detectable by our instrument. To test this interpretation further, a similar experiment was carried out on cells of a human patient with XXX syndrome. Exactly as was expected, all three of the X chromosomes were in the periphery of the nucleus; two of the chromosomes were condensed and intensely fluorescent, while the third was diffused with bright patches alternating with dark patches, indicating regions of exposure and activity alternating with inactive regions (FIG. 7).[23] Similarly, at least one of each pair of chromosomes 2 and 3 were found in the periphery of the nucleus of normal human fibroblasts.

On the other hand, FISH experiments with a probe of human chromosome 21, which contains ribosomal genes, revealed this chromosome to be localized at the nucleolus and the interior of the nucleus. Thus, the conclusion may be drawn that dif-

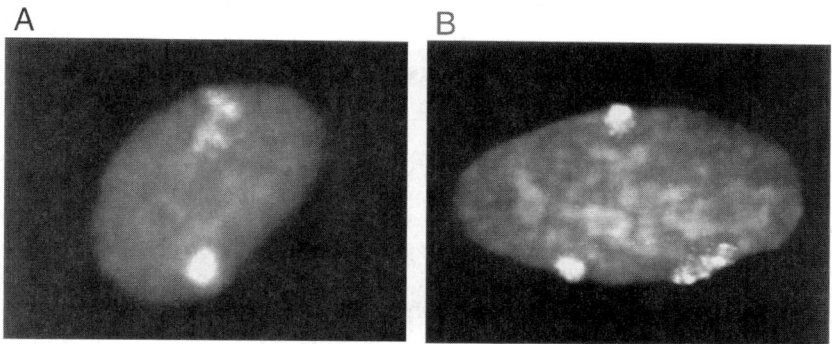

FIGURE 7. A. A cultured G_1-phased normal human female fibroblast treated with X painting probe by FISH procedure. Both chromosomes approximate the nuclear periphery. The inactive X (as confirmed by Barr body staining) is highly condensed, but the central condensed area is surrounded by bright punctate dots. The active X is diffuse, encompassing a greater overall area; is interrupted all through its area by dark spots and also displays bright punctate dots around the periphery. **B.** A cultured G_1-phased human female XXX fibroblast treated with X chromosome painting probe as in A. As expected, two condensed fluorescent spots and one diffuse pattern are obtained, each resembling its prototype in A.

ferent regions of the nucleus are specialized for different gene activations. Therefore, a mechanism must exist for transporting specific chromosomal loci to their particular activation sites within the nucleus. We have proposed that the nuclear fibers participate in this process, as do the noncoding regions of the nucleus and particularly the repetitive regions; and that the nuclear periphery is the site of action of differentiation genes, while at least some of the housekeeping genes are in the interior, associated with nucleoli.

Reports confirming these proposals have appeared from various laboratories. Thus, it has been demonstrated that in PC12 cells reverse transformed by nerve growth factor, the nuclear DNA–sensitive sites migrate to the periphery of the nucleus.[24] Similarly, another paper has been published describing how a nuclear shell of exposed DNA is produced when U937 tumor cells are reverse transformed by transfection with $p21^{waf-1}$. Also in accordance with our theory of genome exposure, these latter authors demonstrated by FISH movement of the centromeres of chromosomes 13, 16, 17, and 21 to the nuclear periphery in the course of the reverse transformation of the tumor cell.[25]

In accordance with this theory, it was also shown by us that the degree of DNA exposure exhibits considerable variation between cells of different normal tissues, a fact to be expected since different tissues would exhibit different activities of different differentiation genes. These findings open up the possibility of using genome exposure methodology to identify the genes of each tissue that participate in its specific differentiation function.

Vimentin is only slightly phosphorylated in transformed CHO cells but is heavily phosphorylated in normal fibroblasts. cAMP addition almost triples the vimentin phosphorylation of CHO cells but does not change that of normal cells. Vimentin phosphorylation is one of the earliest phenomena to occur after addition of cAMP to

CHO cells, and it precedes the cell stretching and other manifestations of reverse transformation.

Vimentin appears as a condensed mass in transformed CHO-K1 cells but has a filamentous structure in the reverse-transformed cell. Thus, it seems likely that phosphorylation of cytoskeletal elements initiates a large-scale genetic regulatory action in which a substantial change in the spectrum of genome exposure and sequestration occurs. Early during reverse transformation of CHO cells by cAMP, a new protein, molecular weight 43,000 and pI of 5.3, was rapidly and specifically induced. This protein is a phosphorylated actin homologue whose induction requires new protein synthesis, and which appears to play a critical role in some cancerous processes.[26]

The gene exposure reaction initiated by cAMP, like other aspects of the reverse transformation reaction, is prevented by agents like colcemid or cytochalasin B, which specifically disorganize components of the cytoskeleton. Thus, cAMP is necessary for organization of the cell cytoskeleton, which in turn is involved in the sequestration or exposure of a large number of specific cell genes. Thus, different states of differentiation appear to be characterized by different specific sets of exposed genes. By this picture, oncogenes would include genes that regulate cytoskeletal structure and genes that control the level of cAMP within the cell. Less than 10% of the human genes are coding genes. The remainder, which includes a large number of repetitive DNA sequences, contain essential DNA structures, which we have suggested regulate the expression of the coding genes. Their constitution is not randomized, as would be expected if they were functionless, but is rigorously maintained; and, indeed, parts of these noncoding regions can be used to establish individual identity in a fashion even more secure than fingerprinting.

If noncoding repetitive sequences are, indeed, involved in exposure, one might well expect to find specific proteins binding to them in order that they may be drawn to their specific region in the nucleus where appropriate activation can be carried out. Such protein attachment to specific DNA-repetitive sites might well be different in cancer and normal cell states. Analysis by gel retardation assay showed that a protein of 66 kDa binds to a human low-repeat repetitive sequence (LRS) in each of the 10 transformed human cell lines examined. This protein-DNA interaction was not observed in 11 normal human cell cultures. In a histiocytic human cell line, U937, which can be induced to differentiate in the presence of phorbol ester, this binding disappeared after cell differentiation.[27]

Our model has proposed that damage to the cytoskeletal and nuclear fiber system can lead to malignant transformation in three ways: by distorting the metabolic patterns necessary for normal biochemical and reproductive regulation; by producing karyotype instability in mitosis, which promotes evolutionary drift toward malignancy through continuous selection of cells that resist signals for reproductive arrest; and by preventing apoptotic death.

This model explains a number of phenomena of malignancy. Transformed cells arising from different tissues may display similar morphological characteristics because they have lost elements of their internal fiber systems that maintain the individual differentiated morphologies and growth behavior. Therefore, they tend toward a similar appearance. The existence of the precancerous state and the frequent development of resistance to therapeutic agents employed in cancer therapy are natural consequences of this picture. We have shown[28] that the pulsating knob behavior exhibited by many transformed cells throughout interphase is readily explained by these considerations, as is the fact that even in normal cells such pulsating membranous knobs exist for an extremely short period after telophase before the organized interphase state of the new microfibrillar system has had time to become established. We have also proposed that the microfibrils are connected, however indirectly, to specific elements in the cell membrane. This is necessary for the maintenance of appropriate cell morphologies and is supported by the fact that antibodies to specific membrane components, lectins, and proteases all cause a transformed CHO cell to become spherical. In all cases the attachment of the fiber system to the membrane is damaged.

The facts of reverse transformation demonstrate a chain of cAMP-triggered events involving the cytoskeleton and extending from the cell membrane to the nuclear chromatin. They imply a continuous channel of information transmission extending from specific sites on the membrane to specific regions of the genome that control expression of normal or transformed behavior in CHO cells.

EFFECT OF cAMP ON BRAIN CELL HYBRIDS

The experiments so far described demonstrate the role of cAMP action in causing a transformed cell like CHO to revert to the normal fibroblast state. These results, implying that cAMP can restore to normal functioning a distorted differentiation process, would lead to the expectation that one could demonstrate cAMP regulation in other kinds of differentiation behavior. A series of experiments was carried out with a hybrid formed from transformed CHO cells and normal cells obtained from a human or Chinese hamster brain. Selected cloned hybrids from such a fusion yielded a very different behavior from that of transformed CHO alone when treated with cAMP derivatives. The main body of the hybrid cell condensed, and neurite-like processes were extended. These processes united with those from adjacent cells to form a continuous network (FIG. 8).[29] Scanning and transmission electron microscopy studies were performed on a hybrid resulting from the Sendai virus fusion of a Chinese hamster ovary glycine auxotrophic mutant cell with a freshly biopsied suspension of Chinese hamster cerebral cortex cells. In normal growth medium, the hybrid differs from the CHO parental cell in displaying a squamous, polygonal, epithelial cell-like appearance with sparse microvilli and filamentous knobs or blebs. Slender filapodia, which sometimes reach a length of 25 μm, extend from interphase cells. Bundles of microfilaments are closely associated with the cell membrane in the perinuclear regions arranged more or less in parallel to the underlying substrate. The untreated hybrid has a relatively unpatterned arrangement of microtubules and reveals desmosomes at points of cell contact. When treated with DBcAMP plus the synergist testololactone, the cells respond by becoming stellate, forming cytoplasmic extensions that radiate from a central microvillus-covered rounded cell body. The arborizing processes number on the average five per cell and form a contiguous network between the cells of a colony.[30] As in reverse transformation, testololactone synergized this process. Extremely small concentrations of colcemid, 3×10^{-7} M, caused complete inhibition of this process formation. However, cytochalasin B, in a concentration of 2 μM, which mimics the action of colcemid in reverse transformation,

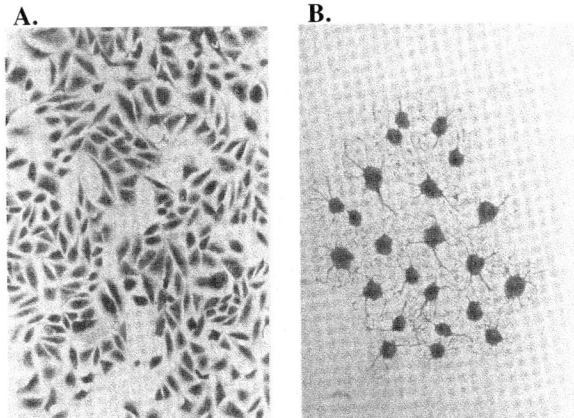

FIGURE 8. Morphology of HB-CHO(+) no. 3 hybrid cells before and after treatment with DBcAMP + testololactone. **A.** Large population of cells cultivated in standard growth medium and then changed to serum-free medium. **B.** Colony of cells as in **A**, but treated with DBcAMP plus testololactone after change to serum-free medium.

showed no effect on process extension in the CHO-brain hybrid. Therefore, while microtubules are required for this action of cAMP, the microfilaments appear not to be involved. This extension of arborized processes, like reverse transformation in the CHO cell itself, is completely reversible when the cAMP is removed. Agents preventing protein synthesis had no effect on the extension of neurite-like processes. It would appear that cAMP is necessary for organization of the microtubular system within various types of cells and that the resulting organized structure is a necessary concomitant of various specific differentiation functions.[29]

ROLE OF PHOSPHORYLATION

The ubiquitous presence of phosphorylation in cellular metabolic pathways raises the question of the role of this moiety in metabolic processes. Phosphorylation is part of the structure of DNA, RNA, a huge number of proteins, and various small molecules. Phosphorylation and dephosphorylation reactions, therefore, constitute one of the most important forms of metabolic regulatory processes.

The phosphate group contains a hydrogen ion that is virtually 100% ionized, thus leaving a very strong and highly localized negative charge at the site of phosphorylation. Electrostatic forces of this kind determine the conformation of macromolecules, since other parts of such molecules that are positively charged will be attracted, and those that are negatively charged will be repelled.

From the phosphorylation site the electrostatic charge distribution as well as the induced multipolar charges determine the closeness of approach of other charged molecular entities. In this connection it should be noted that closeness of approach, which determines reactivity, may also depend on the presence of lipoidal molecules

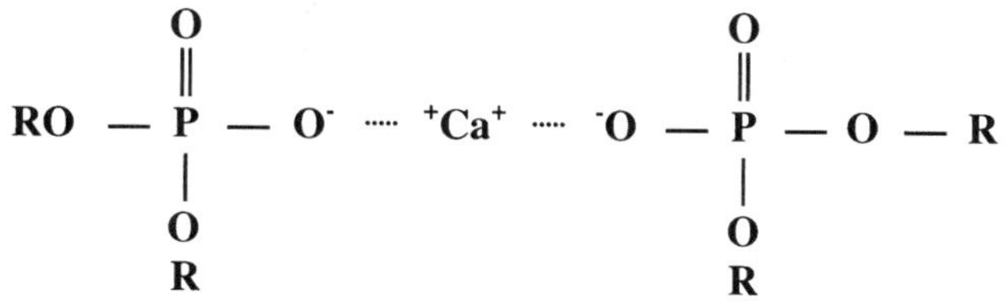

$$\text{RO} - \underset{\underset{\text{R}}{\overset{|}{\text{O}}}}{\overset{\overset{\text{O}}{\|}}{\text{P}}} - \text{O}^- \cdots \text{}^+\text{Ca}^+ \cdots \text{}^-\text{O} - \underset{\underset{\text{R}}{\overset{|}{\text{O}}}}{\overset{\overset{\text{O}}{\|}}{\text{P}}} - \text{O} - \text{R}$$

FIGURE 9. Demonstration of how the divalent Ca^{++} ion could unite two different phosphate groups and so bring about closeness of approach, induced dipoles, and electronic perturbations in the parent molecules which would make possible specific chemical reactions. The *dotted lines* represent electrostatic bonds; the *solid lines* are covalent bands.

in the space between the two moieties. The much lower dielectric constant of lipid as compared to water (2 vs. 80) enormously increases the electrostatic attraction between the two structures and so could strongly promote chemical interaction.[31] In this way phosphate groups placed on particular molecules at particular times during the cell life cycle could determine chemical reactivities of many cellular metabolic changes.

Of similar importance would be ionic components of the medium. For example, calcium ion could form bridges between two particular phosphates, as shown in FIGURE 9. Secondary induced electrostatic forces can arise when two surfaces are brought into close juxtaposition such that activation of specific chemical bonds can be brought about and a highly specific chemical reaction can ensue, as is characteristic of biological, enzymatically catalyzed reactions. Virus activity is also strongly mediated by inorganic cations in the medium and may therefore also involve specific phosphorylation groups.[32]

We have shown that reverse transformation can be elicited by cAMP derivatives in several (though not all) cancer cell lines and that at least some of the phenomena of reverse transformation can be produced in particular malignant cells by chemical agents other than cAMP derivatives.[20]

Sachs and coworkers have demonstrated reverse transformation phenomena by association of malignant cells with other cells or treatment with cell-derived products.[33-52] Waddell has shown that treatment of intraabdominal and abdominal wall desmoid tumors with drugs that affect the metabolism of cAMP can produce gratifying results.[53]

There seems to be good reason for hope that reverse transformation will be an important principle in cancer therapy.

CONCLUSIONS

cAMP, through control of phosphorylation, regulates the structure of the cell cytoskeleton, which in turn strongly influences many differentiation properties including the assumption of normal or cancerous behavior in mammalian cells. A theory of genome regulation involving migration of specific DNA loci to different regions of the nucleus, involving transfer of information from specific molecules in the cell membrane to the DNA in the nucleus, and involving the cell fiber system as a critical element, is presented that explains a variety of different aspects of cell behavior.

ACKNOWLEDGMENTS

This is contribution number 1802 from the Eleanor Roosevelt Institute and the Thomas G. and Mary Vessels Laboratory for Molecular Biology. This work was supported by grants from the Disease Prevention Fund, the Lucille P. Markey Charitable Trust, and The Frost Foundation, Ltd. The authors wish to thank M.H. Puck for editorial assistance.

REFERENCES

1. HSIE, A.W. & T.T. PUCK. 1971. Morphological transformation of Chinese hamster cells by dibutryryl adenosine cyclic 3':5'-monophosphate and testosterone cyclic AMP. Proc. Natl. Acad. Sci. USA **68:** 358–361.
2. PUCK, T.T. 1973. Genetic biochemical studies on the mammalian cell surface. In The Role of Cyclic Nucleotides in Carcinogenesis. J. Schultz & H.G. Gratzner, Eds.: 283–302. Academic Press. New York.
3. PORTER, K.R., T.T. PUCK, A.W. HSIE & D. KELLEY. 1974. An electron microscope study of the effects of dibutryryl cyclic AMP on Chinese hamster ovary cells. Cell **2:** 145–162.
4. HSIE, A.W., C. JONES & T.T. PUCK. 1971. Further changes in differentiation state accompanying the conversion of Chinese hamster cells to fibroblastic form by dibutyryl adenosine cyclic 3':5'-monophosphate and hormones cyclic AMP. Proc. Natl. Acad. Sci. USA **68:** 1648–1652.
5. PUCK, T.T., C.A. WALDREN & A.W. HSIE. 1972. Membrane dynamics in the action of dibutyryl adenosine 3':5'-cyclic monophosphate and testosterone on mammalian cells, cyclic AMP. Proc. Natl. Acad. Sci. USA **69:** 1943–1947.
6. HSIE, A.W. & T.T. PUCK. 1972. Production of variants with respect to reverse transformation in cultured Chinese hamster cells. American Society for Cell Biology Meeting, St. Louis, MO, November 8–11, 1972. J. Cell Biol. **55:** 118A.
7. PATTERSON, D. & C.A. WALDREN. 1973. The effect of inhibitors of RNA and protein synthesis on dibutyryl cyclic AMP mediated morphological transformations of Chinese hamster ovary cells in vitro. Biochem. Biophys. Res. Commun. **50:** 566–573.
8. PUCK, T.T. 1977. Cyclic AMP, the microtubule-microfilament system, and cancer. Proc. Natl. Acad. Sci. USA **74:** 4491–4495.
9. PUCK, T.T. & A. KRYSTOSEK. 1992. The role of the cytoskeleton in genome regulation and cancer. Int. Rev. Cytol. **132:** 75–108.
10. MEEK, W. & T.T. PUCK. 1979. Role of the microfibrillar system in knob action of transformed cells. J. Supramol. Struct. **12:** 335–354.
11. BUNN, P.A., D.G. DIENHART, D. CHAN, et al. 1990. Neuropeptide stimulation of calcium flux in human lung cancer cells: delineation of alternative pathways. Proc. Natl. Acad. Sci. USA **87:** 2162–2166.
12. PUCK, T.T. & A. KRYSTOSEK. 1993. Reverse transformation, genome exposure, and cancer. In Advances in Cancer Research. G.F. Vande & G. Klein, Eds.: 125–151. Academic Press. San Diego, CA.
13. SCHONBERG, S., D. PATTERSON & T.T. PUCK. 1983. Resistance of Chinese hamster ovary cell chromatin to endonuclease digestion. Exp. Cell Res. **145:** 57–62.
14. NIELSON, S.E. & T.T. PUCK. 1980. Deposition of fibronectin in the course of reverse transformation of Chinese hamster ovary cells by cyclic AMP. Proc. Natl. Acad. Sci. USA **77:** 985–989.
15. RUMSBY, G. & T.T. PUCK. 1982.. Ornithine decarboxylase induction and the cytoskeleton in normal and transformed cells. J. Cell. Physiol. **111:** 133–139.

16. PUCK, T.T. 1984. Transformation and reverse transformation in mammalian cells. Adv. Viral Oncol. **4:** 197–216.
17. GURLEY, L.R., A.L. JANDACEK, J.G. VALDEZ, et al. 1998. Br-cAMP induction of apoptosis in synchronized CHO cells. Somatic Cell Mol. Genet. **24:** 173–190.
18. GABRIELSON, E.G., C. SCOGGIN, & T.T. PUCK. 1982. Phosphorylation changes induced by cAMP derivatives in the CHO cell and selected mutants, cyclic AMP. Exp. Cell Res. **142:** 63–68.
19. PUCK, T.T., R.L. ERIKSON, W.D. MEEK & S.E. NIELSON. 1981. Reverse transformation of vole cells transformed by avian sarcoma virus containing the src gene. J. Cell. Physiol. **107:** 399–412.
20. PUCK, T.T., P. WEBB & R. JOHNSON. 1998. Genome exposure and regulation in mammalian cells. Somatic Cell Mol. Genet. **24:** 291–302.
21. ASHALL, F., N. SULLIVAN & T.T. PUCK. 1988. Specificity of the cAMP-induced gene exposure reaction in CHO cells. Proc. Natl. Acad. Sci. USA **85:** 3908–3912.
22. PUCK, T.T., M. BARTHOLDI, A. KRYSTOSEK, et al. 1991. Confocal microscopy of genome exposure in normal, cancer, and reverse-transformed cells. Somatic Cell Mol. Genet. **17:** 489–503.
23. PUCK, T.T. & R. JOHNSON. 1996. DNA exposure and condensation in the X and 21 chromosome. Stem Cells **14:** 548–557.
24. PARK, P.C. & U. DE BONI. 1996. Transposition of DNase hypersensitive chromatin to the nuclear periphery coincides temporally with nerve growth factor-induced regulation of gene expression in PC12 cells. Proc. Natl. Acad. Sci. USA **93:** 11646–11651.
25. LINARES-CRUZ, G., H. BRUZZONI-GIOVANELLI, V. ALVARO, et al. 1998. p21WAF-1 reorganizes the nucleus in tumor suppression. Proc. Natl. Acad. Sci. USA **95:** 1131–1135.
26. MIRANTI, C. & T.T. PUCK. 1990. Gene regulation in reverse transformation: cyclic AMP-induced actin homolog in CHO cells. Somatic Cell Mol. Genet. **16:** 67–78.
27. LAW, M.L., J. GAO & T.T. PUCK. 1989. A nuclear protein associated with human cancer cells binds preferentially to a human repetitive DNA sequence. Proc. Natl. Acad. Sci. USA **86:** 8472–8476.
28. PUCK, T.T. 1977. Cyclic AMP, the microtubule-microfilament system, and cancer. Proc. Natl. Acad. Sci. USA **74:** 4491–4495.
29. KAO, F.T., P. FAIK & T.T. PUCK. 1979. Extension of branching processes from hybrids of brain and Chinese hamster ovary cells. Exp. Cell Res. **122:** 83–91.
30. MEEK, W.D., R.R. PORTER & T.T. PUCK. 1980. The ultrastructure of process formation following treatment with db-cAMP of a Chinese hamster ovary x Chinese hamster brain cell hybrid, cyclic AMP. Exp. Cell Res. **126:** 359–374.
31. PUCK, T.T. 1994. On the role of lipids in cell regulatory economy. Somatic Cell Mol. Genet. **20:** 437–438.
32. PUCK, T.T., A. GAREN & J. CLINE. 1951. The mechanism of virus attachment to host cells. I. The role of ions in the primary reaction. J. Exp. Med. **93:** 65–88.
33. FIBACH, E., T. LANDAU & L. SACHS. 1972. Normal differentiation of myeloid leukaemic cells induced by a differentiation-inducing protein. Nat. New Biol. **237:** 276–278.
34. VLODAVSKY, I., E. FIBACH & L. SACHS. 1975. Control of normal differentiation of myeloid leukemic cells. X. Glucose utilization, cellular ATP and associated membrane changes in D+ and D– cells. J. Cell Physiol. **87:** 167–177.
35. FIBACH, E. & L. SACHS. 1976. Control of normal differentiation of myeloid leukemic cells. XI. Induction of a specific requirement for cell viability and growth during the differentiation of myeloid leukemic cells. J. Cell Physiol. **89:** 259–266.
36. LOTEM, J. & L. SACHS. 1983. Control of in vivo differentiation of myeloid leukemic cells. III. Regulation by T lymphocytes and inflammation. Int. J. Cancer **32:** 781–791.
37. LOTEM, J. & L. SACHS. 1996. Control of apoptosis in hematopoiesis and leukemia by cytokines, tumor suppressor and oncogenes. Leukemia **10:** 925–931.
38. SACHS, L. 1997. Cytokine and apoptosis gene networks that control development and cancer. J. Cell. Physiol. **173:** 126–127.
39. SACHS, L. 1993. The molecular control of hemopoiesis and leukemia. C R Acad. Sci. Ser. III **361:** 871–891.

40. SACHS, L. 1989. The molecular control of normal and leukemic hematopoiesis: myeloid cells as a model system. Ann. N.Y. Acad. Sci. **567:** 141–153.
41. LOTEM, J. & L. SACHS. 1997. Cytokine suppression of protease activation in wild-type p53-dependent and p53-independent apoptosis. Proc. Natl. Acad. Sci. USA **94:** 9349–9353.
42. RABINOWITZ, Z. & L. SACHS. 1970. Control of the reversion of properties in transformed cells. Nature **225:** 136–139.
43. PARAN, M., L. SACHS, Y. BARAK & P. RESNITZKY. 1970. In vitro induction of granulocyte differentiation in hematopoietic cells from leukemic and non-leukemic patients. Proc. Natl. Acad. Sci. USA **67:** 1542–1549.
44. SYMONDS, G. & L. SACHS. 1982. Cell competence for induction of differentiation by insulin and other compounds in myeloid leukemic clones continuously cultured in serum-free medium. Blood **60:** 208–212.
45. LI, Q., L. SACHS, Y.B. SHI & A.P. WOLFFE. 1999. Modification of chromatin structure by the thyroid hormone receptor. Trends Endocrinol. Metab. **10:** 157–164.
46. LOTEM, J. & L. SACHS. 1996. Differential suppression by protease inhibitors and cytokines of apoptosis induced by wild-type p53 and cytotoxic agents. Proc. Natl. Acad. Sci. **93:** 12507–12512.
47. SACHS, L. 1996. Molecular control of development in normal and leukemic myeloid cells by cytokines, tumor suppressor and oncogenes. Curr. Topics Microbiol. Immunol. **211:** 3–5.
48. SACHS, L. & J. LOTEM. 1993. Control of programmed cell death in normal and leukemic cells: new implications for therapy. Blood **82:** 15–21.
49. SACHS, L. 1993. Regulators of normal development and tumor suppression. Int. J. Dev. Biol. **37:** 51–59.
50. LOTEM, J. & L. SACHS. 1992. Regulation of leukaemic cells by interleukin 6 and leukaemia inhibitory factor. Ciba Found. Symp. **167:** 88–99.
51. SACHS, L. 1989. The molecular control of normal and leukemic hematopoiesis: myeloid cells as a model system. Ann. N.Y. Acad. Sci. **567:** 141–153.
52. LOTEM, J. & L. SACHS. 1988. In vivo control of differentiation of myeloid leukemic cells by recombinant granulocyte-macrophage colony-stimulating factor and interleukin 3. Blood **71:** 375–382.
53. WADDELL, W.R. 1975. Treatment of intra-abdominal and abdominal wall desmoid tumors with drugs that affect the metabolism of 3′, 5′-adenosine monophosphate, cyclic AMP. Ann. Surg. **181:** 299–302.

Protein Kinase A as Target for Novel Integrated Strategies of Cancer Therapy

GIAMPAOLO TORTORA AND FORTUNATO CIARDIELLO

Cattedra di Oncologia Medica, Dipartimento di Endocrinologia e Oncologia Molecolare e Clinica, Università di Napoli "Federico II, 80131 Napoli, Italy

> ABSTRACT: We have studied the role of protein kinase A (PKA) in neoplastic transformation, apoptosis, and angiogenesis and its relationship with other signaling molecules, as a basis for developing novel therapeutic strategies. We demonstrated the involvement of PKA type I (PKA-I) in the transduction of mitogenic signals from different sources and demonstrated functional and structural interactions between PKA-I and the activated epidermal growth factor receptor (EGFR). We contributed to the identification and development of several selective inhibitors of PKA-I, such as 8-Cl-cAMP and a hybrid DNA/RNA antisense oligonucleotide of a novel class (AS-PKA-I) and of EGFR, including mAbC225 and ZD1839 (Iressa). All these agents have been investigated in cancer patients. We demonstrated the therapeutic potential of the combined blockade of PKA-I and EGFR, reporting a synergistic antitumor effect when their inhibitors are used in combination. We have also shown that PKA-I and EGFR inhibitors are able to cooperate with selected class of cytotoxic drugs and with ionizing radiation, causing a synergistic inhibition of tumor growth *in vitro* and *in vivo*, accompanied by inhibition of expression of growth and angiogenic factors and by suppression of vessel production. Moreover, PKA-I is implicated in a bcl-2–dependent apoptotic pathway, and we have recently reported a cooperative antitumor and proapoptotic effect of AS-PKA-I in combination with an AS-bcl-2. Finally, we have shown that AS-PKA-I also has antitumor and antiangiogenic effects following oral administration and that they can be greatly enhanced in combination with oral ZD1839 and oral taxanes.
>
> KEYWORDS: cancer therapy; PKA; EGFR; antisense oligonucleotides; apoptosis; angiogenesis; taxanes

We have focused our studies on the design of novel therapeutic strategies based on the development of agents targeting molecular pathways relevant to cancer pathogenesis and progression. In particular, we have studied the role of protein kinase A (PKA) and its connection with other signaling molecules, including the growth factors of the epidermal growth factor (EGF) family and their receptors, in the control of cancer cell proliferation, angiogenesis, and apoptosis and as potential therapeutic targets.

Address for correspondence: Giampaolo Tortora, M.D., Ph.D., Cattedra di Oncologia Medica, Dipart. Endocrinologia e Oncologia Molecolare e Clinica, Università di Napoli Federico II, Via S. Pansini 5, 80131 Napoli, Italy. Voice: +39-081-7462061; fax: +39-081-746206.
gtortora@unina.it

PKA is present in mammalian cells with two distinct isoforms, defined as PKA-I and PKA-II, which differ only in their regulatory subunits (termed RI in PKA-I and RII in PKA-II, respectively). Such differences affect the affinity for cAMP, the subcellular distribution, and, finally, the function of PKA-I and PKA-II.[1] Differential expression of PKA-I and PKA-II has been correlated with cell differentiation and neoplastic transformation. In fact, preferential expression of PKA-II is found in normal nonproliferating tissues and in growth-arrested cells, while PKA-I and/or its regulatory subunit RIα is generally overexpressed in human cancer cell lines and in primary tumors and induced following transformation by certain growth factors, such as TGF-α, or oncogenes, such as *ras* and *erb*B-2 (reviewed in refs. 1 and 2). Moreover, RIα/PKA-I overexpression is associated with a worse prognosis in patients affected by different types of cancer,[3,4] and PKA-I has been directly implicated in the acquisition of the multidrug-resistant (MDR) phenotype.[5] For all the above reasons, PKA-I has been proposed as a potentially relevant target for cancer therapy.

ANTITUMOR EFFECT OF SELECTIVE PKA-I INHIBITORS AND COOPERATION WITH CYTOTOXIC AGENTS

In the past years several PKA-I–selective agents have been developed and tested for their ability to inhibit tumor growth. The most interesting among them are the site-selective cAMP analogue 8-Cl-cAMP and a series of modified antisense oligonucleotides targeting the RIα subunit of PKA-I.[1,2,6–8] More recently, a DNA/RNA hybrid antisense RIα with mixed backbone (MBO AS-PKA-I) has been developed, which exhibits significant improvement of pharmacokinetic properties and target interaction *in vivo*, as compared to first-generation phosphorothioate oligonucleotides.[9]

Both 8-Cl-cAMP and AS-PKA-I are able to inhibit PKA-I expression and function and to promote PKA-II formation, causing cancer cell growth arrest, *in vitro* and *in vivo*, in a wide variety of cancer cell types.[1,10] Moreover, the antiproliferative effect is often preceded by inhibition of the expression of growth factors of the EGF family and of different oncogenes, such as *myc*, *ras* and *erb*B-2.[1,2,11]

PKA-I seems also to affect the cellular sensitivity to cytotoxic drugs. In fact, we have demonstrated that overexpression of PKA-I increases the sensitivity to topoisomerase II drugs, such as doxorubicin and etoposide, and is responsible for the hypersensitivity to these agents observed in human cancer cells overexpressing *ras*. Interestingly, PKA-I does not affect the expression and catalytic activity of topoisomerase II enzyme.[12,13] We have shown that overexpression of *erb*B-2, but not of *ras*, confers resistance to taxanes in human breast cells and that the effect of *erb*B-2 is dominant on *ras*. Treatment with an antisense targeting PKA-I inhibits the growth of cells overexpressing *ras*, *erb*B-2 or both; moreover, down-regulation of PKA-I is able to revert *erb*B-2–dependent resistance to taxanes.[14] These results suggest that the PKA-I signaling pathway also involves targets hit by a certain class of cytotoxic drugs. This hypothesis is supported by a recent study showing that PKA-I is associated with microtubules in the mitotic spindle and is disrupted by taxanes.[15] Consistent with these findings, we have shown that 8-Cl-cAMP, as well as the MBO AS-PKA-I, are able to cooperate with selected anticancer drugs, such as taxanes, topoisomerase II inhibitors, and platinum derivatives, causing a synergistic antitumor activity associated with increased apoptosis in a wide variety of human cancer types *in vitro* and in nude mice bearing human cancer xenografts.[16,17]

More recently, we and others have shown that the MBO AS-PKA-I, owing to its peculiar pharmacokinetic properties, also preserves its antitumor activity and cooperative effects with cytotoxic drugs following oral administration to nude mice.[18,19] Both 8-Cl-cAMP and AS-PKA-I have been evaluated in phase I clinical trials in cancer patients, and biological and antitumor activities have been reported.[2,20] The MBO AS-PKA-I is currently under investigation in combination with taxanes in cancer patients.

THE PKA-I–EGF RECEPTOR CONNECTION

EGF-related growth factors, such as transforming growth factor α (TGF-α), are potent mitogens for human epithelial cell types and have been implicated in cancer development and progression. They bind to the EGF receptor (EGFR) inducing its dimerization and autophosphorylation on tyrosine residues, thereby creating a series of high-affinity binding sites for various molecules that are involved in mitogenic signal transduction mainly through the *ras/raf*/MAPK pathway.[21] Moreover, the various EGF-like ligands may induce preferential activation of different heterodimers between the EGFR and the other three known EGFR-related receptors, *erb*B-2, *erb*B-3, and *erb*B-4. Enhanced expression of TGF-α and/or EGFR has been detected in the majority of human carcinomas and has been associated with poor prognosis in several human tumor types, such as breast cancer. TGF-α and/or EGFR overexpression have been generally found in human cancer cells resistant to chemotherapy, radiotherapy, or hormonotherapy. Therefore, the blockade of the TGF-α–EGFR pathway is a relevant target in a mechanism-based cancer therapy.[22]

PKA-I plays also a physiological role in mitogenic signaling, since increased RIα/PKA-I levels are induced in normal cells by mitogenic stimuli of different hormones and growth factors.[1,2] We have also provided the first experimental evidence of a functional and structural interaction between ligand-induced EGFR and PKA-I. Addition of EGF or TGF-α to quiescent human normal breast cells induces PKA-I expression and cell membrane translocation before cells enter S phase, while pharmacologic PKA-I inhibition prevents S phase entry. On the other hand, constitutive overexpression of PKA-I confers to cells the ability to grow in serum-free medium, bypassing EGF requirement.[23] Furthermore, we have demonstrated a structural interaction of PKA-I with the ligand-activated EGFR, which occurs within 5 minutes following EGF or TGF-α addition—through the binding of the RIα subunit to the SH3 domain(s) of Grb2 adaptor protein—and allows the recruitment of the PKA-I holoenzyme to the activated EGFR.[24] PKA-I overexpression induces mitogen-activated protein kinase (MAPK) activity mimicking the effect of EGFR activation, while PKA-I inhibitors antagonize EGFR-dependent MAPK activation.[24] PKA-I inhibitors are also able to antagonize the transforming effect of EGF or TGF-α and/or to inhibit the expression and function of growth factors of the EGF family in human cancer cells *in vitro* and *in vivo*.[2,15,25] Conversely, constitutive inhibition of EGFR leads to down-regulation of PKA-I.[26] Altogether, these studies have provided the experimental demonstration that a serine-threonine kinase, like PKA-I, is directly involved in the propagation of mitogenic signals triggered by a tyrosine kinase receptor, like EGFR.

COMBINED INHIBITION OF PKA-I AND EGFR

In past years, along with our research on PKA-I inhibitors, we have conducted a parallel effort to develop selective EGFR inhibitors, including an anti-EGFR blocking chimeric human-mouse monoclonal antibody, mAb C225, and a small molecule acting as a selective inhibitor of EGFR tyrosine kinase activity, ZD1839, which is also active following oral administration. Both agents are currently under clinical evaluation in cancer patients.

We have proposed that the mitogenic signaling properties of EGFR and PKA-I and their functional interactions may provide a rational basis for developing a therapeutic strategy based on the combination of their selective inhibitors.

The combination of mAb C225 with either 8-Cl-cAMP or AS-PKA-I has provided the first demonstration in a preclinical model of both the feasibility and the antitumor activity of the simultaneous, combined blockade of biochemical pathways that are critical for cancer cell proliferation, survival, and progression.[27,28] The combination of low doses of these agents causes a marked antitumor cooperative effect *in vitro* and *in vivo* in nude mice. The combined treatment caused a significant and prolonged survival of mice bearing human tumors with no signs of toxicity, accompanied by inhibition of autocrine growth factors, such as TGF-α, AR and CRIPTO, and of angiogenic factors, such as vascular endothelial growth factor (VEGF) and basic fibroblast growth factor (bFGF), that are important for the development of tumor-induced host neoangiogenesis.[27,28] The relevance of such strategy has been demonstrated also in human renal cancer, a disease refractory to chemo- and radiotherapy. Interestingly, renal cancer cells overexpress both PKA-I and EGFR, representing a potential candidate for a combined blockade of these signaling pathways. We have shown that the combination of AS-PKA-I and mAb C225 inhibits PKA-I and EGFR expression and the growth of different human renal cancer cell lines in soft agar. Moreover, they are able to inhibit the growth and induce the regression of tumor xenografts also *in vivo* in nude mice.[29]

EFFECT OF PKA-I INHIBITION ON TUMOR ANGIOGENESIS

The critical role of tumor-induced neovascularization in neoplastic development, progression, and metastasis has been elucidated in recent years as a complex process, requiring the production and local release of various endothelial cell growth factors, such as VEGF and bFGF, that are synthesized by both tumor and normal host cells.[30]

Several studies have demonstrated that molecules involved in mitogenic signaling may provide a major contribution to the development of neoangiogenesis. EGF and TGF-α can up-regulate the production of VEGF in human cancer cells at the transcriptional level,[31,32] while inhibitors of EGFR antagonize VEGF expression.[33,34]

PKA-I seems to be involved in the production of angiogenic factors. In fact, we have previously shown in different human cancer cell lines that treatment with the

PKA-I inhibitor 8-Cl-cAMP determines a dose- and time-dependent down-regulation in the production of VEGF and bFGF, as well of growth factors of the EGF family, and an inhibition of the ability to invade the basement membrane matrix.[18,35]

The involvement of both PKA-I and EGFR in the production of angiogenic factors and vessel formation may provide an explanation for the observation of marked cooperative inhibition associated *in vitro* and *in vivo* with antitumor activity when PKA-I and EGFR inhibitors are used in combination.

INVOLVEMENT OF PKA IN THE APOPTOTIC PATHWAY AND THERAPEUTIC IMPLICATIONS

Bcl-2 is the prominent member of a family of proteins responsible for dysregulation of apoptosis, prevention of death in cancer cells, and resistance to chemo- and radiotherapy.[36] A 18-mer PS-oligonucleotide antisense protein—the human bcl-2 (AS-bcl-2) named G3139—has shown ability to block bcl-2 expression and inhibit tumor growth, both in animal models and in cancer patients, alone and in combination with cytotoxic drugs.[37,38] Different signaling proteins seem to play a role in the control of bcl-2–dependent apoptotic events. For instance, PKA is involved in bcl-2 phosphorylation at a specific consensus sequence and activates the apoptotic cascade following treatment with paclitaxel and other microtubule-damaging agents.[39] A first-generation AS-PKA-I is able to inhibit bcl-2 expression and function and to induce cleavage of PARP, caspase 3 activation, and, finally, apoptosis.[39] Moreover, it has been shown that the PKA-I subunit RIα is directly bound to cytochrome c oxydase and that PKA-I inhibition causes cytochrome c release and apoptosis.[40] These data may provide an explanation for the enhanced apoptotic activity observed when PKA-I inibitors are associated with cytotoxic drugs.

For the above reasons we have investigated whether the combined blockade of PKA and bcl-2 by an antisense strategy may represent a potential therapeutic approach. We have demonstrated that combining the novel MBO AS-PKA-I with the AS bcl-2 cooperatively inhibits bcl-2 expression and soft agar growth and induces apoptosis in different human cancer cell lines. Oral administration of AS-PKA-I in combination with intraperitoneal AS bcl-2 causes a marked antitumor effect and a significant prolongation of survival in nude mice bearing human colon cancer xenografts.[41] The histochemical analysis of tumor specimens has shown inhibition of RIα and Ki67 expression, inhibition of angiogenesis, and parallel induction of apoptosis *in vivo*.[41] These results represent further evidence of an interaction between PKA and bcl-2 signaling pathways, but also provide the rationale for translating this novel therapeutic strategy into a clinical setting.

INTEGRATION OF PKA-I AND EGFR SIGNALING INHIBITORS WITH CONVENTIONAL THERAPY

We have integrated both PKA-I and EGFR inhibitors with conventional treatments, such as radiotherapy and chemotherapy.

We have demonstrated that AS-PKA-I and mAb C225 markedly cooperate with ionizing radiation in different human cancer cells, exhibiting a greater cooperative effect when the three agents are used together. This cooperative antitumor effect was reproduced *in vivo* in nude mice bearing human cancer xenografts, providing a rationale for evaluating the combination of ionizing radiation and selective drugs that block the EGFR and PKA-I pathways in cancer patients.[42]

As described above, we have demonstrated a cooperative effect of the PKA-I inhibitors 8-Cl-cAMP and AS-PKA-I, as well as the EGFR inhibitor mAbC225, with chemotherapy drugs, expecially taxanes. More recently, we and others have shown also that the novel tyrosine kinase inhibitor ZD1839 has a similar cooperative antitumor effect with taxanes, *in vitro* and *in vivo*.[43] Since a physical connection between PKA-I and EGFR and between PKA-I and the microtubules targeted by taxanes has been demonstrate, we have combined the AS-PKA-I with the anti-EGFR antibody mAb C225 and with docetaxel. We have shown that these agents are able to inhibit in a cooperative fashion the *in vitro* growth of human breast cancer cells, achieving maximum antiproliferative effect with low noninhibiting doses of each drug.[44] More recently, we have extended these observations in different human cancer cell models, *in vitro* and *in vivo*, combining the MBO AS-PKA-I, ZD1839, and the novel taxane, IDN5109, which are all active by oral administration. We demonstrated a cooperative growth-inhibitory and proapoptotic effect and inhibition of VEGF expression with any combination of two drugs, and a marked synergistic effect when all three agents were combined. Oral administration of AS-PKA-I in combination with ZD1839 and IDN5109, caused a remarkable antitumor effect accompanied by complete suppression of vessel formation and VEGF expression.[45] These results represent the first demonstration of the cooperative antitumor and antiangiogenic activity of three novel agents that block multiple signaling pathways following oral administration.

CONCLUSIONS

A large body of studies demonstrates that PKA-I plays a role in the transduction of mitogenic signaling and is directly involved in cell proliferation, apoptosis, and angiogenesis. These events are also common to the EGFR-dependent pathway. It has been proposed that future cancer therapy strategies must be based on the integration of conventional therapies with novel inhibitors of signals involved in cell proliferation, apoptosis, and angiogenesis. It has been demonstrated that down-regulation of PKA-I by different tools inhibits cancer cell proliferation and expression of growth and angiogenic factors as well as inducing apoptosis. These effects are enhanced on a cooperative basis by the combination of selective PKA-I and EGFR inhibitors, based on the structural and functional link of these two signaling molecules. Furthermore, we have demonstrated that treatment with both EGFR and PKA-I inhibitors significantly potentiates the proapoptotic and antitumor activity of conventional anticancer treatments, including cytotoxic drugs acting with different mechanisms and ionizing radiation. Taken together, these studies support the hypothesis that cellular damage by chemotherapy or by radiotherapy can convert EGFR ligands from GFs into survival factors for cancer cells that express functional EGFR. In this context, the blockade of PKA-I and EGFR mitogenic signals in combination with cytotoxic drugs or with ionizing radiation could cause irreparable cancer cell damage leading to programmed cell death. This therapeutic approach would allow the use of low doses of drugs and more selective and long-term control of cancer with moderate toxicity.

ACKNOWLEDGMENTS

We wish to thank Dr. Y.S. Cho-Chung, Dr. J. Mendelsohn, and Dr. S. Agrawal for their precious collaboration in years past.

REFERENCES

1. CHO-CHUNG, Y.S. et al. 1995. cAMP-dependent protein kinase: role in normal and malignant growth. Crit. Rev. Oncol. Hematol. **21:** 33–61.
2. TORTORA, G. & F. CIARDIELLO. 2000. Targeting of epidermal growth factor receptor and protein kinase A: molecular basis and therapeutic applications. Ann. Oncol. **11:** 777–783.
3. BRADBURY, A.W. et al. 1994. Protein kinase A (PK-A) regulatory subunit expression in colorectal cancer and related mucosa. Br. J. Cancer **69:** 738–742.
4. MILLER, W.R. et al. 1993. Tumor cyclic AMP binding proteins: an independent prognostic factor for disease recurrence and survival in breast cancer. Breast Cancer Res. Treat. **26:** 89–94.
5. SCALA, S. et al. 1995. Downregulation of mdr-1 expression by 8-Cl-cAMP in multidrug resistant MCF-7 human breast cancer cells. J. Clin. Invest. **96:** 1026–1034.
6. TORTORA, G. 1991. Differentiation of HL-60 leukemia by type I regulatory subunit antisense oligodeoxynucleotide of cAMP-dependent protein kinase. Proc. Natl. Acad. Sci. USA **88:** 2011–2015.
7. NESTEROVA, M. & Y.S. CHO CHUNG. 1997. A single-injection protein kinase A-directed antisense treatment to inhibit tumour growth. Nat. Med. **1:** 528–533.
8. CHO-CHUNG, Y.S. 1999. Antisense oligonucleotide inhibition of serine/threonine kinases: an innovative approach to cancer treatment. Pharmacol. Ther. **82:** 437–449.
9. AKHTAR, S. & S. AGRAWAL. 1997. In vivo studies with antisense oligonucleotides. Trends Pharmacol. Sci. **18:** 12–18.
10. ROHLFF, C. et al. 1993. 8-Cl-cAMP induces truncation and down-regulation of the RIα subunit and up-regulation of the RIIβ subunit of cAMP-dependent protein kinase leading to type II holoenzyme-dependent growth inhibition and differentiation of HL-60 leukemia cells. J. Biol. Chem. **268:** 5774–5782.
11. CIARDIELLO, F. et al. 1993. Downregulation of RIα subunit of the cAMP-dependent protein kinase induces growth inhibition of human mammary epithelial cells transformed by c-Ha-ras and c-erbB2 protooncogenes. Int. J. Cancer **53:** 438–443.
12. NORTH, P. et al. 1994. Overexpression of the RIα regulatory subunit of protein kinase A confers hypersensitivity to topoisomerase II inhibitors and 8-chloro-cyclic adenosine 3′, 5′ monophosphate in Chinese hamster ovary cells. Cancer Res. **54:** 4123–4128.
13. TORTORA, G. et al. 1995. The cAMP-dependent protein kinase type I is involved in hypersensitivity of human breast cells to topoisomerase II inhibitors. Clin. Cancer Res. **1:** 49–56.
14. CIARDIELLO, F. et al. 2000. Resistance to taxanes is induced by c-erbB-2 overexpression in human MCF-10A mammary epithelial cells and is blocked by combined treatment with an antisense oligonucleotide targeting protein kinase A. Int. J. Cancer **85:** 710–715.
15. IMAIZUMI-SCHERRER, T. 2001. Type I protein kinase a is localized to interphase microtubules and strongly associated with the mitotic spindle. Exp. Cell Res. **264:** 250–265.
16. TORTORA, G. et al. 1997. Synergistic inhibition of growth and induction of apoptosis by 8-chloro-cAMP and paclitaxel or cisplatin in human cancer cells. Cancer Res. **57:** 5107–5111.

17. TORTORA, G. 1997. Synergistic inhibition of human cancer cell growth by cytotoxic drugs and mixed backbone antisense oligonucleotide targeting protein kinase A. Proc. Natl. Acad. Sci. USA **94:** 12586–12591.
18. TORTORA, G. 2000. Oral antisense targeting protein kinase A cooperates with taxol and inhibits tumor growth, angiogenesis and growth factors production. Clin. Cancer Res. **6:** 2506–2512.
19. WANG, H. *et al.* 1999. Antitumor activity and pharmacokinetics of a mixed-backbone oligonucleotide targeted to the RIα subunit of protein kinase A following oral administration. Proc. Natl. Acad. Sci. USA **96:** 13989–13994.
20. CHEN, H.X. *et al.* 2000. A safety and pharmacokinetic study of a mixed-backbone oligonucleotide (GEM 231) targeting the type I protein kinase A by 2-hour infusions in patients with refractory solid tumors. Clin. Cancer Res. **6:** 1259–1266.
21. SALOMON, D.S. *et al.* 1995. Epidermal growth factor-related peptides and their receptors in human malignancies. Crit. Rev. Oncol. Hematol. **19:** 183–232.
22. MENDELSOHN, J. & J. BASELGA. 2000. The EGF receptor family as targets for cancer therapy. Oncogene **19:** 6550–6565.
23. TORTORA, G. *et al.* 1994. The RIα subunit of protein kinase A controls serum dependency and entry into cell cycle of human mammary epithelial cells. Oncogene **9:** 3233–3240.
24. TORTORA, G. *et al.* 1997. The RIα subunit of protein kinase A (PKA) binds to Grb2 and allows PKA interaction with the activated EGF-receptor. Oncogene **14:** 923–928.
25. TORTORA, G. *et al.* 1989. Site-selective 8-chloro-adenosine 3′,5′ monophosphate inhibits transformation and transforming growth factor α production in Ki-ras-transformed rat fibroblasts. FEBS Lett. **242:** 363–367.
26. CIARDIELLO, F. *et al.* 1998. Down-regulation of type I protein kinase A by transfection of human breast cancer cells with an epidermal growth factor receptor antisense expression vector. Breast Cancer Res. Treat. **47:** 57–62.
27. CIARDIELLO, F. *et al.* 1995. Cooperative antiproliferative effects of 8-Cl-cAMP and 528 anti-epidermal growth factor receptor monoclonal antibody on human cancer cells. Clin. Cancer Res. **1:** 161–167.
28. CIARDIELLO, F. *et al.* 1996. Antitumor activity of combined blockade of epidermal growth factor receptor and protein kinase A. J. Natl. Cancer Inst. **88:** 1770–1776.
29. CIARDIELLO, F. *et al.* 1998. Cooperative inhibition of renal cancer growth by anti-EGF receptor antibody and protein kinase A antisense oligonucleotide. J. Natl. Cancer Inst. **90:** 1087–1094.
30. HAHNFELDT, P. *et al.* 1999. Tumor development under angiogenic signaling: a dynamical theory of tumor growth, treatment response, and postvascular dormancy. Cancer Res. **59:** 4770–4775.
31. GOLDMAN, C.K. *et al.* 1993. Epidermal growth factor stimulates vascular endothelial growth factor production by human malignant glioma cells: a model of glioblastoma multiforme pathophysiology. Mol. Biol. Cell **4:** 121–133.
32. GILLE, J. *et al.* 1997. Transforming growth factor alpha-induced transcriptional activation of the vascular permeability factor (VPF/VEGF) gene requires AP2-dependent DNA binding and transactivation. EMBO J. **16:** 750–759.
33. PETIT, A.M.V. *et al.* 1997. Neutralizing antibodies against epidermal growth factor and erbB-2/*neu* receptor tyrosine kinases down-regulate vascular endothelial growth factor production by tumor cells *in vitro* and *in vivo*. Am. J. Pathol. **151:** 1523–1530.
34. CIARDIELLO, F. *et al.* 2001. Inhibition of growth factors production and angiogenesis in human cancer cells by ZD1839 (Iressa), a selective epidermal growth factor receptor tyrosine kinase inhibitor. Clin. Cancer Res. **7:** 1459–1465.
35. BIANCO, C. *et al.* 1997. 8-Chloro-cAMP inhibits autocrine and angiogenic growth factors production in human colorectal and breast cancer. Clin. Cancer Res. **3:** 439–448.
36. REED, J. 1999. Dysregulation of apoptosis in cancer. J. Clin. Oncol. **17:** 2941–2953.
37. JANSEN, B. *et al.* 2000. Chemosensitisation of malignant melanoma by *BCL2* antisense therapy. Lancet **356:** 1728–1733.
38. WATERS, J.S. *et al.* 2000. Phase I clinical and pharmacokinetic study of bcl-2 antisense oligonucleotide therapy in patients with non-Hodgkin's lymphoma. J. Clin. Oncol. **18:** 1812–1823.

39. SRIVASTAVA, RK. *et al*. 1998. Involvement of microtubules in the regulation of Bcl2 phosphorylation and apoptosis through cAMP-dependent protein kinase. Mol. Cell. Biol. **18:** 3509–3517.
40. YANG, W.L. *et al*. 1998. Novel function of the regulatory subunit of protein kinase A: regulation of cytochrome c oxydase activity and release of cytochrome c release. Biochemistry **37:** 14175–14180.
41. TORTORA, G. *et al*. 2001. Combined blockade of PKA and bcl-2 by antisense strategy induces apoptosis and inhibits tumor growth and angiogenesis. Clin. Cancer Res. **7:** 2537–2544.
42. BIANCO, C. *et al*. 2000. Antitumor activity of combined treatment of human cancer cells with ionizing radiations and anti-epidermal growth factor receptor monoclonal antibody C225 plus type I protein kinase A antisense oligonucleotide. Clin. Cancer Res. **6:** 4343–4350.
43. CIARDIELLO, F. *et al*. 2000. Antitumor effect and potentiation of cytotoxic drugs activity in human cancer cells by ZD-1839 (Iressa), an EGFR-selective tyrosine kinase inhibitor. Clin. Cancer Res. **6:** 2053–2063.
44. TORTORA. G. *et al*. 1999. Cooperative inhibitory effect of novel mixed backbone oligonucleotide targeting protein kinase A in combination with docetaxel and anti-epidermal growth factor-receptor antibody on human breast cancer cell growth. Clin. Cancer Res. **5:** 875–881.
45. TORTORA, G. *et al*. 2001. Oral administration of a novel taxane, an antisense oligonucleotide targeting protein kinase A, and the epidermal growth factor receptor inhibitor Iressa causes cooperative antitumor and antiangiogenic activity. Clin. Cancer Res. **7:** 4156–4163.

Protein Kinase A and Chromosomal Stability

LUDMILA MATYAKHINA, SARA M. LENHERR, AND
CONSTANTINE A. STRATAKIS

Unit on Genetics & Endocrinology (UGEN), Developmental Endocrinology Branch (DEB), National Institute of Child Health and Human Development (NICHD), National Institutes of Health (NIH), Bethesda, Maryland 20892, USA

ABSTRACT: All malignant human tumors contain chromosomal rearrangements. Among them, the majority of solid tumors show chromosomal instability, caused by abberations in chromosomal segregation during cell division. Chromosomal instability, defined as increased probability of formation of novel chromosomal mutations compared to that of normal or control cells, appears to be a feature of tumorigenesis *in vivo* and *in vitro* (in cancer cell lines). Several enzymatic kinases are involved in maintaining proper chromosomal segregation and regulating cell cycle progression. One such kinase, cAMP-dependent protein kinase A (PKA), has a functional role in many aspects of cell signaling, metabolism, and proliferation. In this review, we will discuss the potential participation of PKA in chromosomal stability. This role includes the association of PKA with the centrosome, microtubules, and the anaphase-promoting complex/cyclosome (ACP/C), all key aspects of proper chromosomal segregation.

KEYWORDS: protein kinase A; regulatory subunits; chromosomes; tumor suppressor gene; genomic instability; cell cycle; mitosis; chromosomal stability

INTRODUCTION

Protein cAMP-dependent protein kinase (PKA) is a serine-threonine kinase that controls many key cellular processes such as cell growth and metabolism, DNA replication, cell division, and actin cytoskeleton rearrangements by catalyzing phosphorylation in response to hormonal stimuli.[1–3] The PKA holoenzyme is a tetramer consisting of a regulatory subunit dimer and two inactive catalytic subunits. When two cAMP molecules bind with one regulatory dimer, the holoenzyme disassociates into a regulatory dimer and two active catalytic subunits. These catalytic subunits are able to phosphorylate a wide variety of substrate proteins.[4] There are four genes encoding the different regulatory subunits (RIα, RIβ, RIIα, RIIβ) and three encoding the catalytic subunits (Cα, Cβ, Cγ).[5] The four types of regulatory subunits have different expression patterns in mammals. While RIα has ubiquitous distribution, RIβ is expressed primarily in brain, testis and B- and T-lymphocytes.[5–8] Similarly, RIIα has ubiquitous distribution, while RIIβ is expressed in brain, adipose, and some endocrine tissues.[9]

Address for correspondence: Constantine A. Stratakis, M.D., D.Sc., Chief, Unit on Genetics and Endocrinology, DEB, NICHD, NIH, Building 10, Room 10N262, 10 Center Dr. MSC1862, Bethesda, MD 20892-1862. Voice: 301-496-4686/402-1998; fax 301-435-4358.
stratakc@cc1.nichd.nih.gov

Protein kinase A types I (PKA-I) and II (PKA-II) are the different forms of PKA composed of catalytic subunits and RI or RII regulatory isoforms, respectively. The different regulatory subunit isoforms play specific roles in response to certain agonists. RI is primarily involved in the control of cell proliferation and neoplastic transformation. It also plays an important role in the transition from G1 to S phase of the cell cycle.[10,11] RII mainly participates in the control of differentiation, growth arrest, and induction of apoptosis.[12] Overexpression of RI is frequently detected in several cancer cell lines.[13]

The PKA holoenzymes are targeted to specific subcellular locations, such as cytoplasm, the cytoskeleton, plasma membrane, nucleus, Golgi apparatus, endoplasmic reticulum, and other organelles through interactions of their R subunits with PKA–anchoring proteins (AKAPs).[14,15] AKAPs target PKA molecules close to their substrates. The compartmentalization of the PKA holoenzyme by AKAPs may mediate *in vivo*, at least in part, some of the tissue specificity of cAMP-activated signal transduction. Initially, only PKA-II was observed to be targeted by AKAPs, but recent evidence also demonstrates the association of some AKAPs with PKA-I.[16,17]

Mutations of the RIα regulatory subunit of PKA are responsible for Carney complex (CNC) disease in half of CNC patients.[18,19] Tumor cell lines from these patients have contained dicentric chromosomes, telomeric associations, translocations, and polyploidy, indicative of chromosomal instability.[20–22] These findings led to the suggestion that changes in PKA activity caused by mutations of RIα could lead to chromosomal instability observed in cell lines from CNC tumors.

DEFINITION OF CHROMOSOMAL INSTABILITY

The term *chromosomal instability* describes a variety of karyotypic alterations frequently observed in cancers, which include both changes in chromosome number (aneuploidy) and structural chromosomal rearrangements. Chromosomal instability may be determined as an increased probability of formation of novel chromosomal mutations as compared to normal or other control cells. Numerical changes can be caused by abnormal chromosomal segregation at metaphase/anaphase transition resulting in the gain or loss of chromosomes or by multipolar divisions associated with abnormal number or structure of centrosomes.[23–26] Structural rearrangements can be initiated by breaks in chromosomes or telomeric dysfunction. Chromosome breaks can lead to inversions, deletions, translocations, or dicentric chromosomes. Telomeric dysfunction can lead to telomeric associations, which can give rise to unstable dicentric or ring chromosomes as well as aneuploidy.[27–31] Chromosomal instability has been considered to be a necessary step for tumorigenesis because it increases the accumulation of mutations responsible for malignant phenotypes.[32,33]

Usually, the solid tumors contain both types of abnormalities; the number of changes increases with tumor progression. Additionally, most solid tumors demonstrate considerable intratumor variability in the character of their genomic rearrangements.[34–41]

Studies of model organisms such as yeast showed that chromosomal instability could be caused by alterations of many genes. Among them are genes involved in chromosome condensation, sister-chromatid cohesion, kinetochore structure and function, and centrosome/microtubule formation and dynamics, as well as checkpoint genes that regulate the proper progression of the cell cycle.[42,43]

Since protein phosphorylation has a central role in the orderly progression of the cell cycle, it is not unexpected that several kinases, including PKA, are involved in these processes. Although several studies have been performed on the role of cAMP/PKA in cell cycle progression, very little is known regarding the direct involvement of PKA in the regulation of chromosomal stability. In this review, we will discuss the potential participation of PKA in chromosomal stability. This role includes the association of PKA with the centrosome and microtubules, the anaphase-promoting complex/cyclosome (ACP/C)—all key aspects of proper segregation of chromosomes.

PKA LOCALIZATION IN THE CELL

It has been shown that RIα and RIIα of PKA may be associated with microtubules,[44,45] RIIα and RIIβ subunits with pericentriolar matrix of the centrosome during interphase,[46] and the catalytic subunits with microtubules or mitotic spindles.[45] Association with these structures, which are major components of the cytoskeleton and mitotic apparatus, suggests that PKA can play a very important role in different phases of chromosomal replication and/or division.

PKA AND THE CENTROSOME

The cell cycle is composed of two phases: interphase and M phase. Interphase is divided into G1, S, and G2 phases. M phase includes mitosis and cytokinesis. Each of them is very important for proper cell progression. The centrosome plays an important role during the entire course of the cell cycle, functioning as the major organizer of the microtubule network in interphase nuclei and of the mitotic spindle in mitosis. The centrosome consists of a pair of centrioles surrounded by pericentriolar material from which microtubules are nucleated. During S phase, prior to mitotic cell division, the centrosome duplicates. As prophase begins, the two daughter centrosomes move to opposite poles of the cell, where they induce microtubule elongation and initiate the formation of the mitotic spindle.[43,47–49]

Thus, alterations in normal centrosome number, function, or regulation could lead to the failure of cell division and the missegregation of chromosomes, resulting in chromosomal instability.[42,48] This idea has been supported by several studies showing a correlation between changes in the centrosomes and the presence of chromosomal instability in cancer.[24,25,50,51]

Currently, it is well known that many regulatory proteins or protein complexes that play important roles in the orderly progression of the cell cycle are present in the centrosome. Examples include γ-tubulin, pericentrin, Nm23-H1, p53, BRCA1, and BRCA2.[49,52,53] Additionally, numerous kinases that participate in the regulation of cell cycle progression have been shown to localize at the centrosome and mitotic spindle poles. In addition to PKA, these include Cdk2, Cdk4/6, polo-like and aurora kinases, pEg2 and others.[47,54,55]

Association of the PKA regulatory and catalytic subunits with centrosomes suggests that PKA could also be involved in the phosphorylation of proteins that are important for the proper cell cycle progression. PKA is involved in phosphorylation of centrin, a component of the centrioles and the pericentriolar matrix; phosphorylation of centrin is necessary for normal centrosome division and for formation of the mitotic spindle during the G2/M phase of the cell cycle.[56]

It has been shown that tethering of PKA-II to the centrosome is performed by centrosomal proteins AKAP350/450/CG-NAP and pericentrin (the latter is also an AKAP; see below).[57–60] It is interesting that both these AKAPs target PKA to the same subcellular site, the centrosome, but probably spatially segregate independent signaling events within it.[61]

Pericentrin, an integral component of the pericentriolar material of the centrosome, interacts with γ-tubulin and the motor protein dynein and is involved in the organization of the mitotic spindle.[62–64] Diviani and colleagues have demonstrated that PKA anchoring at the centrosome through pericentrin is an important step in normal centrosome function; disruption of PKA-II anchoring results in spindle abnormalities, apparently due to deficient dynein function.[60,61]

The above suggest that changes in PKA activity may induce chromosomal instability through deficient centrosomal structure and/or function. This is certainly true with mutations of polo-like kinase (PLK-1) and overexpression of STK15/BTAK, all components of the centrosome, which have been shown to be responsible for centrosome abnormalities and, consequently, chromosomal instability in rodent cells.[65,66]

PKA AND MICROTUBULES

Microtubules, a main component of the cytoskeleton, participate in many processes that take place during the cell cycle. They are necessary for intracellular transport of molecules and organelles, establishment of cell polarity and morphology, and cell division. Accomplishment of these functions occurs through the rapid, numerous microtubule interactions that are based on the high dynamic nature of microtubule ends and the activity of microtubule motor proteins.[67–69] Association of RIα and RIIα as well as the catalytic subunits of PKA with microtubules suggests that PKA could be involved in these processes.

Stathmin, a protein that is necessary for the stability of microtubules, is phosphorylated by PKA;[70] overexpression of stathmin mutants that cannot be phosphorylated by PKA prevents the assembly of the mitotic spindle *in vitro*.[71,72] Accordingly, Imaizumi-Scherrer *et al.* showed that mouse hepatoma cells transfected with wild-type or mutant RIα subunits demonstrated aberrant mitosis with multipolar spindles and mono- or multinucleated giant cells.[45] They postulated that association of the RIα with microtubules during the mitotic process plays an important role in centrosome duplication, segregation, sister chromatid separation, or cytokinesis. Additionally, the same investigators observed an asymmetric distribution of both RIα and C subunits on microtubules, a finding that suggested that PKA could be also associated with the motor proteins regulating cytokinesis during cellular division.

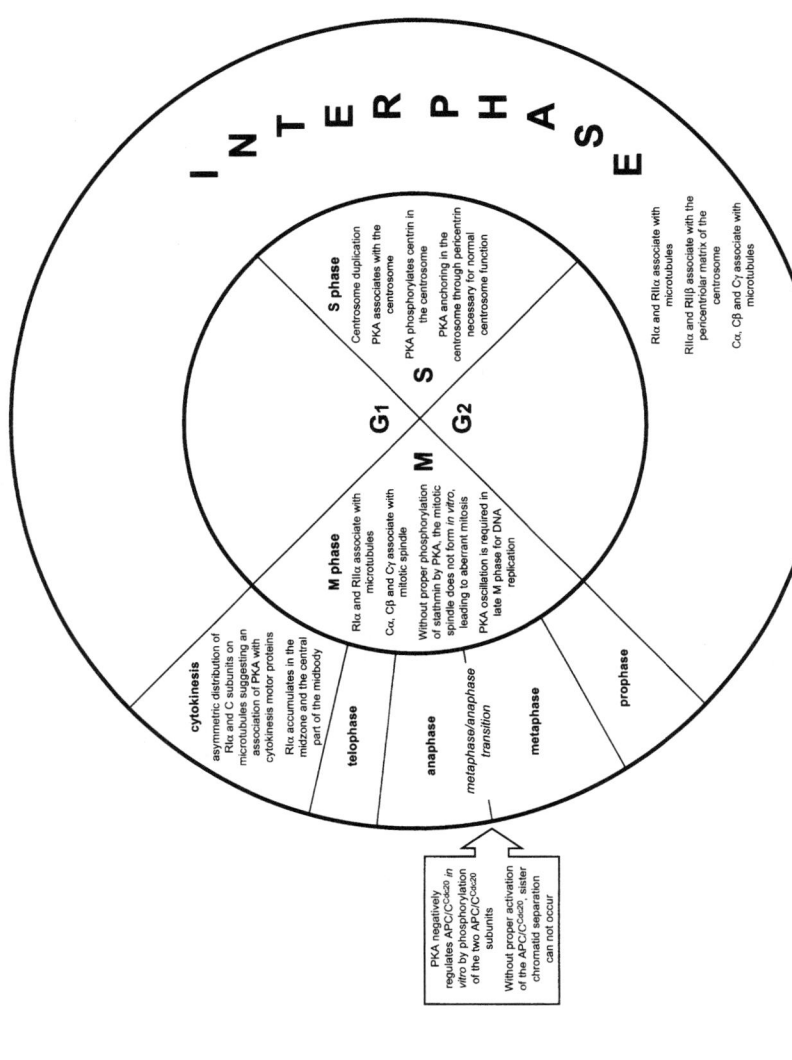

FIGURE 1. Schematic diagram of the cell cycle and possible PKA roles, as suggested by the references reviewed in the present report.

PKA AND APC: REGULATION OF METAPHASE-ANAPHASE TRANSITION

One potential target for PKA regulation is the metaphase-anaphase transition. Errors that occur during this transition lead to numerical chromosomal aberrations as a result of asymmetric distribution of chromosomes. Many epithelial tumors exhibit chromosomal instability caused by abnormalities during this transition.[26] The primary regulatory protein complex allowing for the initiation of anaphase is the anaphase-promoting complex, otherwise known as the cyclosome (APC/C). The APC/C, a highly regulated ubiquitin ligase, promotes the transition from metaphase to anaphase by targeting key mitotic proteins for proteolysis. If chromosomes are properly attached and aligned, the APC/C is activated by Cdc20. APC/C^{Cdc20} activation is inhibited by the PKA pathway in both fission yeast and mammalian cell models.[73,74] APC/C control over mitosis progression is modulated by the phosphorylation of both PKA and Cdc2-CyclinB–activated polo-like kinase. These proteins are involved in the correct timing of substrate-specific ubiquitination through binding of activators and cell cycle–dependent phosphorylation.[73,75] Finally, sister chromatid separation is the main event in metaphase-anaphase transition; it is also regulated by APC/C and, thus, indirectly by PKA and other kinases.[73,75,76] Mutations in PKA could affect chromatid separation and, thus, cause chromosomal instability by yet another mechanism.

PKA AND CYTOKINESIS

Imaizumi-Scherrer et al. observed the accumulation of RIα in the midzone and in the central part of the midbody during cytokinesis similar to the situation described for other kinases, such as the polo-like, citron, and aurora-related kinases, which are involved in cytokinesis.[45,77–79] It is noteworthy that mutations in polo-like kinases or changes in activity of aurora-related kinases have been shown to cause polyploidy and aneuploidy.[77–79]

CONCLUSION

Over the last decade, an emerging body of literature showed ample evidence that PKA is involved in the regulation of chromosomal stability in both interphase and metaphase nuclei, through its association with the centrosome, mitotic spindle and microtubules, and, perhaps, cytokinesis (FIG. 1). However, a lot of work remains to be done in this field: what are the exact mechanisms through which PKA mediates its functions in the regulation of chromosomal stability? What are the target proteins? Does this regulation involve mainly PKA-II, which is what the current literature seems to focus on, or does it also involve PKA-I? The answers to these questions are expected to shed light on some of the most fundamental links between signaling pathways and chromosomal role in tumorigenesis.

REFERENCES

1. FRANCIS, S.H. & J.D. CORBIN. 1994. Structure and function of cyclic nucleotide-dependent protein kinases. Annu. Rev. Physiol **56:** 237–272.
2. COSTANZO, V., E.V. AVVEDIMENTO, M.E. GOTTESMAN, et al. 1999. Protein kinase A is required for chromosomal DNA replication. Curr. Biol. **9:** 903–906.
3. BEEBE, S.J. 1994. The cAMP-dependent protein kinases and cAMP signal transduction. Semin. Cancer Biol. **5:** 285–294.
4. MONTMINY, M. 1997. Transcriptional regulation by cyclic AMP. Annu. Rev. Biochem. **66:** 807–822.
5. TASKEN, K., B.S. SKALHEGG, R. SOLBERG, et al. 1993. Novel isozymes of cAMP-dependent protein kinase exist in human cells due to formation of RI alpha-RI beta heterodimeric complexes. J. Biol. Chem. **268:** 21276–21283.
6. CLEGG, C.H., G.G. CADD & G.S. MCKNIGHT. 1988. Genetic characterization of a brain-specific form of the type I regulatory subunit of cAMP-dependent protein kinase. Proc. Natl. Acad. Sci. USA **85:** 3703–3707.
7. SCOTT, J.D. 1991. Cyclic nucleotide-dependent protein kinases. Pharmacol. Ther. **50:** 123–145.
8. SKALHEGG, B.S. & K. TASKEN. 1997. Specificity in the cAMP/PKA signaling pathway. differential expression, regulation, and subcellular localization of subunits of PKA. Front. Biosci. **2:** d331–d342.
9. FOSS, K.B., B. LANDMARK, B.S. SKALHEGG, et al. 1994. Characterization of in-vitro-translated human regulatory and catalytic subunits of cAMP-dependent protein kinases. Eur. J. Biochem. **220:** 217–223.
10. SEWING, A. & R. MULLER. 1994. Protein kinase A phosphorylates cyclin D1 at three distinct sites within the cyclin box and at the C-terminus. Oncogene **9:** 2733–2736.
11. TORTORA, G., V. DAMIANO, C. BIANCO, et al. 1997. The RIalpha subunit of protein kinase A (PKA) binds to Grb2 and allows PKA interaction with the activated EGF-receptor. Oncogene **14:** 923–928.
12. CHO-CHUNG, Y.S., S. PEPE, T. CLAIR, et al. 1995. cAMP-dependent protein kinase: role in normal and malignant growth. Crit. Rev. Oncol. Hematol. **21:** 33–61.
13. CIARDIELLO, F. & G. TORTORA. 1998. Interactions between the epidermal growth factor receptor and type I protein kinase A: biological significance and therapeutic implications. Clin. Cancer Res. **4:** 821–828.
14. DELL'ACQUA, M.L. & J.D. SCOTT. 1997. Protein kinase A anchoring. J. Biol. Chem. **272:** 12881–12884.
15. FAUX, M.C. & J.D. SCOTT. 1996. More on target with protein phosphorylation: conferring specificity by location. Trends Biochem. Sci. **21:** 312–315.
16. ANGELO, R. & C.S. RUBIN. 1998. Molecular characterization of an anchor protein (AKAPCE) that binds the RI subunit (RCE) of type I protein kinase A from *Caenorhabditis elegans*. J. Biol. Chem. **273:** 14633–14643.
17. KUSSEL-ANDERMANN, P., A. EL AMRAOUI, S. SAFIEDDINE, et al. 2000. Unconventional myosin VIIA is a novel A-kinase-anchoring protein. J. Biol. Chem. **275:** 29654–29659.
18. KIRSCHNER, L.S., J.A. CARNEY, S.D. PACK, et al. 2000. Mutations of the gene encoding the protein kinase A type I-alpha regulatory subunit in patients with the Carney complex. Nat. Genet. **26:** 89–92.
19. KIRSCHNER, L.S., F. SANDRINI, J. MONBO, et al. 2000. Genetic heterogeneity and spectrum of mutations of the PRKAR1A gene in patients with the Carney complex. Hum. Mol. Genet. **9:** 3037–3046.
20. DEWALD, G.W., R.J. DAHL, J.L. SPURBECK, et al. 1987. Chromosomally abnormal clones and nonrandom telomeric translocations in cardiac myxomas. Mayo Clin. Proc. **62:** 558–567.
21. STRATAKIS, C.A., R.B. JENKINS, E. PRAS, et al. 1996. Cytogenetic and microsatellite alterations in tumors from patients with the syndrome of myxomas, spotty skin pigmentation, and endocrine overactivity (Carney complex). J. Clin. Endocrinol. Metab **81:** 3607–3614.
22. DIJKHUIZEN, T., B.E. VAN DEN BERG, W.M. MOLENAAR, et al. 1992. Cytogenetics of a case of cardiac myxoma. Cancer Genet. Cytogenet. **63:** 73–75.

23. SAUNDERS, W.S., M. SHUSTER, X. HUANG, et al. 2000. Chromosomal instability and cytoskeletal defects in oral cancer cells. Proc. Natl. Acad. Sci. USA **97:** 303–308.
24. SATO, N., K. MIZUMOTO, M. NAKAMURA, et al. 2001. Correlation between centrosome abnormalities and chromosomal instability in human pancreatic cancer cells. Cancer Genet. Cytogenet. **126:** 13–19.
25. GHADIMI, B.M., D.L. SACKETT, M.J. DIFILIPPANTONIO, et al. 2000. Centrosome amplification and instability occurs exclusively in aneuploid, but not in diploid colorectal cancer cell lines, and correlates with numerical chromosomal aberrations. Genes Chromosomes Cancer **27:** 183–190.
26. STEINBECK, R.G. 1998. Chromosome division figures reveal genomic instability in tumorigenesis of human colon mucosa. Br. J. Cancer **77:** 1027–1033.
27. SCHWARTZ, J.L., R. JORDAN & H.H. EVANS. 2001. Characteristics of chromosome instability in the human lymphoblast cell line WTK1. Cancer Genet. Cytogenet. **129:** 124–130.
28. GISSELSSON, D., T. JONSON, A. PETERSEN, et al. 2001. Telomere dysfunction triggers extensive DNA fragmentation and evolution of complex chromosome abnormalities in human malignant tumors. Proc. Natl. Acad. Sci. USA **98:** 12683–12688.
29. GISSELSSON, D., L. PETTERSSON, M. HOGLUND, et al. 2000. Chromosomal breakage-fusion-bridge events cause genetic intratumor heterogeneity. Proc. Natl. Acad. Sci. USA **97:** 5357–5362.
30. HACKETT, J.A., D.M. FELDSER & C.W. GREIDER. 2001. Telomere dysfunction increases mutation rate and genomic instability. Cell **106:** 275–286.
31. COUNTER, C.M., A.A. AVILION, C.E. LEFEUVRE, et al. 1992. Telomere shortening associated with chromosome instability is arrested in immortal cells which express telomerase activity. EMBO J. **11:** 1921–1929.
32. NOWELL, P.C. 1976. The clonal evolution of tumor cell populations. Science **194:** 23–28.
33. SOLOMON, E., J. BORROW & A.D. GODDARD. 1991. Chromosome aberrations and cancer. Science **254:** 1153–1160.
34. LENGAUER, C., K.W. KINZLER & B. VOGELSTEIN. 1997. Genetic instability in colorectal cancers. Nature **386:** 623–627.
35. FADL-ELMULA, I., L. GORUNOVA, N. MANDAHL, et al. 2000. Karyotypic characterization of urinary bladder transitional cell carcinomas. Genes Chromosomes Cancer **29:** 256–265.
36. GORUNOVA, L., S. DAWISKIBA, A. ANDREN-SANDBERG, et al. 2001. Extensive cytogenetic heterogeneity in a benign retroperitoneal schwannoma. Cancer Genet. Cytogenet. **127:** 148–154.
37. MITELMAN, F. 2002. http://cgap.nci.nih.gov/Chromosomes/Mitelman.
38. GORUNOVA, L., B. JOHANSSON, S. DAWISKIBA, et al. 1995. Massive cytogenetic heterogeneity in a pancreatic carcinoma: fifty-four karyotypically unrelated clones. Genes Chromosomes Cancer **14:** 259–266.
39. GORUNOVA, L., B. JOHANSSON, S. DAWISKIBA, et al. 1995. Cytogenetically detected clonal heterogeneity in a duodenal adenocarcinoma. Cancer Genet. Cytogenet. **82:** 146–150.
40. GORUNOVA, L., L.A. PARADA, J. LIMON, et al. 1999. Nonrandom chromosomal aberrations and cytogenetic heterogeneity in gallbladder carcinomas. Genes Chromosomes Cancer **26:** 312–321.
41. ROSAI, J., M. AKERMAN, P. DAL CIN, et al. 1996. Combined morphologic and karyotypic study of 59 atypical lipomatous tumors. Evaluation of their relationship and differential diagnosis with other adipose tissue tumors (a report of the CHAMP Study Group). Am. J. Surg. Pathol. **20:** 1182–1189.
42. LENGAUER, C., K.W. KINZLER & B. VOGELSTEIN. 1998. Genetic instabilities in human cancers. Nature **396:** 643–649.
43. PIHAN, G.A. & S.J. DOXSEY. 1999. The mitotic machinery as a source of genetic instability in cancer. Semin. Cancer Biol. **9:** 289–302.
44. VALLEE, R.B., M.J. DIBARTOLOMEIS & W.E. THEURKAUF. 1981. A protein kinase bound to the projection portion of MAP 2 (microtubule-associated protein 2). J. Cell Biol. **90:** 568–576.

45. IMAIZUMI-SCHERRER, T., D.M. FAUST, S. BARRADEAU, *et al.* 2001. Type I protein kinase A is localized to interphase microtubules and strongly associated with the mitotic spindle. Exp. Cell Res. **264:** 250–265.
46. KERYER, G., B.S. SKALHEGG, B.F. LANDMARK, *et al.* 1999. Differential localization of protein kinase A type II isozymes in the Golgi-centrosomal area. Exp. Cell Res. **249:** 131–146.
47. BALCZON, R. 1996. The centrosome in animal cells and its functional homologs in plant and yeast cells. Int. Rev. Cytol. **169:** 25–82.
48. DOXSEY, S. 2001. Re-evaluating centrosome function. Nat. Rev. Mol. Cell Biol. **2:** 688–698.
49. OU, Y. & J.B. RATTNER. 2000. A subset of centrosomal proteins are arranged in a tubular conformation that is reproduced during centrosome duplication. Cell Motil. Cytoskeleton **47:** 13–24.
50. PIHAN, G.A., A. PUROHIT, J. WALLACE, *et al.* 1998. Centrosome defects and genetic instability in malignant tumors. Cancer Res. **58:** 3974–3985.
51. GUSTAFSON, L.M., L.L. GLEICH, K. FUKASAWA, *et al.* 2000. Centrosome hyperamplification in head and neck squamous cell carcinoma: a potential phenotypic marker of tumor aggressiveness. Laryngoscope **110:** 1798–1801.
52. MACK, G.J., Y. OU & J.B. RATTNER. 2000. Integrating centrosome structure with protein composition and function in animal cells. Microsc. Res. Tech. **49:** 409–419.
53. ROYMANS, D., K. VISSENBERG, C. DE JONGHE, *et al.* 2001. Identification of the tumor metastasis suppressor Nm23-H1/Nm23-R1 as a constituent of the centrosome. Exp. Cell Res. **262:** 145–153.
54. NIGG, E.A. 2001. Mitotic kinases as regulators of cell division and its checkpoints. Nat. Rev. Mol. Cell Biol. **2:** 21–32.
55. WHITEHEAD, C.M. & J.L. SALISBURY. 1999. Regulation and regulatory activities of centrosomes. J. Cell Biochem. Suppl. **32–33:** 192–199.
56. LUTZ, W., W.L. LINGLE, D. MCCORMICK, *et al.* 2001. Phosphorylation of centrin during the cell cycle and its role in centriole separation preceding centrosome duplication. J. Biol. Chem. **276:** 20774–20780.
57. SCHMIDT, P.H., D.T. DRANSFIELD, J.O. CLAUDIO, *et al.* 1999. AKAP350, a multiply spliced protein kinase A-anchoring protein associated with centrosomes. J. Biol. Chem. **274:** 3055–3066.
58. WITCZAK, O., B.S. SKALHEGG, G. KERYER, *et al.* 1999. Cloning and characterization of a cDNA encoding an A-kinase anchoring protein located in the centrosome, AKAP450. EMBO J. **18:** 1858–1868.
59. TAKAHASHI, M., H. SHIBATA, M. SHIMAKAWA, *et al.* 1999. Characterization of a novel giant scaffolding protein, CG-NAP, that anchors multiple signaling enzymes to centrosome and the Golgi apparatus. J. Biol. Chem. **274:** 17267–17274.
60. DIVIANI, D., L.K. LANGEBERG, S.J. DOXSEY & J.D. SCOTT. 2000. Pericentrin anchors protein kinase A at the centrosome through a newly identified RII-binding domain. Curr. Biol. **10:** 417–420.
61. DIVIANI, D. & J.D. SCOTT. 2001. AKAP signaling complexes at the cytoskeleton. J. Cell Sci. **114:** 1431–1437.
62. DOXSEY, S.J., P. STEIN, L. EVANS, *et al.* 1994. Pericentrin, a highly conserved centrosome protein involved in microtubule organization. Cell **76:** 639–650.
63. DICTENBERG, J.B., W. ZIMMERMAN, C.A. SPARKS, *et al.* 1998. Pericentrin and gamma-tubulin form a protein complex and are organized into a novel lattice at the centrosome. J. Cell Biol. **141:** 163–174.
64. PUROHIT, A., S.H. TYNAN, R. VALLEE & S.J. DOXSEY. 1999. Direct interaction of pericentrin with cytoplasmic dynein light intermediate chain contributes to mitotic spindle organization. J. Cell Biol. **147:** 481–492.
65. SMITH, M.R., M.L. WILSON, R. HAMANAKA, *et al.* 1997. Malignant transformation of mammalian cells initiated by constitutive expression of the polo-like kinase. Biochem. Biophys. Res. Commun. **234:** 397–405.
66. ZHOU, H., J. KUANG, L. ZHONG, *et al.* 1998. Tumour amplified kinase STK15/BTAK induces centrosome amplification, aneuploidy and transformation. Nat. Genet. **20:** 189–193.

67. Cassimeris, L. 1999. Accessory protein regulation of microtubule dynamics throughout the cell cycle. Curr. Opin. Cell Biol. **11:** 134–141.
68. Desai, A. & T.J. Mitchison. 1997. Microtubule polymerization dynamics. Annu. Rev. Cell Dev. Biol. **13:** 83–117.
69. Andersen, S.S. 1999. Balanced regulation of microtubule dynamics during the cell cycle: a contemporary view. Bioessays **21:** 53–60.
70. Larsson, N., U. Marklund, H.M. Gradin, *et al.* 1997. Control of microtubule dynamics by oncoprotein 18: dissection of the regulatory role of multisite phosphorylation during mitosis. Mol. Cell Biol. **17:** 5530–5539.
71. Howell, B., N. Larsson, M. Gullberg & L. Cassimeris. 1999. Dissociation of the tubulin-sequestering and microtubule catastrophe-promoting activities of oncoprotein 18/stathmin. Mol. Biol. Cell **10:** 105–118.
72. Gradin, H.M., N. Larsson, U. Marklund & M. Gullberg. 1998. Regulation of microtubule dynamics by extracellular signals: cAMP-dependent protein kinase switches off the activity of oncoprotein 18 in intact cells. J. Cell Biol. **140:** 131–141.
73. Yanagida, M., Y.M. Yamashita, H. Tatebe, *et al.* 1999. Control of metaphase-anaphase progression by proteolysis: cyclosome function regulated by the protein kinase A pathway, ubiquitination and localization. Philos. Trans. R. Soc. Lond. B Biol. Sci. **354:** 1559–1569.
74. Kotani, S., S. Tugendreich, M. Fujii, *et al.* 1998. PKA and MPF-activated polo-like kinase regulate anaphase-promoting complex activity and mitosis progression. Mol. Cell **1:** 371–380.
75. Zachariae, W. 1999. Progression into and out of mitosis. Curr. Opin. Cell Biol. **11:** 708–716.
76. Kotani, S., H. Tanaka, H. Yasuda & K. Todokoro. 1999. Regulation of APC activity by phosphorylation and regulatory factors. J. Cell Biol. **146:** 791–800.
77. Glover, D.M., I.M. Hagan & A.A. Tavares. 1998. Polo-like kinases: a team that plays throughout mitosis. Genes Dev. **12:** 3777–3787.
78. Madaule, P., M. Eda, N. Watanabe, *et al.* 1998. Role of citron kinase as a target of the small GTPase Rho in cytokinesis. Nature **394:** 491–494.
79. Giet, R. & C. Prigent. 1999. Aurora/Ipl1p-related kinases, a new oncogenic family of mitotic serine-threonine kinases. J. Cell Sci. **112** (Pt. 21): 3591–3601.

Protein Kinase A: Regulation and Receptor-Mediated Delivery of Antisense Oligonucleotides and Cytotoxic Drugs

P. G. SVESHNIKOV, I. D. GROZDOVA, M. V. NESTEROVA, AND E. S. SEVERIN

Russian Research Center for Molecular Diagnostics and Therapy, 113149 Moscow, Russia

ABSTRACT: Protein kinases help regulate eukaryotic cell division. We investigated the regulation of cAMP-dependent protein kinase A (PKA) and casein kinase (CK) type I activity in normal cells and in cancer. To assess this activity in biopsies we suggest a new parameter—the ratio of CK activity and total PKA activity divided by cAMP concentration: CK/PKA/cAMP. In 98 samples of colon mucosa in normal, inflamed, polyp, and adenocarcinoma cells, we found this parameter to be fairly constant in normal conditions and increased 10-fold in colon cancer; the ratio does not depend on the place of biopsy or the patient's age or sex. Experiments with model systems of concanavalin A–stimulated lymphocytes and regenerating rat liver showed that in normal cell proliferation the parameter increases 2–3-fold, as compared with a 30-fold increase in cancer. Unlike normal cells, malignant cells show CK activation and decrease of cAMP; therefore, PKA activity decreases. This suggests a correlation of CK and PKA activity and significant damage to their regulation at pathological changes of tissue proliferation. To further study concerted CK and PKA regulation we used monoclonal antibodies (mAbs) against cAMP-dependent protein kinase regulatory subunit RKIIβ. We produced 11 antibodies in three groups: inhibiting, which block cAMP binding with RIIβ and inhibit holoenzyme formation (RS6); activating, which enhance cAMP binding and do not affect holoenzyme formation (RS28); and neutral (RS17). To investigate mAb influence on protein kinase regulation in live cells we used permeabilized pheochromocytoma PC12 by digitonin. When used at 5-μM concentration for 5 min, digitonin allowed us to deliver mAb into PC12 cells at 30–34-nM concentration, leaving 68–75% viable cells. Protein kinase activity was measured within 0.5 and 4 h after incorporation of mAbs into cells. After 30 min incorporation, mAb RS6 blocked PKA activation in PC12 cells under the influence of cAMP; other mAbs showed no effect. mAb RS6 caused a 4-fold increase of free C subunit activity 4 h after incorporation. mAb RS38 decreased R2C2 activity and did not influence C subunit activity. The change of free C subunit activity caused by mAb incorporation was followed by a synchronized, well-balanced change of CK type I activity, which suggests a correlation between the two phosphorylation systems of cell proteins.

KEYWORDS: protein kinase A (PKA); monoclonal antibodies; adenocarcinoma; colon cancer

Address for correspondence: Professor E. S. Severin, General Director, Russian Research Center for Molecular Diagnostics and Therapy, Sympheropolsky Blvd., 8, 113149 Moscow, Russia. Voice: 7-095-113-23-51; fax: 7-095-113-26-33.
bob@aha.ru

TABLE 1. Basic features of protein kinases from pig brain and *Physarum polycephalum*

Characteristics	Protein kinases	
	Pig brain	*P. polycephalum*
Holoenzyme formation	$R_2C_2 \rightarrow R_2 + 2C$	$C_4 \rightarrow 4C$
MW of the holoenzyme	186.000	286.000
MW of catalytic subunit	40.000	67.000
MW of regulatory subunit	53.000	–
Preferential protein substrates	Histone H1	Casein, histone H1
A.A. residues target for phosphorylation	Ser^{37} histone H1	Thr
Specific activity	470 units/min	70 units/min
K_m^{app} ATP	1.2×10^{-5} M	1.5×10^{-5} M
K_m^{app} histone H1	3×10^{-5} M	0.2 mg/mL
Degree of activation by cAMP	2.5–4 fold	–

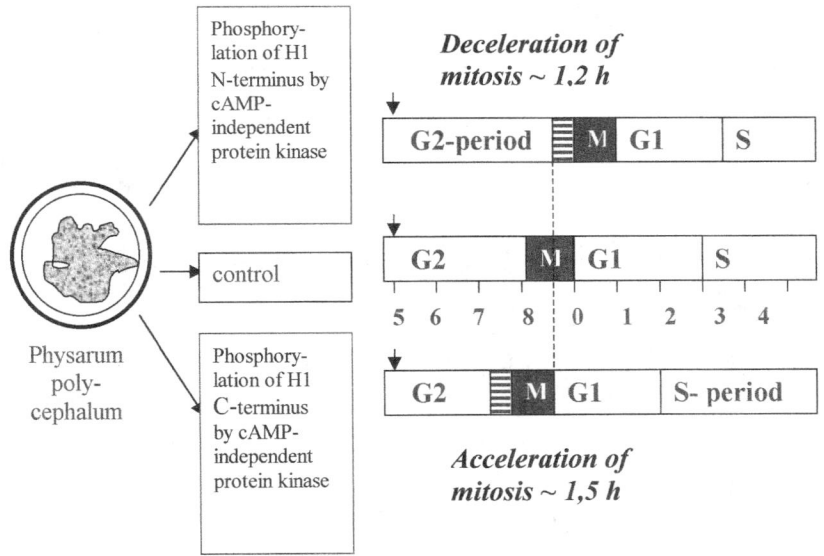

FIGURE 1. Influence of histone H1 phosphorylation by protein kinases on the *Physarum polycephalum* cell cycle progression.

For many years our attention has been focused primarily on the study of cAMP-dependent and -independent protein kinases. The main characteristics of these enzymes are shown in TABLE 1. cAMP-dependent protein kinase was purified from pig brain and tested using histone H1 as substrate. It is interesting to note that in the literature this enzyme is commonly referred to as "cAMP-dependent histone kinase." The isolated peptide fragments of histones H1, H2A, and H2B subject to phospho-

TABLE 2. Inhibitory action of ATP analogues on the activity of the catalytic subunit of histone kinase

№ n/п	Structural formulae	Compound title	Inhibitory action
1	AdH$_2$'C-O-P(O)(O$^-$)-O-P(O)(O$^-$)-O-P(O)(O$^-$)-OH	ATP	Substrate $K_m = 1.2 \times 10^{-5}$ M
2	AdH$_2$'C-O-P(O)(O$^-$)-O-P(O)(O$^-$)-CH$_2$-Cl	Adenosine-5'-chloromethanepyrophosphonate	Irreversible inhibitor $K_1 = 1.6 \times 10^{-3}$ M $k_2 = 0.05$ min^{-1}
3	AdH$_2$'C-O-P(O)(O$^-$)-CH$_2$-CH$_2$-Cl	Adenosine-5'-(β-chloroethylphosphate)	Irreversible inhibitor $K_1 = 4.7 \times 10^{-3}$ M $k_2 = 0.03$ min^{-1}
4	AdH$_2$'C-O-P(O)(O$^-$)-O-P(O)(O$^-$)-CH$_2$-CH$_2$-Br	Adenosine-5'-(β-bromoethanepyrophosphonate)	Competitive inhibitor $K_1 = 1.7 \times 10^{-4}$ M
5	AdH$_2$'C-O-P(O)(O$^-$)-CH$_2$-CH$_2$-Br	Adenosine-5'-bromomethanephosphonate	Competitive inhibitor $K_1 = 2 \times 10^{-4}$ M
6	AdH$_2$'C-O-P(O)(O$^-$)-CH$_2$-Cl	Adenosine-5'-chloromethanephosphonate	Competitive inhibitor $K_1 = 4 \times 10^{-4}$ M
7	AdH$_2$'C-O-P(O)(O$^-$)-CH$_2$-NH-C(O)-CH$_2$-Cl	Adenosine-5'-chloroacetylaminomethanephosphonate	Competitive inhibitor $K_1 = 2.3 \times 10^{-3}$ M
8	AdH$_2$'C-O-P(O)(O$^-$)-O-C$_6$H$_4$-S(O)$_2$-F	Adenosine-5'-(p-fluorosulphonylphenylphosphate)	Irreversible inhibitor $K_1 = 2.3 \times 10^{-3}$ M $k_2 = 0.067$ min^{-1}
9	AdH$_2$'C-O-P(O)(O$^-$)-O-C$_6$H$_4$-NH-C(O)-CHCl-CH$_2$	Adenosine-5'-(p-chloroacetylphenylphosphate)	Competitive inhibitor $K_1 = 2.6 \times 10^{-4}$ M
10	AdH$_2$'C-O-P(O)(O$^-$)-CH$_2$-NH-C(O)-C$_6$H$_4$-S(O)$_2$-F	Adenosine-5'-(p-fluorosulphonylbenzoylaminomethanephosphonate)	Competitive inhibitor $K_1 = 4.2 \times 10^{-4}$ M

rylation contained identical amino acid sequences—namely, Lys(Arg)-Ala-Ser. The phosphorylated Ser-37 residue of histone H1 was localized in the N terminal fragment, whereas cAMP-independent protein kinases phosphorylated serine residues in the C terminal fragment of histone H1. cAMP-dependent and -independent protein kinases had fundamentally different effects on the cell cycle of the synchronous culture of *Physarum polycephalum*: whereas the former decelerated mitosis, the latter accelerated it (FIG. 1).[1-9]

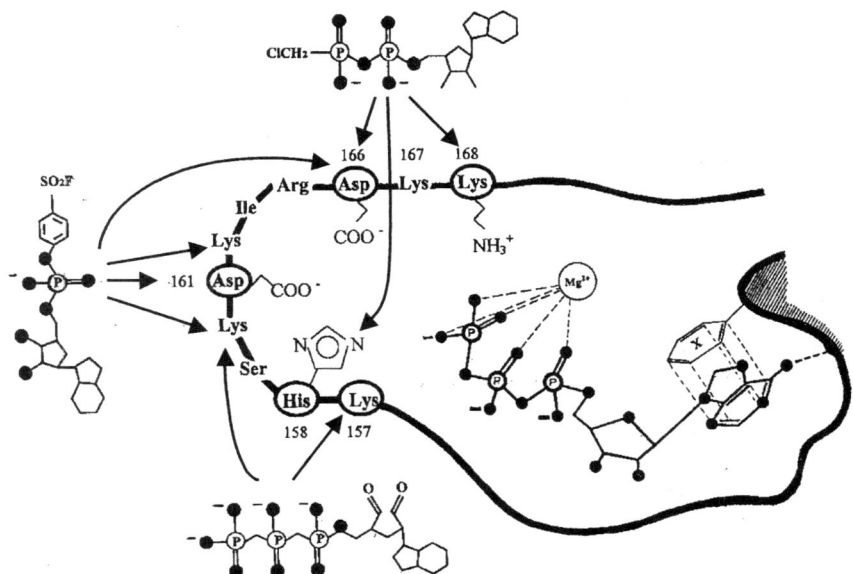

FIGURE 2. Structure of the active site of the PKA catalytic subunit.

As far as we know, these results are considered the first evidence of the essential biological role of the protein kinases in the regulation of the cell cycle progression, which is now described in all cell biology textbooks.

Our next step was to study the active center of the catalytic subunit of cAMP-dependent protein kinase. To this end, we synthesized a series of structural analogues of ATP containing active alkylating groups.[10,11]

As can be seen from TABLE 2, three of these analogues (compounds 2, 3 and 8) have the ability to induce covalent blocking of functional groups in the active center of the ATP binding site of the catalytic subunit. It is of note that compound 2 manifested the highest affinity ($K_i = 1.6 \cdot 10^{-3}$ M) and a high rate of irreversible alkylation of the functional group in the enzyme active center ($K_2 = 0.05$ min^{-1}).

The results of kinetic studies of inhibitors and analysis of peptide fragments in the ATP site of the active center of the catalytic subunit give a very general idea of the functional topography of the active center of the catalytic subunit (FIG. 2). This picture illustrates the role of the cationic cluster (Lys, Arg) in the binding of negative charges in the side chain of the ATP molecule and possible involvement of the carboxy group of glutamate and the imidazole ring of histidine in the transfer of terminal phosphate from the ATP molecule to the protein substrate (the Ser-37 residue of histone H1) in the course of the phosphotransferase reaction.[12]

In the next series of experiments, we studied the kinetic mechanism of the phosphotransferase reaction catalyzed by cAMP-dependent protein kinase from pig brain.[13–15]

Based on the two feasible kinetic mechanisms of this reaction—namely, the random bi-bi and ping-pong mechanisms—we provided a rationale for the ping-pong

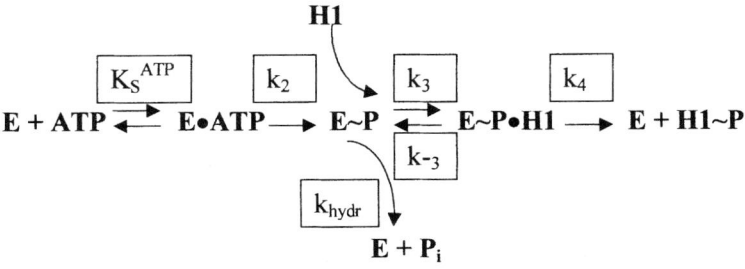

K_S^{ATP} = 50 μM K_M^{ATP} = 12 μM K_S^{H1} = k_{-3}/k_3 = 83 μM
k_2 = 25 sec.$^{-1}$ k_3 = 50 sec.$^{-1}$ k_4 = 20 sec.$^{-1}$
k_{cat} = 18 sec.$^{-1}$ k_{hydr} = 7.6 sec.$^{-1}$

1. k_2 (25 sec.$^{-1}$) and k_3 (50 sec.$^{-1}$) > k_{cat} (18 sec.$^{-1}$)
   E~P – true kinetic intermediate

2. k_3 (50 sec.$^{-1}$) >> k_{hydr} (7.6 sec.$^{-1}$)
   ATP hydrolysis cannot be measured in the presence of the protein substrate

3. k_3 (50 sec.$^{-1}$) > k_2 (25 sec.$^{-1}$)
   The formation of the phosphoenzyme cannot be measured under steady-state conditions

4. k_2 (25 sec.$^{-1}$) and k_4 (20 sec.$^{-1}$) **have the lowest values**
   k_2 - formation of the phosphoenzyme (rate – limiting step)
   k_4 – the enzyme – phosphohistone complex is highly stable due to protein – protein interactions

FIGURE 3. Kinetic mechanism of the PKA phosphotransferase reaction.

mechanism, which includes the intermediate formation of the phosphoryl-enzyme complex (FIG. 3). The values of rate constants for ATP binding to the enzyme active center (K_5 = 5·10^{-5} M), phosphorylation (K_2 = 27 s^{-1}), and degradation (K_3 = 7.6 s^{-1} [H$_2$O]) constants of the phosphoryl enzyme and the rate of transfer of the phosphate residue from the phosphoryl enzyme to the protein substrate (histone H1) (K_4 = 42 s^{-1} [H1]) were determined by fast kinetics methods. The data obtained made it possible to draw three main conclusions concerning the mechanism of the phosphotransferase reaction:

- Assuming that K_2 = 25 s^{-1} and K_3 = 50 s^{-1}, which is greater than K_{cat} (18 s^{-1}), the phosphoryl-enzyme (E ~ P) is a true kinetic intermediate product.
- The formation of the phosphoenzyme (K_2 = 25 s^{-1}) is a rate-limiting step of the phosphotransferase reaction.
- The enzyme-phosphohistone complex is rather stable due to protein-protein interactions.

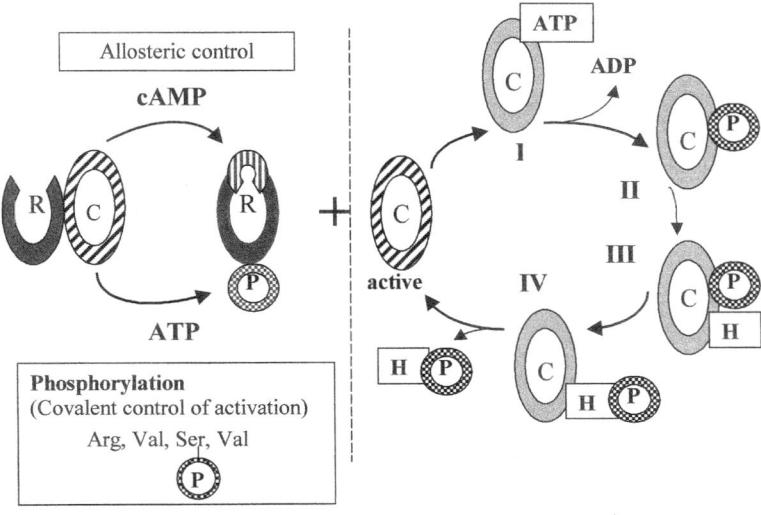

FIGURE 4. General scheme of the action of cAMP-dependent PKA.

A general scheme of the phosphotransferase reaction is shown in FIGURE 4, which illustrates the formation of a complex between the catalytic subunit and ATP (stage I), the formation of the phosphoryl enzyme (stage II) concomitantly with dissociation of ADP, the formation of a complex between the phosphoenzyme and the protein substrate (histone H1) (stage III), and phosphorylation of histone H1 with subsequent dissociation of the phosphohistone from the active center of the enzyme (stage IV).[16–19]

Since the 1980s our attention has been focused on the study of physiological roles of cAMP-binding proteins and phosphorylation reactions in the regulation of various cell-mediated processes.[20,21] This was exactly the time when Dr. I. Pastan, of the National Institutes of Health (USA), published a series of elegant studies devoted to immunotoxins. We in our laboratory combined our efforts and experience gained in the field of activation mechanisms of cAMP-dependent protein kinases in order to design principally new selective-action toxins as receptors for specifically screened (protected) toxins, respecrins. A scheme of dissociation of the holoenzyme (cAMP-dependent protein kinase) used in the design of toxins activated on the surface of target cells during their interaction with the corresponding antigens is shown in FIGURE 5.[22,23]

The inactive form of the toxin was synthesized by covalent binding of the epitope fragment of the antigen to the toxin and its interaction with an antibody or a protein specifically recognizing this epitope. In this case, the antibody has to play the role of a regulatory subunit of cAMP-dependent protein kinase. Dissociation of the toxin-epitope-antibody complex should take place on the surface of target cells carrying surface antigens and manifesting greater affinity for the antibody. Thus, activation of respecrins and, correspondingly, their toxic effects were manifested with respect to definite cells only. The mechanisms of action of respecrins are shown in FIGURE

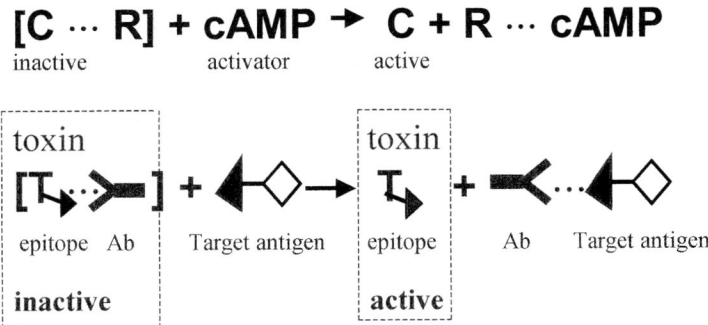

FIGURE 5. Activation of cAMP-dependent protein kinase as a background for the respecrin development principle.

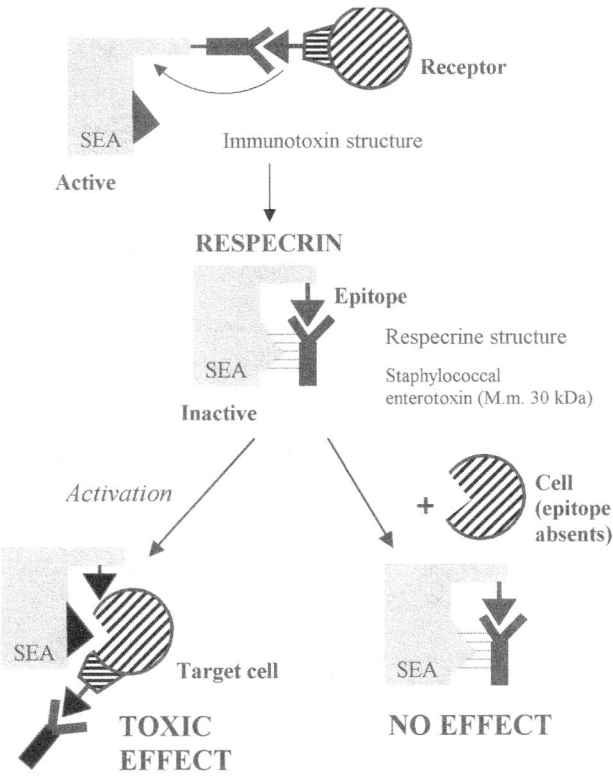

FIGURE 6. Mechanism of respecrin activation in the presence of the target antigen.

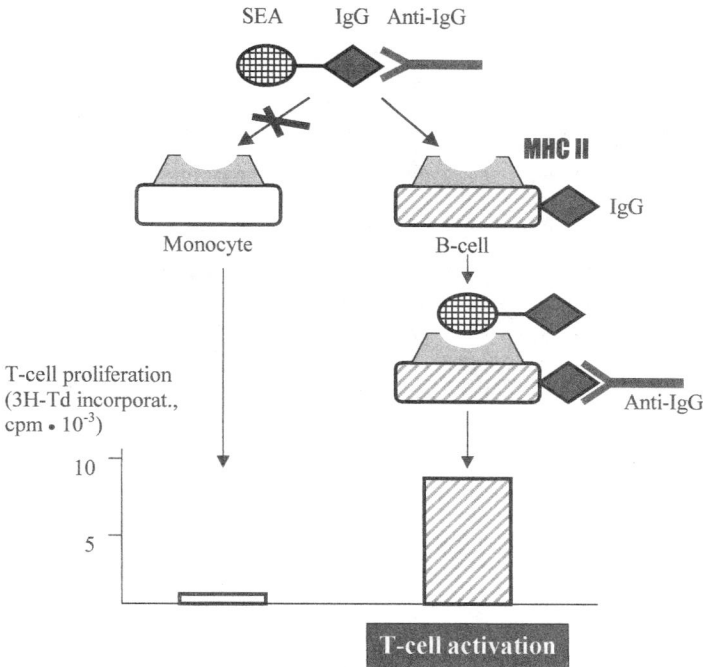

FIGURE 7. Effect of anti-IgG–based respecrin on *S. aureus* enterotoxin A (SEA)–dependent T cell activation.

6. In FIGURE 7 a model of the selective action of *Staphylococcus aureus* enterotoxin A on B cells is shown.

As can be seen from FIGURE 7, a pronounced mitogenic effect of the *S. aureus* enterotoxin A– IgG complex was observed only in the case of B cells carrying IgG molecules on their surface. Under these conditions, respecrin was inactive with respect to monocytes having no surface IgG molecules.

Another series of our experiments was aimed at the study of translocation mechanisms of the regulatory subunit of cAMP-dependent protein kinase.[24–28] FIGURE 8 shows a general scheme of translocation of the regulatory subunit into the nucleus and subsequent activation of synthesis of the low–molecular weight protein factor P-15. It is important to note that according to our data, the phospho- form of the regulatory subunit is devoid of the ability to be translocated into the nucleus. Hence, cAMP level, phosphatase activity, and accessibility of protein substrates for phosphorylation are the factors that strongly affect manifestations of biological activities of cAMP-dependent protein kinase subunits.

The essential biological role of protein kinases in metabolic regulation has been clearly demonstrated in our experiments designed to investigate these enzymes' activities in different human tissues under both normal and pathological conditions.[29,30] Toward this end, we measured the activities of cAMP-dependent (PKA) and cAMP-independent (casein kinase, CK) protein kinases, determined cAMP lev-

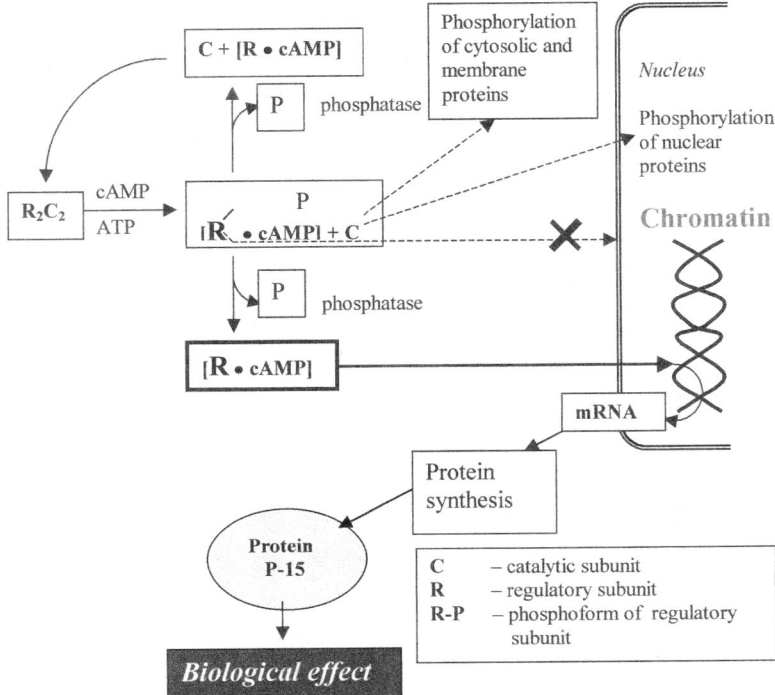

FIGURE 8. Hypothetical scheme of the action of cAMP-dependent protein kinase II.

els, and calculated CK/PKA/cAMP ratios in each sample. The latter parameter is especially important, since it reflects the correlation between casein kinase activities and the true endogenous activities of cAMP-dependent protein kinases, which, in turn, are determined by the cAMP level in body tissues.

We established CK/PKA/cAMP ratios for two types of normal human tissues—large intestinal mucosa and gastric mucosa—as well as for resting human lymphocytes and rat hepatocytes and studied their dynamics during stimulation of benign proliferation and of cancer.(TABLE 3). The total number of biopsies examined in this

TABLE 3. Mean values of CK/PKA/cAMP ratios in different tissues under normal conditions, during proliferation, and in cancer

Tissue	Normal	Proliferation	Cancer
Colon mucosa	0.066 ± 0.028	0.16 ± 0.1 (polypi)	0.69 ± 0.57
Intestine mucosa	0.046 ± 0.018	0.106 ± 0.050 (ulcer)	0.25 ± 0.15
Human lymphocytes	0.034 ± 0.020	0.083 ± 0.015 (Con A)	0.88 ± 0.2 (Jurkat cells)
Rat liver	0.046 ± 0.005	0.116 ± 0.013	

TABLE 4. Comparative activities of PKA and casein kinase, cAMP levels, and CK/PKA/cAMP ratios in various pathologies

Biopsy/number colon mucosa	Casein kinase activity	Protein kinase activity	cAMP	CK / (PKA • cAMP)
Normal/25	257 ± 23	308 ± 25	14.4 ± 1.6	0.066 ± 0.006
Inflammation/27	268 ± 42	306 ± 45	13.4 ± 2.1	0.08 ± 0.01
Polypi/12	517 ± 69	228 ± 19	16.8 ± 2.4	0.16 ± 0.03
Adenocarcinoma/24	988 ± 149	367 ± 35	5.6 ± 0.6	0.69 ± 0.13

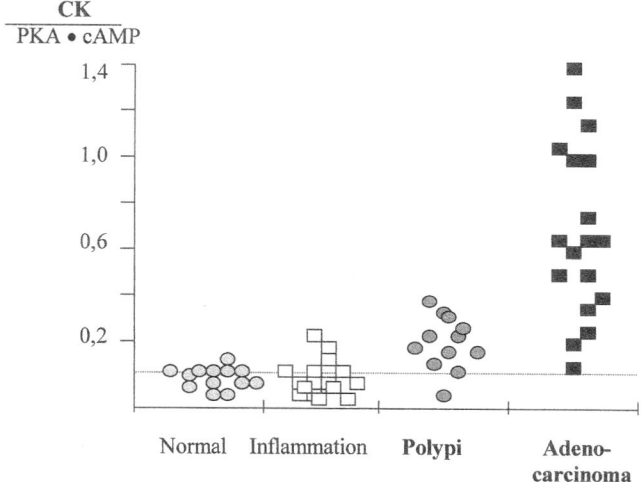

FIGURE 9. Comparison of protein kinase casein kinase activities and cAMP level in colon mucosa at different pathologies.

study was 600. We did not observe any significant changes in CK/PKA/cAMP ratios in those pathologies that are unrelated to activation of proliferative processes in body tissues (FIG. 9, inflammation). In case of normal (benign) cell proliferation, the ratios of enzyme activities changed only insignificantly and reversibly—that is, they returned to normal level after inhibition of cell proliferation, as was observed in experiments with regenerating rat liver tissues. In contrast to normal proliferation, malignant transformation was concomitant with an increase in the CK/PKA/cAMP ratio, on the average, by one order of magnitude. This phenomenon was observed inside tumor cells and at some distance from them, in visually and histologically unchanged tissues. The increase in the malignization index was characteristic of all three types of cells examined in this study (TABLE 3). At the same time, the drastic change in the activity ratios of two protein kinase systems was a characteristic sign of malignant transformation of intact cells.

These effects can be attributed to an increase in casein kinase activity and a decrease in endogenous activities of protein kinase A with a decline of cAMP (TABLE

TABLE 5. Monoclonal antibodies against the regulatory subunit of PKA (RIIβ) and their effects on cAMP binding and inhibition of the catalytic subunit

	Mab	Kd	cAMP binding	Holoenzyme formation
Control			100%	71–76%
Inhibiting				
	RS 1	n.d.	45%	40%
	RS 2	50 nM	56%	0%
	RS 4	9 nM	46%	60–70%
	RS 6	3.8 nM	0–10%	0%
Activating				
	RS 24	n.d.	160%	74%
	RS 26	10 nM	250%	76%
	RS 28	4.7 nM	183%	78%
	RS 30	1.4 nM	160%	80%
Neutral				
	RS 13	19 nM	100%	84%
	RS 17	38.4 nM	100%	67%
	RS 21	n.d.	n.d.	83%

4) as well as to a decrease in the total amount of these enzymes in gastric tumors. Therefore, reliable differentiation of tumor and normal tissues, which is critical for the correct diagnosis of cancer, is possible only when these three parameters are determined in tissue samples simultaneously and after calculation of their ratio (CK/PKA/cAMP).

Also, we have developed a procedure for more detailed elucidation of the role of cAMP-dependent protein kinases (of intracellular cAMP-dependent protein kinase, in particular) in the regulation of cell metabolism—a procedure that makes it possible to modulate these enzyme activities in live cells. To this end, we obtained monoclonal antibodies to the regulatory subunit RIIβ isolated from bovine brain.[31,32] Depending on their effects on RIIβ, these antibodies can be classified into three groups: inhibiting, activating, and so-called neutral antibodies—those that do not interfere in the reaction of RIIβ with cAMP or the catalytic subunit of protein kinase (TABLE 5). Attachement of activating antibodies to RIIβ enhanced the ability of the latter to bind cAMP. Inhibiting antibodies, such as RS6, suppressed the interaction of RIIβ with cAMP or the catalytic subunit of protein kinase nearly completely. At the same time, these antibodies could only prevent the formation of the holoenzyme without causing dissociation of the already-formed protein kinase. The latter circumstance is especially important for the correct understanding of the effects produced by inhibitory antibodies inside the cells.

We have developed a procedure for incorporating antibodies into cells with full preservation of cell viability (FIG. 10).[23] Prior to antibody incorporation, PC12 cells were permeabilized by the nonionic detergent digitonin. The latter is widely used for permeabilizing cells, but its damaging effect is irreversible. The main distinctive fea-

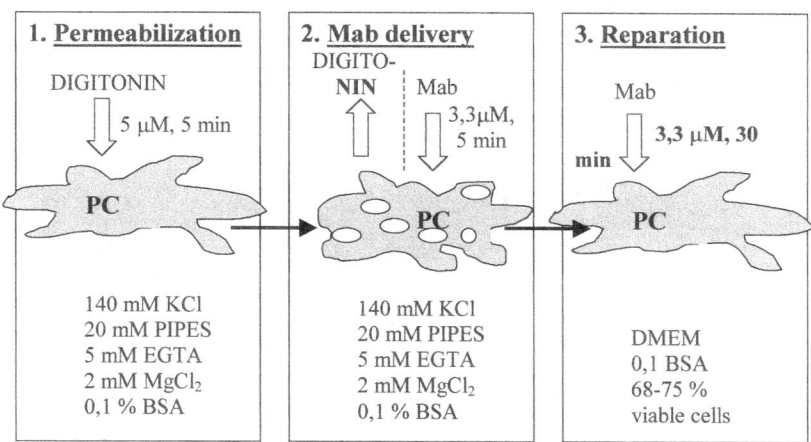

FIGURE 10. Delivery of monoclonal antibodies against the regulatory subunit of PKA (RIIβ) to digitonin-permeabilized live cells (PC12).

ture of our method in comparison with those described previously is that we used very low concentrations of digitonin, and the reaction time was very short (5–8 min), but sufficient to induce perforation of cellular membranes (FIG. 10.1); at the end of the treatment complete removal of the detergent was achieved immediately. Under these conditions, about 70% of PC12 cells retained their ability to repair plasma membrane and remained viable (FIG. 10.3). At the same time, the size of perforations formed in the cellular membrane allowed penetration of large protein molecules, such as antibodies.

The effect of monoclonal antibodies on cAMP-dependent protein kinase in PC12 cells (neuronal cells) is shown in FIGURE 11. After repair, the cells were homogenized and examined for histone kinase activity in the absence and presence of cAMP. The phosphotransferase activity increased after addition of cAMP to all control samples. However, in experimental samples—that is, in cells loaded with inhibiting antibodies (RS6)—histone kinase activity was the same both in the presence and absence of cAMP. Incorporated into cells, antibodies inhibited activation of intracellular cAMP-dependent protein kinase by cAMP. These results suggest that antibodies that enter into cells by simple diffusion are able to recognize RIIβ inside the cells and bind to it, as a result of which the properties of RIIβ change as they do *in vitro*, which is manifested as inhibition of RIIβ binding to cAMP and activation of cAMP-dependent protein kinase.

This route of delivery of antibodies to body cells can be used for purposeful modification of various properties of cAMP-dependent protein kinases inside live cells.

In conclusion, we would like to cite some of our most recent findings concerning selective delivery of biologically active compounds by receptor-mediated endocytosis of certain growth factors. We examined the possibility of using neural and epidermal growth factors as well as alpha-fetoprotein for this purpose. These studies showed that the rate of endocytosis was the highest in the case of AFP. Moreover, the AFP receptor is expressed on the surface of the majority of tumor and fetal cells,

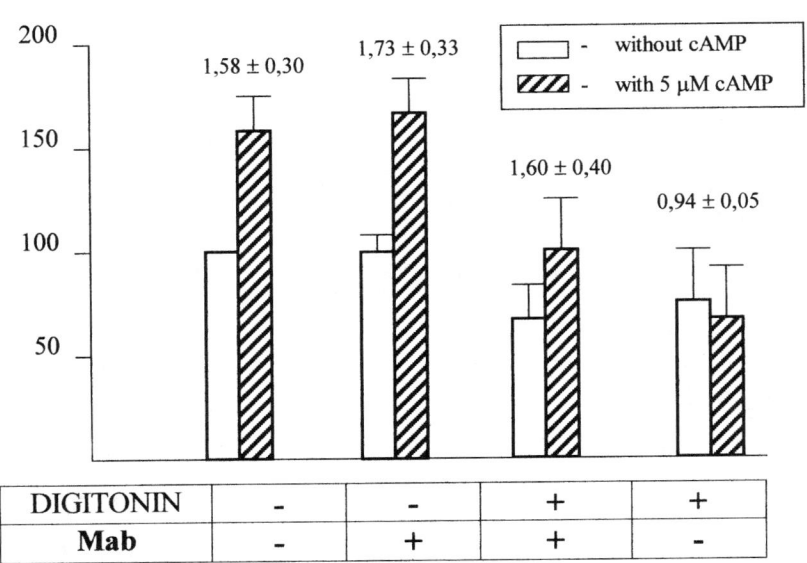

FIGURE 11. Relative PKA activity in digitonin-permeabilized PC12 cells after RS6 antibody treatment.

is less presented in monocytes, and is practically absent on the surface of normal cells. It follows from these data that AFP is an ideal ligand for targeted delivery of cytostatic drugs to tumor cells.[35,36] We performed the synthesis of AFP conjugates with some popular cytostatic drugs. These studies demonstrated that our conjugates produced more potent selective effects in comparison with free cytostatic agents.[37,38] This approach appeared to be useful during AFP-vehicled transport of antisense oligonucleotides to target cells. We performed covalent cross-linking of a series of antisense oligonucleotides (AS c-myc, Astel, AS bcl-2) to AFP via the disulfide bridge. After reduction of the disulfide bond, the antisense oligonucleotide was bound to a specific fragment (RNA or DNA). The required concentrations of AFP-antisense conjugates were approximately 10–100 times lower in comparison with those used in control experiments with free antisense oligonucleotides.

REFERENCES

1. SEVERIN, E.S., S.N. KOCHETKOV, M.V. NESTEROVA & N.N. GULYAEV. 1974. Isolation of the regulatory subunit of pig brain histone kinase by affinity chromatography on cAMP-containing adsorbent. FEBS Lett. **49:** 61–64.
2. SEVERIN, E.S., M.V. NESTEROVA, N.N. GULYAEV & S.V. SHLYAPNIKOV. 1975. Brain histone kinase: the structure, substrate specificity and mechanism of action. Adv. Enzyme Regul. **14:** 407–444.

3. SHLYAPNIKOV, S.V., A.A. ARUTYUNYAN, L.V. MEMELOVA, et al. 1975. Investigation of the sites phosphorylated in lysine-rich histones by protein kinase from pig brain. FEBS Lett. **53:** 316–319.
4. ZELENIN, A.V., E.A. KIRIANOVA, M.V. NESTEROVA, et al. 1975. Influence of specific histone kinase on the physico-chemical properties of chromatin in situ. Mol. Biol. Rep. **2:** 241–245.
5. BUSHUEV, V.N., S.N. KUROCHKIN, V.A. KAROL, et al. 1976. Structure of histone H1 and its phosphorylated by Ser-38 derivative in complexes with DNA. Rep. USSR Acad. Sci. **227:** 489–492.
6. SEVERIN, E.S., S.N. KUROCHKIN, I.N. TRAKHT & R.D. COLE. 1977. Investigations of major sites phosphorylated in histone H1 by kinase from diffetent stages of the cell cycle. FEBS Lett. **84:** 163–166.
7. GLOTOV, B.O., L.G. NIKOLAYEV, S.N. KUROCHKIN & E.S. SEVERIN. 1977. Histone H1-DNA interaction. Influence of phosphorylation on the interaction of histone H1 with linear fragmented DNA. Nucleic Acids Res. **4:** 1065–1082.
8. TRAKHT, I.N., I.D. GROZDOVA & E.S. SEVERIN. 1979. Investigation of histone H1 phosphorylation role in regulation of synchronous culture *Physarum polycephalum* cell division. Rep. USSR Acad. Sci **244:** 763–767.
9. TRAKHT, I.N., I.D. GROZDOVA, N.N. GULYAEV, et al. 1980. Effects of some protein kinases, cyclic nucleotides and specific inhibitors of phosphorylation on the mitotic cycle of *Physarum polycephalum*. Biokhimiya **45:** 788–793.
10. GULYAEV, N.N., V.L. TUNITSKAYA, L.A. BARANOVA, et al. 1976. Investigation of the active sity of catalytic histonekonase subunit. Biokhimiya **41:** 1241–1249.
11. SEVERIN, E.S., N.N. GULYAEV, T.V. BULARGINA & M.N. KOCHETKOVA. 1978. Specific inhibition of cyclic AMP-dependent protein kinase, adenylate cyclase and phosphodiesterase by ATP and cAMP analogs. Adv. Enzyme Regul. **17:** 251–282.
12. KOCHETKOV, S.N., T.V. BULARGINA, L.P. SASHCHENKO & E.S. SEVERIN. 1977. Studies on the mechanism of action of histone kinase dependent on adenosine 3′,5′-monophosphate. Evidence for involvement of histidine and lysine residues in the phosphotransferase reaction. Eur. J. Biochem. **81:** 111–118.
13. GABIBOV, A.G., S.N. KOCHETKOV, L.P. SASHCHENKO, et al. 1981. Studies on the mechanism of action of the histone kinase dependent on cAMP. Interation of ATP with the catalytic subunit of the pig-brain enzyme. Application of the quenched-flow technique. Eur. J. Biochem. **115:** 297–301.
14. GABIBOV, A.G. S.N. KOCHETKOV, L.P. SACHCHENKO, et al. 1983. Studies on the mechanism of action of the histone kinase dependent on cAMP. Investigation of protein-protein interaction by electron spin-resonance spectroscopy and stopped-flow methods. Eur. J. Biochem. **132:** 339–344.
15. GABIBOV, A.G., S.N. KOCHETKOV, L.P. SASHCHENKO, et al. 1983. Studies on the mechanism of action of the histone kinase dependent on cAMP. Fast kinetics of histone H1 phosphorylation. Eur. J. Biochem. **135:** 491–495.
16. KOCHETKOV, S.N., A.G. GABIBOV & E.S. SEVERIN. 1984. Mechanisms of phosphoryl transferin enzmatic reactions. Bioorg. Khim. **10:** 1301–1325.
17. KOCHETKOV, S.N., E.M. BAGIROV, A.G. GABIBOV & E.S. SEVERIN. 1984. Mechanism of action of cAMP-dependent protein kinase. III. A suggested role of the enzyme carboxyl group. Mol. Biol. (Mosc.) 18: 704–711.
18. KOCHETKOV, S.N., A.G. GABIBOV, L.I. MARYASH, et al. 1984. Mechanism of action of cAMP-dependent protein kinase. IV. Interaction of the enzyme catalytic subunit with structural analogs of histone H1. Mol. Biol. (Mosc.) **18:** 901–906.
19. KOCHETKOV, S.N., A.G. GABIBOV, T.N. LUKACHINA & E.S. SEVERIN. 1984. Physico-chemical principles of cAMP-dependent protein phosphorylation. Catalysis of phosphoryl group transfer to nucleophilic agents. FEBS Lett. **173:** 179–184.
20. SEVERIN, E.S. & M.V. NESTEROVA. 1981. Effect of cyclic AMP-dependent protein kinase on gene expression. Adv. Enzyme Regul. **20:** 167–193.
21. KAFIANI, C.A., A.V. ITKES, O.N. KARTACHEVA, et al. 1982. A study on the relationship between the interferon enzyme system and the system of cyclic nucleotide metabolism. Adv. Enzyme Regul. **21:** 353–365.

22. ALAKHOV, V.YU., S.A. ARZHAKOV, et al. 1988. A new principle of construction of immunotherapeutic compounds of directed action. Physiologically active substances reversibly screened by target-recognizing macromolecules. Dokl. Akad. Nauk SSR **303**(6): 1494–1497.
23. ALAKHOV, V.YU., S.A. ARZHAKOV, O.V. VASILENKO, et al. 1990. Respectins—a new type of compound with targeted action. Biomed. Sci. **1:** 155–159.
24. NESTEROVA, M.V., S.F. BARBASHEV, A.A. ARIPDZHANOV, et al. 1980. Nuclear translocation and effect of cAMP-dependent protein kinase on transcription. Biokhimiya **45:** 979–991.
25. NESTEROVA, M.V., KH.A. ULMASOV, A. ABDUKARIMOV, et al. 1981. Nuclear translocation of cAMP-dependent protein kinase. Exp. Cell Res. **132:** 367–373.
26. NESTEROVA, M.V., A.I. GLUKHOV & E.S. SEVERIN. 1982. Effect of the regulatory subunit of cAMP-dependent protein kinase on the genetic activity of eukaryotic cells. Mol. Cell. Biochem. **49:** 53–61.
27. GLUKHOV, A.I. M.V. NESTEROVA, V.L. BUKHMAN & E.S. SEVERIN. 1986. Study of interaction between cAMP-dependent protein kinase subunits and structural components of the nucleus. Biokhimiya **51:** 103–111.
28. NESTEROVA, M.V., A.I. GLUKHOV, A.G. APRIKIAN & E.S. SEVERIN. 1987. Nuclear proteins: substrates of cAMP-dependent protein kinase. Biokhimiya **52:** 1150–1153.
29. GROZDOVA, I.D., A.V. MIKHAILOVSKY, I.R. ESHBA, et al. 1986. Alteration in actitvity of protein kinases in skin under conditions of malignancy.Vopr. Med. Khim. **32:** 4–8.
30. GROZDOVA, I.D., N.P. KLIMOV, E.G. MAMAEVA, et al. 1989. Activity of protein kinases and content of cAMP in gastric musocal membrane under conditions of nonmalignant discases. Vopr. Med. Khim. **6:** 83–87.
31. GROZDOVA, I.D., E.V. NYUPENKO, E.V. SVESHNIKOVA & E.S. SEVERIN. 1990. Immunochemical properties of the cAMP-dependent protein kinase regulatory subunit type II. Biokhimiya **55:** 1244–1250.
32. SVESHNIKOVA, E.V., N.A. ALEXANDROVA, I.D. GROZDOVA, et al. 1996. Immunochemical studies on human, bovine and pig brain regulatory subunits of cAMP-dependent protein kinase type II. Biochem. Mol. Biol. Int. **39:** 1063–1070.
33. GROZDOVA, I.D., N.A. ALEXANDROVA, E.V. SVESHNIKOVA, et al. 1996. Properties of the regulatory subunit of cAMP-dependent protein kinase type II from human brain. Biochem. Mol. Biol. Int. **40:** 1159–1166.
34. ALEXANDROVA, N.A., P.G. SVESHNIKOV, N.K. NAGRADOVA & I.D. GROZDOVA. 1998. Incorporation of monoclonal antibodies in living rat pheochromocytoma PC12 cells. Evidence for intracellular formation of immune complex between the incorporated antibody and a target protein. FEBS Lett. **432:** 187–190.
35. SEVERIN, S.E., E.YU. MOSKALEVA, I.I. SHMYREV, et al. 1995. Alpha-fetoprotein-mediated targeting of anti-cancer drugs to tumor cells in vitro. Biochem. Mol. Biol. Int. **37:** 385–392.
36. SEVERIN, S.E., E.YU. MOSKALEVA, G.A. POSYPANOVA, et al. 1996. In vivo antitumor activity of cytotoxic drugs conjugated with human alpha-fetoprotein. Tumor Targeting **2:** 299–306.
37. KANEVSKY, V. YU., L.P. POZDNYAKOVA, O.A. AKSENOVA, et al. 1997. Isolation and characterization of AFP-binding proteins from tumor and fetal human tissues. Biochem. Mol. Biol. Int. **41:** 1143–1151.
38. MOSKALEVA, E.YU., G.A. POSYPANOVA, I.I. SHMYREV, et al. 1997. Alpha-fetoprotein-mediated targeting—a new strategy to overcome multidrug resistance of tumour cells in vitro. Cell Biol. Int. **21:** 793–799.

Gs$_\alpha$ Mutations and Imprinting Defects in Human Disease

LEE S. WEINSTEIN, MIN CHEN, AND JIE LIU

Metabolic Diseases Branch, National Institute of Diabetes, Digestive, and Kidney Diseases, National Institutes of Health, Bethesda, Maryland 20892, USA

ABSTRACT: Gs is the ubiquitously expressed heterotrimeric G protein that couples receptors to the effector enzyme adenylyl cyclase and is required for receptor-stimulated intracellular cAMP generation. Activated receptors promote the exchange of GTP for GDP on the Gs α-subunit (Gs$_\alpha$), resulting in Gs activation; an intrinsic GTPase activity of Gs$_\alpha$ deactivates Gs by hydrolyzing bound GTP to GDP. Mutations of Gs$_\alpha$ residues involved in the GTPase reaction that lead to constitutive activation are present in endocrine tumors, fibrous dysplasia of bone, and McCune-Albright syndrome. Heterozygous loss-of-function mutations lead to Albright hereditary osteodystrophy (AHO), a disease characterized by short stature, obesity, and skeletal defects, and are sometimes associated with progressive osseous heteroplasia. Maternal transmission of Gs$_\alpha$ mutations leads to AHO plus resistance to several hormones (e.g., parathyroid hormone) that activate Gs in their target tissues (pseudohypoparathyroidism type IA), while paternal transmission leads only to the AHO phenotype (pseudopseudohypoparathyroidism). Studies in both mice and humans demonstrate that Gs$_\alpha$ is imprinted in a tissue-specific manner, being expressed primarily from the maternal allele in some tissues and biallelically expressed in most other tissues. This likely explains why multihormone resistance occurs only when Gs$_\alpha$ mutations are inherited maternally. The Gs$_\alpha$ gene *GNAS1* has at least four alternative promoters and first exons, leading to the production of alternative gene products including Gs$_\alpha$, XLαs (a novel Gs$_\alpha$ isoform expressed only from the paternal allele), and NESP55 (a chromogranin-like protein expressed only from the maternal allele). The fourth alternative promoter and first exon (exon 1A) located just upstream of the Gs$_\alpha$ promoter is normally methylated on the maternal allele and is transcriptionally active on the paternal allele. In patients with parathyroid hormone resistance but without AHO (pseudohypoparathyroidism type IB), the exon 1A promoter region is unmethylated and transcriptionally active on both alleles. This *GNAS1* imprinting defect is predicted to decrease Gs$_\alpha$ expression in tissues where Gs$_\alpha$ is normally imprinted and therefore to lead to renal parathyroid hormone resistance.

KEYWORDS: G proteins; cAMP; Albright hereditary osteodystrophy; McCune-Albright syndrome; fibrous dysplasia; progressive osseous heteroplasia; genomic imprinting

Address for correspondence: Dr. Lee S. Weinstein, Metabolic Diseases Branch, NIDDK/NIH, Bethesda, MD 20892-1752. Voice: 301-402-2923; fax: 301-402-0374.
leew@amb.niddk.nih.gov

Gs_α STRUCTURE AND FUNCTION

Gs is one member of a large family of heterotrimeric G proteins that are integral components of diverse signaling pathways. G proteins transmit signals from cell surface receptors (which classically have seven transmembrane domains) to intracellular enzymes or ion channels that generate second messengers, such as cyclic AMP (cAMP) or other metabolites. Each G protein is defined by its specific α subunit, which binds guanine nucleotide and in the inactive state is bound to a βγ dimer that is attached to the plasma membrane through lipid modifications of the γ subunit. Gα's also undergo lipid modifications that are important for membrane targeting.[1]

The Gs α subunit (Gs_α) is expressed in virtually all cell types and couples receptors to adenylyl cyclase, leading to receptor-stimulated intracellular cAMP generation. cAMP activates cAMP-dependent protein kinase (protein kinase A [PKA]) and has more recently been shown to interact with ion channels and other proteins. Gs_α might also activate other effectors, such as cardiac Ca^{2+} channels[2,3] and src tyrosine kinases,[4] and has been implicated in the regulation of intracellular membrane trafficking.[5] Evidence suggests that Gs_α may also be activated by various tyrosine kinase receptors, including the epidermal growth factor (EGF)[6,7] and basic fibroblast growth factor receptors,[8] perhaps by phosphorylation of specific Gs_α tyrosine residues.[9,10]

In the inactive state Gs exists as a Gs_α-βγ heterotrimer with GDP bound to Gs_α. Activated receptors promote the release of GDP, which is replaced with ambient GTP. GTP binding causes Gs_α to switch into an active conformation, and activated GTP-bound Gs_α dissociates from βγ, allowing the GTP-bound Gs_α to interact directly with and activate its effectors. Over time GTP is hydrolyzed to GDP, leading to reassociation with βγ. There is no evidence to suggest that Gs_α is affected by RGS (regulator of G protein signaling) proteins, although interaction with adenylyl cyclase may directly promote GTP hydrolysis.[11]

Gs_α has two domains, a *ras*-like GTPase domain, which includes the sites for guanine nucleotide binding and effector interaction, and a more variable helical domain.[12,13] Alternative exon splicing produces several Gs_α isoforms with helical domains of slightly differing length.[14,15] Guanine nucleotide gains access to its binding site via a cleft between the two domains, and the helical domain may therefore be important for maintaining GDP binding.[16,17] Three regions within the GTPase domain (named switch 1, 2, and 3) undergo conformational changes upon GTP binding. Upon activation, switches 2 and 3 approach one another and stabilize the active conformation by forming multiple interactions between acidic and basic amino acid residues.[18–21] Interactions between switch 3 and the helical domain may be important for maintaining guanine nucleotide binding.[16,17] Two residues in the GTPase domain (Arg^{201}, Gln^{227}) are catalytically important for hydrolysis of bound GTP, and mutation of these residues leads to constitutive activation.[22–25] The Gs_α carboxyl terminus is important for receptor interactions.[26–30]

THE Gs_α GENE (*GNAS1*)

The human Gs_α gene (*GNAS1*) is located at 20q13.2-13.3 (refs. 31–33) while its mouse ortholog (*Gnas*) is located in a syntenic region in chromosome 2.[34,35] The

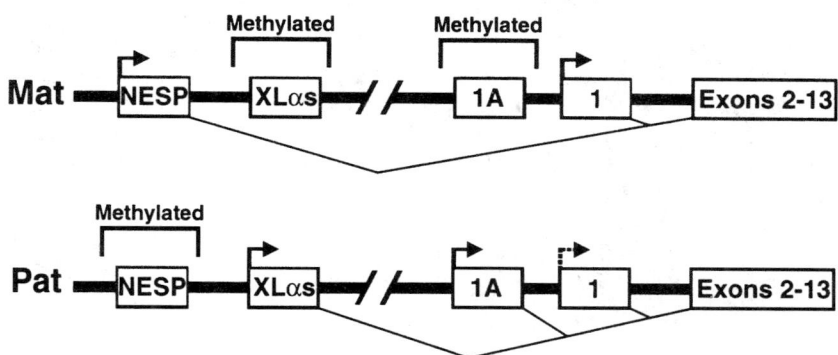

FIGURE 1. The methylation and expression patterns of the four alternative promoters and first exons of *GNAS1*. Alternative upstream exons labeled NESP55, XLαs, 1A, and 1 splice onto a common set of downstream exons (2–13) to generate NESP55, XLαs, an unknown gene product, and $G_s\alpha$, respectively. Transcriptionally active promoters are designated by *horizontal arrows* with the splicing onto exon 2 designated below. Differentially methylated regions are outlined above. The maternal allele is shown above and the paternal allele is below. The *dashed horizontal arrow* for exon 1 in the paternal allele indicates that Gs_α is biallelically expressed in most tissues but is silenced on the paternal allele in some tissues. A paternally expressed antisense transcript that traverses the NESP55 exon is not shown in the figure. (Adapted from Liu *et al.*[38])

overall structure of the mouse and human genes is very similar. These genes were originally defined by the 13 exons that encode Gs_α.[15] Four functionally similar Gs_α isoforms result from alternative splicing involving exon 3.[14,15] The Gs_α promoter and first exon are within a CpG island that is unmethylated in both human and mouse.[36–38]

GNAS1 is now known to have at least three additional alternative promoters and first exons located upstream of the Gs_α promoter that all splice onto a common exon and that are all imprinted (FIG. 1). Imprinting is a phenomenon affecting a small number of genes, which leads to transcriptional silencing of one parental allele and is often, although not always, associated with methylation of the inactive promoter. The most upstream *GNAS1* promoter generates transcripts for the chromogranin-like protein NESP55.[37,39,40] The entire NESP55 coding region is within the upstream exon, and Gs_α exons 2–13 contribute to the 3' untranslated region of the NESP55 transcripts.[39,41] The NESP55 promoter region is methylated only on the paternal allele, and NESP55 is transcribed only from the maternal allele.[37,39,42] An alternative promoter, located ~11 kb downstream of the NESP55 promoter, generates transcripts for XLαs, an isoform of Gs_α with a long amino-terminal extension.[36,37,40,43] The XLαs promoter region is methylated on the maternal allele and XLαs is expressed only from the paternal allele.[36,37,42] Like Gs_α, XLαs can activate adenylyl cyclase, although it remains to be seen whether it can be activated by receptors.[44] Both NESP55 and XLαs are expressed primarily in neuroendocrine tissues,[41,43,45,46] and little is known about their biological functions. Recently, paternally expressed antisense transcripts, which originate from a promoter upstream of the XLαs promoter and traverse the NESP55 promoter region, have been identified.[42,47]

A fourth promoter and first exon (exon 1A) located ~2.5 kb upstream of Gs_α exon 1 generates ubiquitously expressed transcripts that are presumed to be untranslated.[38,48] The exon 1A region is methylated on the maternal allele, and exon 1A transcripts are expressed only from the paternal allele.[38,49] Two observations suggest that this region may be important for *GNAS1* imprinting. First, in mouse the methylation of exon 1A is established in the oocyte and maintained throughout pre- and postimplantation development,[38] a feature characteristic of so-called methylation imprint marks that are critical for the establishment of imprinting.[50,51] Second, as discussed below, imprinting of this region is lost in patients with an isolated form of parathyroid hormone (PTH) resistance and this most likely affects Gs_α expression in specific tissues.[49]

Gs_α ACTIVATING MUTATIONS

Endocrine Tumors

Activating Gs_α mutations produce functional endocrine tumors by increasing intracellular cAMP, which is known to stimulate both cell proliferation and hormone secretion in many endocrine glands (see TABLE 1). Activating mutations encoding substitutions of Gs_α residues Arg^{201} or Gln^{227} were initially identified in a subset of human pituitary growth hormone (GH)–secreting tumors that had high levels of membrane-associated adenylyl cyclase activity.[52] Subsequent studies confirmed that these mutations occur in about one-third of GH-secreting tumors.[22,53,54] These are somatic mutations, as they are not found in peripheral blood samples from the same patients. Arg^{201} or Gln^{227} are catalytically important for the intrinsic GTPase activity of Gs_α, and mutations of these residues leads to constitutive activation by allowing the G protein to remain in the GTP-bound active state for a long period of time.[22,23,55] The covalent modification of Arg^{201} that is catalyzed by cholera toxin also leads to Gs_α activation. The severe secretory diarrhea that characterizes intestinal cholera infection results from elevated cAMP levels in intestinal lining cells. In somatotrophs excess cAMP induces the tissue-specific transcription factor GHF1, which eventually leads to excess cell proliferation and GH secretion.[56] Transgenic mice expressing cholera toxin in their somatotrophs developed pituitary hyperplasia and excess GH secretion.[57]

Because they have been identified in GH-secreting tumors, these mutations, which are dominant acting, have been called *gsp* mutations.[22,53] *Gsp* mutations are present in about 10% of clinically nonfunctional pituitary tumors[53,58,59] and in a very small number of corticotroph adenomas.[53,60] Gs_α is imprinted in human pituitaries, where it is expressed only from the maternal allele.[61] In one study examining *gsp*+ GH-secreting tumors, the mutation was present on the maternal allele in 21 of 22 tumors, suggesting that tumor growth depends on expression of the activated protein.[61] This study also provided evidence that imprinting of Gs_α is relaxed in these tumors. No *gsp* mutations have been identified in lactotroph or thyrotroph tumors.[53,62]

Gsp mutations are also present in a small subset of autonomously functioning thyroid tumors[53,63] and are rarely present in differentiated thyroid cancers.[63–65] Cell culture[66] and transgenic mouse studies[67] show that activated Gs_α leads to thyroid

TABLE 1. Human diseases associated with *GNAS1* defects

Activating mutations (somatic)	Inactivating mutations (germline)	Imprinting defect
GH-secreting pituitary tumors	Albright hereditary osteodystrophy (AHO)	PHPIB
Thyroid adenomas, carcinomas	PHPIA (maternal mutation)	
Fibrous dysplasia	PPHP (paternal mutation)	
McCune-Albright syndrome (MAS)	Progressive osseous heteroplasia (POH)	
Leydig cell tumors	PHPIA + testotoxicosis	
Myxomas	(maternal A366S mutation)	
Pheochromocytoma (very rare)	PHPIB (maternal ΔI382 mutation)	
Parathyroid adenoma (very rare)		
ACTH-secreting pituitary tumor (very rare)		

cell proliferation and increased thyroid hormone secretion, perhaps through activation of mitogen-activated protein kinase pathways.[68] *Gsp* mutations are rarely found in other endocrine tumors.[53,69–72] Several compensatory mechanisms might mitigate the effects of *gsp* mutations in endocrine cells, including the rapid turnover of activated Gs_α proteins[73] and the induction of phosphodiesterases that degrade cAMP.[74]

McCune-Albright Syndrome

McCune-Albright syndrome (MAS) is classically defined as the co-occurrence of hyperpigmented (café-au-lait) skin lesions, sexual precocity, and fibrous dysplasia of bone (FD).[75–77] However, MAS patients may present with only one or two of these manifestations or may also present with other endocrine or nonendocrine manifestations.[78–80] MAS is a sporadic disease and is not inherited. The café-au-lait spots are tan-brown hyperpigmented flat macules that are often arranged in a segmental pattern, which follows the developmental lines of Blaschko.[81] Melanocytes cultured from these lesions have increased levels of tyrosinase, the rate-limiting enzyme for the production of melanin, as a result of increased intracellular cAMP.[82]

In female MAS patients sexual precocity presents as premature menses associated with fluctuating serum estrogen levels and enlarged ovarian follicles.[83] In males sexual precocity presents with testicular enlargement and/or premature onset of secondary sex characteristics and spermatogenesis.[80,84,85] MAS patients may also develop hyperfunctional thyroid nodules,[80,84,86–88] macronodular adrenal hyperplasia or adrenal adenomas associated with hypercortisolism,[80,84,89,90] somatomammotroph adenomas,[80,84,91] or hypophosphatemic rickets or osteomalacia.[80,84,92] In each case the peripheral endocrine organs (thyroid, adrenal cortex, gonads) appear activated despite low circulating levels of their respective stimulatory pituitary hormone (e.g., thyrotropin, adrenocorticotropin, gonadotropins).[80,84,86,87,93–96] Hypophosphatemia is associated with a more general renal tubular dysfunction[97] and is likely to be caused by excess secretion of a phosphaturic factor from fibrous dysplasia lesions similar to the factor presumed to be oversecreted by tumors in tumor-induced osteomalacia.

MAS patients may also develop one or more nonendocrine abnormalities affecting the liver, heart, thymus, spleen, bone marrow, gastrointestinal tract, or the brain.[98–102] Some abnormalities reported in MAS patients include cholestatic jaundice, atypical cardiomyocyte hypertrophy, tachycardia, sudden death, thymic hyperplasia, myelofibrosis, intestinal polyps, and microcephaly.

On the basis of the specific distribution pattern of café-au-lait lesions, Happle proposed that MAS is likely to be caused by a dominant somatic mutation occurring early in development or a gametic half chromatid mutation, producing a mosaic with a widespead distribution of mutant cells.[81] The extent of clinical manifestations within each patient is determined by the specific distribution of mutant-bearing cells. Presumably the mutation would be lethal in the germline, as MAS is never inherited. The endocrine and skin manifestations of MAS provided the clue that activating Gs_α mutations might be the underlying mutation, as cAMP was known to stimulate endocrine gland proliferation, hormone secretion, and pigment granule formation; and cAMP levels were shown to be elevated in various tissues isolated from MAS patients.[82,103–105] This was further supported by the fact that circulating levels of trophic hormones, which stimulate these tissues, are suppressed in MAS patients.[84,95,106,107]

Many studies show that MAS is caused by the widespread distribution of somatic *gsp* mutations affecting Arg^{201} (refs. 98,100,108–110). A possible germline Arg^{201} to Leu mutation was identified in a patient with severe skeletal, endocrine, and developmental abnormalities.[111] *Gsp* mutations coding for Gs_α Gln^{227} substitutions have not been found in MAS, perhaps because these mutations are more activating and therefore more lethal.[22] As in sporadic endocrine tumors, the mutation is always heterozygous, consistent with its being dominantly acting. *Gsp* mutations have been identified in both normal and abnormal nonendocrine tissues.[98,100,101] It seems plausible that increased chronotropic and ionotropic stimulation due to Gs activation could lead to the cardiac hypertrophy and sudden death that is occasionally associated with MAS.[100,112] It remains to be determined how *gsp* mutations lead to the other nonendocrine manifestations of MAS.

Fibrous Dysplasia of Bone

FD is a nonfamilial bone disorder characterized by focal benign lesions that most often occur in the long bones of the extremities, the ribs, and craniofacial bones.[76,77] Most FD patients (~70%) have a single bone lesion (monoostotic FD), while the remainder have multiple lesions (polyostotic FD).[113] A small minority of polyostotic FD patients have other features of MAS. In some patients FD is associated with intramuscular myxomas (Mazabraud's syndrome).[114–116] FD lesions may be asymptomatic or may produce bone deformity, pain, pathological fractures, or cranial nerve compression.[80,99,117]

FD lesions originate in the medullary cavity and expand concentrically outward into the surrounding cortical bone. Most of the cells are immature mesenchymal cells with a spindle-shaped fibroblastic appearance[118] that express various osteoblast-specific markers.[119] Embedded within the fibroblast-like cells are spicules of immature woven bone that are lined by flat cells with retracted cell bodies, which form pseudolacunar spaces.[119,120] The observed cell retraction is likely to be the result of increased intracellular cAMP.[119] Sometimes islands of hyaline cartilage are also present.[118,121] Osteoclasts are present at the periphery where they actively resorb surrounding normal bone.[104,119] Bisphosphonates provide therapeutic benefit due to their antiresorptive activity.[122] Rarely these lesions undergo malignant degeneration.[123,124]

FD appears to result from abnormal proliferation and differentiation of bone marrow stromal cells.[125] Cells isolated from FD lesions have an increased proliferation rate and are poorly differentiated, expressing early, but not late, osteoblast differentiation markers.[105,119] Transplantation of these cells into immunocompromised mice produces lesions that are similar to human FD.[126,127]

As in MAS, *gsp* mutations have also been found in FD lesions from patients both with and without other MAS manifestations.[105,108,127–131] Similar mutations are also found in intramuscular myxomas, both those that occur alone and those that are present in patients with Mazabraud's syndrome.[132] Elevated cAMP in osteoprogenitor cells may lead to FD by stimulating proliferation and inhibiting differentiation. In cell culture models both PTH and forskolin, agents that stimulate cAMP production, inhibit osteoblast differentiation.[133–135] cAMP induces c-*fos* and other genes in a PKA-dependent manner.[136] PKA phosphorylates nuclear transcription factors, such as cAMP–response element binding protein (CREB), which activates the transcription of genes such as c-*fos*.[137] Fos overexpression in osteogenic cells in response to increased cAMP appears to be an important common pathway in the development of FD.[138,139] Fos is one component of AP-1, transcriptional factors that are present at high levels in proliferating osteoblasts[140] and that suppress the expression of osteoblast-specific genes, such as osteocalcin.[140] cAMP, through the induction of Fos, also stimulates the expression of the interleukin IL-6, which stimulates bone resorption by recruiting osteoclasts.[104,141] Studies suggest that local production of other factors, such as PTH-related protein (PTHrP) and platelet-derived growth factor B may also play a role in the development of FD.[142,143]

Gs_α LOSS-OF-FUNCTION MUTATIONS

Albright Hereditary Osteodystrophy

Albright hereditary osteodystrophy (AHO) is characterized by the following features: short stature, obesity, brachydactyly, subcutaneous ossifications, and mental deficits or developmental delay.[144,145] Brachydactyly refers to shortening and widening of long bones in the hands and feet, which may or may not be symmetric and most often involves the distal thumb and fourth and fifth metacarpals.[146–148] Brachydactyly becomes clinically obvious within the first decade and is associated with premature closure and coning of the epiphyses. Ectopic intramembranous ossifications are generally confined to the superficial subcutaneous tissues. Decreased lipolysis in adipose tissue may contribute to the obesity that is commonly associated with AHO.[149,150] The severity of the AHO phenotype varies widely between patients, with some patients having few, or no, features of the syndrome.[151,152]

Pseudohypoparathyroidism Type IA

Patients who inherit the AHO trait from their mother also develop multihormone resistance, a condition referred to as pseudohypoparathyroidism type IA (PHPIA).[153] The multihormone resistance in PHPIA involves hormones that activate Gs-coupled pathways in their target tissues. Renal parathyroid hormone (PTH) resistance develops over the first several years of life, initially presenting with hyperphosphatemia and elevated PTH levels and eventually with hypocalcemia.[154–157] Some PHPIA patients remain eucalcemic with elevated PTH levels.[158–160] PHPIA patients have a markedly reduced urinary cAMP response to administered PTH,[161,162] consistent with a signaling defect proximal to cAMP generation in renal proximal tubules. As would be predicted, phosphate excretion and synthesis of 1,25-dihydroxyvitamin D, two physiological responses to PTH that are mediated by cAMP, are both impaired in PHPIA.[163–169] 1,25-Dihydroxyvitamin D deficiency leads to hypocalcemia by preventing the intestinal absorption of calcium and the mobilization of calcium from bone.[158,170,171]

Almost all PHPIA patients also present with thyrotropin (TSH) resistance,[172,173] typically from the time of birth.[157,174–176] The resistance is generally mild with minimally elevated TSH levels and normal or slightly low thyroid hormone levels. Goiter is usually absent, presumably due to the resistance of the thyroid to the growth-promoting effects of TSH. Low TSH-stimulated adenylyl cyclase activity was demonstrated in thyroid membranes isolated from one PHPIA patient.[177]

Female PHPIA patients often present with hypogonadism, with delayed or incomplete sexual maturation, oligomenorrhea, and/or impaired fertility.[172,178–181] These patients tend to be hypoestrogenic,[181] but gonadotropin levels are not clearly elevated.[172,178–181] It has been proposed that PHPIA patients have a partial resistance to gonadotropins that effects ovulation but not follicular development.[181] Many PHPIA patients also have prolactin deficiency[172,182–186] and some have GH deficiency, which may contribute to short stature.[179,187] Olfactory defects have been reported in PHPIA patients.[188–191]

PHPIA patients do not show resistance to vasopressin[182,192] or adrenocorticotropin,[172,182] hormones that also stimulate Gs-coupled pathways in their target tissues. This may reflect the fact that the levels of intracellular cAMP generated by these hormones are still sufficient to produce a maximal physiological response. For example, although the plasma cAMP responses to glucagon[172,193] and isoproterenol[194] are somewhat reduced in PHPIA, the hyperglycemic response to these hormones is unaffected.

Pseudopseudohypoparathyroidism

Patients who inherit the AHO trait from their father present only with the AHO phenotype and do not have multihormone resistance. This condition has also been called pseudopseudohypoparathyroidism (PPHP).[153] PPHP patients have normal levels of PTH and TSH and have normal urinary cAMP responses to administered PTH.[162] Because many of the clinical features of AHO are nonspecific, the diagnosis of PPHP can only be made in patients within a well-documented AHO kindred or in whom a Gs defect is identified.

Pathogenesis of PHPIA/PPHP

Biochemical reconstitution assays that measure hormone signaling in patient cell membranes showed that Gs bioactivity is decreased by ~50% in membranes isolated from various tissues in PHPIA patients.[162,195–197] Decreased Gs bioactivity is often associated with decreased Gs_α mRNA[198,199] and/or protein[152,200] expression. A similar Gs_α deficiency is also present in PPHP patients,[162] suggesting that both disorders are caused by a common genetic defect.

Consistent with 50% loss of Gs_α expression in PHPIA and PPHP, multiple heterozygous inactivating mutations within the Gs_α coding exons have been identified in these patients (for a compilation of mutations, see refs. 145,201). These mutations are spread throughout the coding exons, except exon 3, which can be spliced out and still produce a functional Gs_α protein.[201] One specific 4-bp deletion within exon 7 has been identified in at least 13 families[201–209] and appears to result from the frequent occurrence of *de novo* mutation due to DNA polymerase pausing and slipped strand mispairing.[203] Two missense mutations (Met^1 to Val, Ala^{366} to Ser) have each been reported in two kindreds.[201,210,211] All other mutations to date are unique to a single kindred.

GNAS1 mutations associated with AHO include deletional and insertional frameshift mutations, nonsense mutations, and splice junction mutations, most of which disrupt Gs_α mRNA expression, as well as missense mutations, which alter Gs_α protein stability or function. Protein stability may be disrupted by steric effects due to the presence of an amino acid with a bulky side change (e.g., Ser^{250} to Arg, ref. 212; Glu^{259} to Val, refs. 21, 207) or by decreased affinity for guanine nucleotide (e.g., Ser^{250} to Arg, ref. 212; Arg^{258} to Trp, ref. 17; Ala^{366} to Ser, ref. 211), as bound guanine nucleotide is required for normal α-subunit stability.

Missense mutations can also produce more specific biochemical defects. One mutation within the carboxy terminus (Arg^{385} to His, ref. 213) leads to global uncoupling of Gs_α from receptors, while a nearby mutation (deletion of Ile^{382} ref. 214) leads to selective uncoupling from PTH receptors, producing an isolated PTH resistance phenotype. Mutations of the basic residue Arg^{231} within the switch 2 region[20,215] or the acidic residue Glu^{259} in switch 3 (ref. 21) lead to a receptor activation defect, presumably by disrupting interactions between the two switch regions that stabilize the active conformation. Mutation of the switch 3 residue Arg^{258} leads to increased GDP dissociation due to loss of an interaction with a helical domain that normally forms a "lid" over the cleft between the two domains.[17,216] This mutation also increases the catalytic rate of the intrinsic GTPase reaction.[216] The Ala^{366} to Ser mutation produces a syndrome of PHPIA plus gonadotropin-independent precocious puberty in males by subtly increasing the rate of GDP dissociation.[211] At ambient body temperature the protein is unstable, leading to PHPIA, while at the slightly lower temperature of the testis, the protein remains stable and constitutively activated due to increased basal GDP release (the rate-limiting step for activation), leading to testotoxicosis.

Identical *GNAS1* mutations are present in both PHPIA and PPHP patients within the same family,[151,157,204,205,208,217,218] confirming that both disorders are caused by the same genetic defect. Many studies now show that PHPIA results from maternal inheritance of *GNAS1* mutations, while PPHP results from paternal inheritance of the same mutations,[151,153,157,204,205,208,217,218] strongly suggesting that *GNAS1* is imprinted. This model would predict that Gs_α is normally expressed primarily from the maternal allele due to silencing (imprinting) of the paternal Gs_α allele in hormone target tissues (e.g., renal proximal tubules, the major site of PTH action in the kidney). Inactivating mutations on the active maternal allele would lead to loss of Gs_α expression and hormone signaling (PHPIA), while mutations on the inactive paternal allele would have no effect on Gs_α expression or hormone signaling (PPHP). This would explain why the urinary cAMP response to administered PTH is markedly reduced in PHPIA patients but is normal in PPHP patients.[161,162] Support for this model comes from a recent study demonstrating maternal-specific expression of Gs_α in human pitutitaries[61] as well as studies on *Gnas* knockout mice that confirm Gs_α imprinting in renal proximal tubules.[219]

The imprinting of Gs_α would have to be tissue specific, as $G_s\alpha$ is not imprinted in several human and mouse tissues,[36,39,219,220] and is supported by the fact that in several tissues Gs_α expression is similarly reduced by ~50% in both PHPIA and PPHP patients.[162] Lack of Gs_α imprinting in certain hormone target tissues (e.g., adrenal cortex, renal collecting ducts) may explain why PHPIA patients fail to demonstrate resistance to other hormones (e.g., adrenocorticotropin, vasopressin), which also activate Gs. The partial loss of Gs_α expression resulting from heterozygous mutations may fail to produce hormone resistance, either because the amounts of Gs_α are still not rate limiting for hormone-stimulated cAMP, or because the reduced cAMP levels are still adequate to elicit the downstream physiological responses.[193,221,222] Lack of Gs_α imprinting in the distal nephron[219,223] may also explain why the anticalciuric effect of PTH in the distal nephron is maintained in PHPIA patients.[224]

The AHO phenotype most likely results from Gs_α deficiency, rather than from loss of NESP55 or XLαs expression. The paternal-specific XLαs gene product should be lost only in PPHP patients, while NESP55 should be lost only in those PHPIA patients whose mutations disrupt mRNA expression. Given that the AHO phenotype is similar in both PHPIA and PPHP, it seems unlikely that loss of either XLαs or NESP55 could explain the phenotype. Moreover, loss of NESP55 expression in some patients is not associated with the AHO phenotype.[49]

Exactly how Gs_α haploinsufficiency might lead to AHO is unclear, although there are potential mechanisms to explore. While activating Gs_α mutations appear to lead to inhibition of osteoblast differentiation, resulting in FD, decreased Gs_α signaling may lead to ectopic ossifications by promoting osteoblast differentiation. Brachydactyly in AHO patients results from premature closure of growth plates. PTHrP, through Gs pathways, normally inhibits chondrocyte differentiation within the growth plate.[225–228] Decreased Gs_α expression would be expected to lead to shortened bone growth by decreasing the proliferative phase of growth plate chondrocytes, and recent evidence for this has been provided by examining the differentiation pattern of growth plate chondrocytes with a *Gnas* knockout allele.[229]

Gs_α mutations could lead to obesity by several mechanisms. Decreased cAMP levels result in decreased lipolysis.[149,150] Mice with increased PKA activity due to loss of a PKA regulatory subunit are lean,[230] while mice with an increased ratio of α2- to β-adrenergic receptors in adipocytes (which lowers the ability of catecholamines to raise intracellular cAMP) are prone to obesity.[231] It is also possible that decreased sympathetic activity may contribute to decreased lipolysis as well.[150,232] In addition to its effects on metabolism, Gs_α deficiency may also promote adipocyte differentiation.[233] Exactly how Gs_α deficiency might lead to the mental deficits in AHO is unknown, although cAMP is known to be important for learning and memory.[234,235]

Progressive Osseous Heteroplasia

Progressive osseous heteroplasia (POH) is a congenital disorder of ectopic intramembranous bone formation characterized by the formation of large dermal and subcutaneous ossifications that invade deep connective tissues and skeletal muscle.[236] In most cases POH is not associated with the other characteristic features of AHO. Recently inactivating Gs_α mutations were identified in two POH patients[237,238] and reduced Gs_α expression was identified in a third case.[238] Two of these patients also presented with other signs of AHO.

GNAS1 IMPRINTING DEFECTS

Pseudohypoparathyroidism Type IB

Pseudohypoparathyroidism type IB (PHPIB) patients have renal PTH resistance, but do not have AHO or resistance to other hormones. PHPIB can occur sporadically or in a familial setting. In PHPIB families, PTH resistance develops only when the trait is inherited maternally, similar to the inheritance pattern for PTH resistance in patients with AHO.[239] As in PHPIA, the PTH-stimulated urinary cAMP responses are markedly reduced in PHPIB,[172] consistent with a proximal signaling defect. Familial PHPIB was mapped to 20q13, in the vicinity of GNAS1.[239] However, Gs function is normal in erythrocyte membranes from PHPIB patients,[172,240] ruling out disrupting mutations within the Gs_α coding exons that are typically present in AHO/PHPIA.

We hypothesized that if both GNAS1 alleles had a paternal-specific imprinting pattern, then Gs_α expression in renal proximal tubules would be markedly reduced due to lack of an "active maternal" allele, resulting in renal PTH resistance. In contrast, abnormal GNAS1 imprinting would not affect Gs_α expression in most other tissues where Gs_α is normally equally expressed from the maternal and paternal allele. This would account for the combination of renal PTH resistance and normal erythrocyte Gs_α expression that is characteristic of PHPIB. Normal Gs_α expression in the vast majority of tissues where Gs_α is normally not imprinted might explain why these patients lack the AHO phenotype.

Recent studies have shown that imprinting of the exon 1A region is lost in virtually all PHPIB patients (40 of 40 patients examined—refs. 49, 241, 242 and L.S.W. and J.L., unpublished data), with both alleles having a paternal-specific imprinting pattern (unmethylated, transcriptionally active) in the exon 1A region. In contrast, imprinting of the NESP55 and XLαs promoter regions is unaffected in most PHPIB patients. Therefore, PHPIB appears to result from abnormal imprinting of the exon 1A region. Paternal uniparental disomy (UPD, the inheritance of both chromosomes or chromosomal regions from the same parent) of chromosome 20 does not appear to be common in PHPIB,[49] although it was reported in one patient with PHPIB and

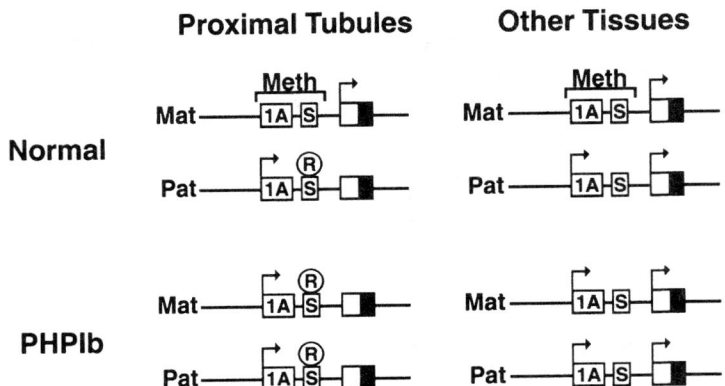

FIGURE 2. A model for the role of a *GNAS1* imprinting defect in PHPIB. In the model shown the exon 1A region has a *cis*-acting silencer element (S) that binds a repressor (R) that is expressed only in specific tissues such as renal proximal tubules. The repressor normally binds to the paternal allele and inhibits Gs_α expression (Gs_α exon 1 shown as *partially filled box*), but is prevented from binding to the maternal allele by methylation of the silencer, allowing Gs_α to be maternally expressed. In most tissues the repressor is not expressed, and, therefore, Gs_α is biallelically expressed despite the fact that the maternal allele is methylated. In PHPIB the exon 1A region is unmethylated on both parental alleles, allowing the repressor to bind to both alleles in proximal tubules, which results in loss of Gs_α expression and PTH resistance. In other tissues Gs_α is biallelically expressed despite the abnormal methylation because the repressor is not expressed. This would explain why Gs_α expression is unaffected in tissues (e.g., erythrocytes) derived from PHPIB patients. (Reproduced from L.S. Weinstein. 2001. J. Clin. Endocrinol. Metab. **86:** 4622–4626).

other associated abnormalities.[243] Mapping of familial PHPIB to 20q13 suggests that the imprinting defect may result from an underlying mutation within the vicinity of *GNAS1*.[239,241] Analysis of one such patient provided evidence that the mutation leading to the imprinting defect may be located at a distance from the exon 1A region.[241] Recently a Gs_α coding mutation producing a protein that is selectively uncoupled from the PTH receptor due to deletion of residue Ile^{382} in the carboxyl-terminus was also shown to result in a PHPIB-like phenotype.[77]

If the imprinted exon 1A region is important for tissue-specific imprinting of Gs_α, how might this occur? Any potential model must account for the fact that imprinting of the exon 1A region is ubiquitous while imprinting of Gs_α is tissue specific. For instance, the exon 1A region could contain a silencer element that binds a tissue-specific repressor on the paternal allele, but that cannot bind the repressor on the maternal allele because the binding site is methylated (FIG. 2). This would lead to silencing of Gs_α in the paternal allele in tissues where the repressor is expressed (e.g., renal proximal tubules). In PHPIB patients the maternal-specific methylation of the exon 1A region is lost, resulting in loss of Gs_α expression from both alleles. In most other tissues, the imprinting status of the exon 1A region has no affect on the allele-specific expression of Gs_α because the repressor is not expressed in these tissues. Alternatively, this region may contain a boundary element that insulates the Gs_α promoter from an upstream enhancer on the paternal allele, but that does not function as an boundary element on the maternal allele because methylation prevents binding of an insulator protein. This is precisely the mechanism by which the *Igf2* gene is imprinted.[244,245]

GNAS KNOCKOUT MICE

Mice with paternal UPD/maternal deletion and maternal UPD/paternal deletion of a distal chromosome 2 region including the *Gnas* locus lead to distinct phenotypes.[35,246] Mice with targeted disruption of *Gnas* exon 2 in both alleles die in early gestation, demonstrating that the gene is necessary for development.[219] Early on, heterozygotes with disruption of the maternal (m–/+) or paternal (+/p–) allele have distinct phenotypes that are very similar to those described for paternal and maternal UPD mice, respectively.[219,246] These findings suggest that the m–/+ and +/p– phenotypes might result from loss of maternal (e.g., Gs_α)– and paternal (XLαs, Gs_α)–specific *Gnas* gene products, respectively.

Similar to what is observed in humans with AHO, maternal, but not paternal, transmission of the knockout allele resulted in renal PTH resistance and reduced PTH-stimulated cAMP production in renal proximal tubules.[219] Consistent with these biochemical observations, maternal, but not paternal, transmission of the knockout allele also resulted in reduced Gs_α expression in proximal tubules,[219] confirming that Gs_α is imprinted in this tissue with expression mostly from the maternal allele. Within the kidney, imprinting of Gs_α is limited to the proximal tubules, as Gs_α is not imprinted in more distal portions of the nephron, such as the thick ascending limbs or collecting ducts.[219,221,223] Lack of Gs_α imprinting in the distal nephron might explain why PHP patients seem to maintain the ability of PTH to stimulate calcium reabsorption in the distal nephron and therefore do not develop hypercalciuria.[224] This may also contribute to the fact that PHPIA patients do not have obvious renal resistance to vasopressin in the collecting ducts.[182,192,223]

M–/+ and +/p– mice also have abnormalities in energy and glucose metabolism.[232,247] M–/+ mice develop obesity in association with low metabolic rates and activity levels. In contrast, +/p– mice are leaner than normal and are both hypermetabolic and hyperactive. Similar effects are also observed in mice with maternal and paternal inheritance of a radiation-induced mutation that maps near *Gnas*.[248] Preliminary results suggest that these opposite effects may be the result of decreased and increased sympathetic activity in m–/+ and +/p– mice, respectively.[232] Circulating levels of norepinephrine were reported to be low in three PHPIA patients, suggesting that reduced sympathetic activity might also contribute to obesity in PHPIA.[150] However, it remains unclear why +/p– mice are lean while PPHP patients who also have paternal mutations are obese. Both m–/+ and +/p– mice have increased sensitivity to insulin *in vivo* and increased insulin-stimulated glucose uptake in skeletal muscle.[247] The similar effects on insulin sensitivity in both lean and obese groups of mice may reflect the fact that Gs_α expression in skeletal muscle is similarly reduced in m–/+ and +/p– mice (Gs_α appears not to be imprinted in this tissue[232]). These findings suggest that Gs_α is a negative regulator of insulin action.

SUMMARY

Activating and inactivating Gs_α mutations produce opposite effects on endocrine function and bone development, presumably secondary to opposite effects on intracellular cAMP. Activating mutations lead to the activation of hormone signaling pathways in the absence of circulating hormone, while inactivating mutations lead to hormone resistance. Activating mutations inhibit osteoblast differentiation, leading to FD, while inactivating mutations promote osteoblast differentiation, leading to ectopic ossifications in AHO and POH. Activating Gs_α mutations are somatic, and the specific manifestations in a given individual are primarily determined by the extent and specific distribution of mutant-bearing cells. In contrast, inactivating Gs_α mutations are germline and therefore lead to global manifestations, which may be altered by other genetic and environmental factors.

The manifestations of Gs_α mutations are further modified by parental inheritance because Gs_α is imprinted in a tissue-specific manner. Maternal inheritance of inactivating Gs_α mutations (PHPIA) or an imprinting defect that leads to a paternal-specific imprinting pattern on both parental alleles leads to renal PTH resistance due to loss of expression of Gs_α from the active maternal allele in renal proximal tubules. In most other tissues Gs_α is not imprinted, and therefore mutations on either the maternal (PHPIA) or paternal (PPHP) allele lead to a similar, ~50%, loss of Gs_α expression. This haploinsufficiency is likely to result in the AHO phenotype that is present in both PHPIA and PPHP patients. Recent findings also suggest that the clinical manifestations in individuals with activating Gs_α mutation are likely to be affected by whether the mutation is present in the maternal or paternal allele.

REFERENCES

1. WEDEGAERTNER, P.B., P.T. WILSON & H.R. BOURNE. 1995. Lipid modifications of trimeric G proteins. J. Biol. Chem. **270:** 503–506.
2. MATTERA, R. *et al.* 1989. Splice variants of the alpha subunit of the G protein G_s activate both adenylyl cyclase and calcium channels. Science **243:** 804–807.
3. YATANI, A. *et al.* 1988. The stimulatory G protein of adenylyl cyclase, G_s, also stimulates dihydropyridine-sensitive Ca^{2+} channels. Evidence for direct regulation independent of phosphorylation by cAMP-dependent protein kinase or stimulation by a dihydropyridine agonist. J. Biol. Chem. **263:** 9887–9895.
4. MA, Y. *et al.* 2000. Src tyrosine kinase is a novel direct effector of G proteins. Cell **102:** 635–646.
5. BOMSEL, M. & K. MOSTOV. 1992. Role of heterotrimeric G proteins in membrane traffic. Mol. Biol. Cell **3:** 1317–1328.
6. NAIR, B.G. *et al.* 1990. $G_{s\alpha}$ mediates epidermal growth factor-elicited stimulation of rat cardiac adenylate cyclase. J. Biol. Chem. **265:** 21317–21322.
7. SUN, H. *et al.* 1997. The juxtamembrane, cytosolic region of the epidermal growth factor receptor is involved in association with α-subunit of G_s. J. Biol. Chem. **272:** 5413–5420.
8. KRIEGER-BRAUER, H.I., P. MEDDA & H. KATHER. 2000. Basic fibroblast growth factor utilized both types of component subunits of G_s for dual signaling in human adipocytes. Stimulation of adenylyl cyclase via $G_{\alpha s}$ and inhibition of NADPH oxidase by Gβγs. J. Biol. Chem. **275:** 35920–35925.
9. NAIR, B.G. & T.B. PATEL. 1993. Regulation of cardiac adenylyl cyclase by epidermal growth factor (EGF). Role of EGF receptor protein tyrosine kinase activity. Biochem. Pharmacol. **46:** 1239–1245.

10. POPPLETON, H. et al. 1996. Activation of $G_s\alpha$ by the epidermal growth factor receptor involves phosphorylation. J. Biol. Chem. **271:** 6947–6951.
11. SCHOLICH, K. et al. 1999. Facilitation of signal onset and termination by adenylyl cyclase. Science **283:** 1328–1331.
12. SUNAHARA, R.K. et al. 1997. Crystal structure of the adenylyl cyclase activator $G_{s\alpha}$. Science **278:** 1943–1947.
13. TESMER, J.J. et al. 1997. Crystal structure of the catalytic domains of adenylyl cyclase in a complex with $G_s\alpha \cdot GTP\gamma S$. Science **278:** 1907–1916.
14. BRAY, P. et al. 1986. Human cDNA clones for four species of $G_{\alpha s}$ signal transduction protein. Proc. Natl. Acad. Sci. USA **83:** 8893–8897.
15. KOZASA, T. et al. 1988. Isolation and characterization of the human Gs alpha gene. Proc. Natl. Acad. Sci. USA **85:** 2081–2085.
16. MIXON, M.B. et al. 1995. Tertiary and quaternary structural changes in $G_{i\alpha 1}$. Science **270:** 954–960.
17. WARNER, D.R. et al. 1998. A novel mutation in the switch 3 region of $G_s\alpha$ in a patient with Albright hereditary osteodystrophy impairs GDP binding and receptor activation. J. Biol. Chem. **273:** 23976–23983.
18. LI, Q.B. & R.A. CERIONE. 1997. Communication between switch II and switch III of the transducin α subunit is essential for target activation. J. Biol. Chem. **272:** 21673–21676.
19. IIRI, T., Z. FARFEL & H.R. BOURNE. 1998. G-protein diseases furnish a model for the turn-on switch. Nature **394:** 35–38.
20. IIRI, T., Z. FARFEL & H.R. BOURNE. 1997. Conditional activation defect of a human $G_{s\alpha}$ mutant. Proc. Natl. Acad. Sci. USA **94:** 5656–5661.
21. WARNER, D.R. et al. 1999. Mutagenesis of the conserved residue Glu^{259} of $G_s\alpha$ demonstrates the importance of interactions between switches 2 and 3 for activation. J. Biol. Chem. **274:** 4977–4984.
22. LANDIS, C.A. et al. 1989. GTPase inhibiting mutations activate the α chain of G_s and stimulate adenylyl cyclase in human pituitary tumours. Nature **340:** 692–696.
23. GRAZIANO, M.P. & A.G. GILMAN. 1989. Synthesis in *Escherichia coli* of GTPase-deficient mutants of G_s alpha. J. Biol. Chem. **264:** 15475–15482.
24. COLEMAN, D. E. et al. 1994. Structures of active conformations of $G_{i\alpha 1}$ and the mechanism of GTP hydrolysis. Science **265:** 1405–1412.
25. SONDEK, J. et al. 1994. GTPase mechanism of G proteins from the 1.7-Å crystal structure of transducin α-GDP-AlF_4^-. Nature **372:** 276–279.
26. RALL, T. & B.A. HARRIS. 1987. Identification of the lesion in the stimulatory GTP-binding protein of the uncoupled S49 lymphoma. FEBS Lett. **224:** 365–371.
27. SULLIVAN, K.A. et al. 1987. Identification of receptor contact site involved in receptor-G protein coupling. Nature **330:** 758–760.
28. WEST, R.E., JR. et al. 1985. Pertussis toxin-catalyzed ADP-ribosylation of transducin: cysteine 347 is the ADP-ribose acceptor site. J. Biol. Chem. **260:** 14428–14430.
29. SIMONDS, W.F. et al. 1989. G_{i2} mediates α_2-adrenergic inhibition of adenylyl cyclase in platelet membranes: *in situ* identification with G_α C-terminal antibodies. Proc. Natl. Acad. Sci. USA **86:** 7809–7813.
30. SIMONDS, W.F. et al. 1989. Receptor and effector interactions of Gs. Functional studies with antibodies to the αs carboxyl-terminal decapeptide. FEBS. Lett. **249:** 189–194.
31. GEJMAN, P.V. et al. 1991. Genetic mapping of the Gs-α subunit gene (*GNAS1*) to the distal long arm of chromosome 20 using a polymorphism detected by denaturing gradient gel electrophoresis. Genomics **9:** 782–783.
32. RAO, V.V., S. SCHNITTGER & I. HANSMANN. 1991. G protein $Gs\alpha$ (*GNAS1*), the probable candidate gene for Albright hereditary osteodystrophy, is assigned to human chromosome 20q12-q13.2. Genomics **10:** 257–261.
33. LEVINE, M.A., W.S. MODI & S.J. O'BRIEN. 1991. Mapping of the gene encoding the α subunit of the stimulatory G protein of adenylyl cyclase (*GNAS1*) to 20q13.2-q13.3 in human by *in situ* hybridization. Genomics **11:** 478–479.
34. BLATT, C. et al. 1988. Chromosomal localization of genes encoding guanine nucleotide-binding protein subunits in mouse and human. Proc. Natl. Acad. Sci. USA **85:** 7642–7646.

35. PETERS, J. et al. 1994. Mapping studies of the distal imprinting region of mouse chromosome 2. Genet. Res. **63:** 169–174.
36. HAYWARD, B.E. et al. 1998. The human *GNAS1* gene is imprinted and encodes distinct paternally and biallelically expressed G proteins. Proc. Natl. Acad. Sci. USA **95:** 10038–10043.
37. PETERS, J. et al. 1999. A cluster of oppositely imprinted transcripts at the *Gnas* locus in the distal imprinting region of mouse chromosome 2. Proc. Natl. Acad. Sci. USA **96:** 3830–3835.
38. LIU, J. et al. 2000. Identification of a methylation imprint mark within the mouse *Gnas* locus. Mol. Cell. Biol. **20:** 5808–5817.
39. HAYWARD, B.E. et al. 1998. Bidirectional imprinting of a single gene: *GNAS1* encodes maternally, paternally, and biallelically derived proteins. Proc. Natl. Acad. Sci. USA **95:** 15475–15480.
40. KELSEY, G. et al. 1999. Identification of imprinted loci by methylation-sensitive representational difference analysis: application to mouse distal chromosome 2. Genomics **62:** 129–138.
41. ISCHIA, R. et al. 1997. Molecular cloning and characterization of NESP55, a novel chromogranin-like precursor of a peptide with $5-HT_{1B}$ receptor antagonist activity. J. Biol. Chem. **272:** 11657–11662.
42. LI, T. et al. 2000. Tissue-specific expression of antisense and sense transcripts at the imprinted *Gnas* locus. Genomics **69:** 295–304.
43. KEHLENBACH, R.H., J. MATTHEY & W.B. HUTTNER. 1994. XLαs is a new type of G protein. Nature **372:** 804–809.
44. KLEMKE, M. et al. 2000. Characterization of the extra-large G protein α-subunit XLαs. II. Signal transduction properties. J. Biol. Chem. **275:** 33633–33640.
45. LOVISETTI-SCAMIFORM, P. et al. 1999. Relative amounts and molecular forms of NESP55 in various bovine tissues. Brain Res. **829:** 99–106.
46. PASOLLI, H.A. et al. 2000. Characterization of the extra-large G protein α-subunit XLαs. I. Tissue distribution and subcellular localization. J Biol. Chem. **275:** 33622–33632.
47. HAYWARD, B.E. & D.T. BONTHRON. 2000. An imprinted antisense transcript at the human *GNAS1* locus. Hum. Mol. Genet. **9:** 835–841.
48. ISHIKAWA, Y. et al. 1990. Alternative promoter and 5′ exon generate a novel $G_s\alpha$ mRNA. J. Biol. Chem. **265:** 8458–8462.
49. LIU, J. et al. 2000. A *GNAS1* imprinting defect in pseudohypoparathyroidism type IB. J. Clin. Invest. **106:** 1167–1174.
50. CONSTANCIA, M. et al. 1998. Imprinting mechanisms. Genome Res. **8:** 881–900.
51. REIK, W. & J. WALTER. 2001. Genomic imprinting: parental influence on the genome. Nat. Rev. Genet. **2:** 21–32.
52. VALLAR, L., A. SPADA & G. GIANNATTASIO. 1987. Altered Gs and adenylate cyclase activity in human GH-secreting pituitary adenomas. Nature **330:** 566–568.
53. LYONS, J. et al. 1990. Two G protein oncogenes in human endocrine tumors. Science **249:** 655–659.
54. YANG, I. et al. 1996. Characteristics of *gsp*-positive growth hormone-secreting pituitary tumors in Korean acromegalic patients. Eur. J. Endocrinol. **134:** 720–726.
55. MASTERS, S.B., C.A. LANDIS & H.R. BOURNE. 1990. GTPase-inhibiting mutations in the α subunit of Gs. Adv. Second Messenger Phosphoprotein Res. **24:** 70–75.
56. CASTRILLO, J-L., L.E. THEILL & M. KARIN. 1991. Function of the homeodomain protein GHF1 in pituitary cell proliferation. Science **253:** 197–199.
57. BURTON, F.H. et al. 1991. Pituitary hyperplasia and gigantism in mice caused by a cholera toxin transgene. Nature **350:** 74–77.
58. TORDJMAN, K. et al. 1993. Activating mutations of the G_s α-gene in nonfunctioning pituitary tumors. J. Clin. Endocrinol. Metab. **77:** 765–769.
59. WILLIAMSON, E.A. et al. 1994. Gsα and Gi2α mutations in clinically non-functioning pituitary tumours. Clin. Endocrinol. (Oxf.) **41:** 815–820.
60. WILLIAMSON, E.A. et al. 1995. G-protein mutations in human pituitary adrenocorticotrophic hormone-secreting adenomas. Eur. J. Clin. Invest. **25:** 128–131.
61. HAYWARD, B.E. et al. 2001. Imprinting of the $G_s\alpha$ gene *GNAS1* in the pathogenesis of acromegaly. J. Clin. Invest. **107:** R31–R36.

62. DONG, Q. *et al.* 1996. Screening of candidate oncogenes in human thyotroph tumors: absence of activating mutations of the Gα-q, Gα-11, Gα-s, or thyrotropin-releasing hormone receptor genes. J. Clin. Endocrinol. Metab. **81:** 1134–1140.
63. O'SULLIVAN, C. *et al.* 1991. Activating point mutations of the gsp oncogene in human thyroid adenomas. Mol. Carcinogenesis **4:** 345–349.
64. SUAREZ, H.G. *et al.* 1991. gsp mutations in human thyroid tumours. Oncogene **6:** 677–679.
65. ESAPA, C. *et al.* 1997. G protein and thyrotropin receptor mutations in thyroid neoplasia. J. Clin. Endocrinol. Metab. **82:** 493–496.
66. MUCA, C. & L. VALLAR. 1994. Expression of mutationally activated Gαs stimulates growth and differentiation of thyroid FRTL5 cells. Oncogene **9:** 3647–3653.
67. ZEIGER, M.A. *et al.* 1997. Thyroid-specific expression of cholera toxin A1 subunit causes thyroid hyperplasia and hyperthyroidism in transgenic mice. Endocrinology **138:** 3133–3140.
68. POMERANCE, M. *et al.* 2000. Thyroid-stimulating hormone and cyclic AMP activate p38 mitogen-activated protein kinase cascade. Involvement of protein kinase A, Rac1, and reactive oxygen species. J. Biol. Chem. **275:** 40539–40546.
69. WILLIAMSON, E.A. *et al.* 1995. G protein gene mutations in patients with multiple endocrinopathies. J. Clin. Endocrinol. Metab. **80:** 1702–1705.
70. YOSHIMOTO, K. *et al.* 1993. Rare mutations of the G$_s\alpha$ subunit gene in human endocrine tumors: mutation detection by polymerase chain reaction-primer-introduced restriction analysis. Cancer **72:** 1386–1393.
71. VESSEY, S.J.R. *et al.* 1994. Absence of mutations in the G$_s\alpha$ and G$_i$2α genes in sporadic parathyroid adenomas and insulinomas. Clin. Sci. **87:** 493–497.
72. REINCKE, M. *et al.* 1993. No evidence for oncogenic mutations in guanine nucleotide-binding proteins of human adrenocortical neoplasms. J. Clin. Endocrinol. Metab. **77:** 1419–1422.
73. BALLARE, E. *et al.* 1998. Activating mutations of the Gsα gene are associated with low levels of Gsα protein in growth hormone-secreting tumors. J Clin. Endocrinol. Metab. **83:** 4386–4390.
74. LANIA, A. *et al.* 1998. Constitutively active Gsα is associated with an increased phosphodiesterase activity in human growth hormone-secreting adenomas. J. Clin. Endocrinol. Metab. **83:** 1624–1628.
75. RINGEL, M.D., W.F. SCHWINDINGER & M.A. LEVINE. 1996. Clinical implications of genetic defects in G proteins. The molecular basis of McCune-Albright syndrome and Albright hereditary osteodystrophy. Medicine (Baltimore) **75:** 171–184.
76. WEINSTEIN, L.S. 1996. Other skeletal diseases resulting from G protein defects—fibrous dysplasia and McCune-Albright syndrome. *In* Principles of Bone Biology. J.P. Bilezikian, L.G. Raisz & G.A. Rodan, Eds.: 877–888. Academic Press. San Diego, CA.
77. WEINSTEIN, L.S. 2000. Fibrous dysplasia and the McCune-Albright syndrome. *In* The Genetics of Osteoporosis and Metabolic Bone Disease. M.J. Econs, Ed.: 163–177. Humana Press. Totowa, NJ.
78. GRANT, D.B. & L. MARTINEZ. 1983. The McCune-Albright syndrome without typical skin pigmentation. Acta Paediatr. Scand. **72:** 477–478.
79. RIETH, K.G. *et al.* 1984. Pituitary and ovarian abnormalities demonstrated by CT and ultrasound in children with features of the McCune-Albright syndrome. Radiology **153:** 389–393.
80. DANON, M. & J.D. CRAWFORD. 1987. The McCune-Albright syndrome. Engeb. Inn. Med. Kinderheilkd. **55:** 81–115.
81. HAPPLE, R. 1986. The McCune-Albright syndrome: a lethal gene surviving by mosaicism. Clin. Genet. **29:** 321-324.
82. KIM, I. *et al.* 1999. Activating mutation of Gsα in McCune-Albright syndrome causes skin pigmentation by tyrosinase gene activation on affected melanocytes. Horm. Res. **52:** 235–240.
83. FEUILLAN, P.P. 1993. Treatment of sexual precocity in girls with the McCune-Albright syndrome. *In* Sexual Precocity: Etiology, Diagnosis and Management. G.D. Grave & G.B. Cutler, Eds.: 243–251. Raven Press. New York.

84. MAURAS, N. & R.M. BLIZZARD. 1986. The McCune-Albright syndrome. Acta Endocrinol. Suppl. (Copenh.) **279:** 207–217.
85. GIOVANELLI, G., S. BERNASCONI & G. BANCHINI. 1978. McCune-Albright syndrome in a male child: a clinical and endocrinologic enigma. J. Pediatr. **92:** 220–226.
86. FEUILLAN, P.P. *et al.* 1990. Thyroid abnormalities in the McCune-Albright syndrome: ultrasonography and hormone studies. J. Clin. Endocrinol. Metab. **71:** 1596–1601.
87. MASTORAKOS, G. *et al.* 1997. Hyperthyroidism in McCune-Albright syndrome with a review of thyroid abnormalities sixty years after the first report. Thyroid **7:** 433–439.
88. LAIR-MILAN, F. *et al.* 1996. Thyroid sonographic abnormalities in McCune-Albright syndrome. Pediatr. Radiol. **26:** 424–426.
89. BENJAMIN, D.R. & J.W. MCROBERTS. 1973. Polyostotic fibrous dysplasia associated with Cushing syndrome. Arch. Pathol. **96:** 175–178.
90. KIRK, J.M.W. *et al.* 1999. Cushing's syndrome caused by nodular adrenal hyperplasia in children with McCune-Albright syndrome. J. Pediatr. **134:** 789–792.
91. KOVACS, K. *et al.* 1984. Mammosomatotroph hyperplasia associated with acromegaly and hyperprolactinemia in a patient with the McCune-Albright syndrome. Virchows Arch. **403:** 77–86.
92. SCHWINDINGER, W.F. & M.A. LEVINE. 1993. McCune-Albright syndrome. Trends. Endocrinol. Metab. **4:** 238–242.
93. FOSTER, C.M. *et al.* 1984. Absence of pubertal gonadotropin secretion in girls with McCune-Albright syndrome. J. Clin. Endocrinol. Metab. **58:** 1161–1165.
94. DANON, M. *et al.* 1975. Cushing syndrome, sexual precocity, and polyostotic fibrous dysplasia (Albright syndrome) in infancy. J. Pediatr. **87:** 917–921.
95. SCULLY, R.E. & B.U. MCNEELY. 1975. Case records of the Massachusetts General Hospital: Case 4-1975. N. Engl. J. Med. **292:** 199–203.
96. D'ARMIENTO, M. *et al.* 1983. McCune-Albright syndrome: evidence for autonomous multiendocrine hyperfunction. J. Pediatr. **102:** 584–586.
97. COLLINS, M.T. *et al.* 2001. Renal phosphate wasting in fibrous dysplasia of bone is part of a generalized renal tubular dysfunction similar to that seen in tumor-induced osteomalacia. J. Bone. Miner. Res. **16:** 806–813.
98. WEINSTEIN, L.S. *et al.* 1991. Activating mutations of the stimulatory G protein in the McCune-Albright syndrome. N. Engl. J. Med. **325:** 1688–1695.
99. HARRIS, W.H., H.R.J. DUDLEY & R.J. BARRY. 1962. The natural history of fibrous dysplasia. J. Bone Joint Surg. **44-A:** 207–233.
100. SHENKER, A. *et al.* 1993. Severe endocrine and nonendocrine manifestations of the McCune-Albright syndrome associated with activating mutations of stimulatory G protein G_s. J. Pediatr. **123:** 509–518.
101. SILVA, E.S. *et al.* 2000. Demonstration of McCune-Albright mutations in the liver of children with high γGT progressive cholestasis. J. Hepatol. **32:** 154–158.
102. MCCUNE, D.J. & H. BRUCH. 1937. Osteodystrophia fibrosa: report of a case in which the condition was combined with precocious puberty, pathologic pigmentation of the skin and hyperthyroidism, with a review of the literature. Am. J. Dis. Child. **54:** 806–848.
103. ZUNG, A. *et al.* 1995. Urinary cyclic adenosine 3′,5′-monophosphate response in McCune-Albright syndrome: clinical evidence for altered renal adenylate cyclase activity. J. Clin. Endocrinol. Metab. **80:** 3576–3581.
104. YAMAMOTO, T. *et al.* 1996. Increased IL-6-production by cells isolated from the fibrous bone dysplasia tissues in patients with McCune-Albright syndrome. J. Clin. Invest. **98:** 30–35.
105. MARIE, P.J. *et al.* 1997. Increased proliferation of osteoblastic cells expressing the activating Gsα mutation in monostotic and polyostotic fibrous dysplasia. Am. J. Path. **150:** 1059–1069.
106. LEE, P.A., C. VAN DOP & C.J. MIGEON. 1986. McCune-Albright syndrome: long term followup. JAMA **256:** 2980–2984.
107. DUMONT, J.E., J-C. JAUNIAUX & P.P. ROGER. 1989. The cyclic AMP-mediated stimulation of cell proliferation. Trends Biochem. Sci. **14:** 67–71.
108. SHENKER, A. *et al.* 1994. An activating $G_s α$ mutation is present in fibrous dysplasia of bone in the McCune-Albright syndrome. J. Clin. Endocrinol. Metab. **79:** 750–755.

109. SCHWINDINGER, W.F., C.A. FRANCOMANO & M.A. LEVINE. 1992. Identification of a mutation in the gene encoding the α subunit of the stimulatory G protein of adenylyl cyclase in McCune-Albright syndrome. Proc. Natl. Acad. Sci. USA **89:** 5152–5156.
110. RIMINUCCI, M. *et al.* 1999. A novel *GNAS1* mutation, R201G, in McCune-Albright syndrome. J. Bone Miner. Res. **14:** 1987–1989.
111. MOCKRIDGE, K.A. *et al.* 1999. Polyendocrinopathy, skeletal dysplasia, organomegaly, and distinctive facies associated with a novel, widely expressed Gsα mutation [abstract]. Am. J. Hum. Genet. **65** (Suppl.): A426.
112. ZHENG, M. *et al.* 2000. β_2-adrenergic receptor-induced p38 MAPK activation is mediated by protein kinase A rather than by G_i or $G\beta\gamma$ in adult mouse cardiomyocytes. J. Biol. Chem. **275:** 40635–40640.
113. NAGER, G.T., D.W. KENNEDY & E. KOPSTEIN. 1982. Fibrous dysplasia: a review of the disease and its manifestations in the temporal bone. Ann. Otol. Rhinol. Laryngol. Suppl. **92:** 1–52.
114. MAZABRAUD, A. & J. GIRARD. 1957. Un cas particulier de dysplasia localisations osseus et tendineuses. Rev. Rhum. Mal. Osteartic. **34:** 652–659.
115. LEVER, E.G. & K.W. PETTINGALE. 1983. Albright's syndrome associated with a soft-tissue myxoma and hypophosphataemic osteomalacia. Report of a case and review of the literature. J. Bone. Jt. Surg. Br. Vol. **65:** 621–626.
116. PRAYSON, M.A. & M.C. LEESON. 1993. Soft-tissue myxomas and fibrous dysplasia of bone: a case report and review of the literature. Clin. Orthop. Related Res. **291:** 222–228.
117. SCHLUMBERGER, H.G. 1946. Fibrous dysplasia of single bones (monostotic fibrous dysplasia). Milit. Surg. **99:** 504–527.
118. GRECO, M.A. & G.C. STEINER. 1986. Ultrastructure of fibrous dysplasia of bone: a study of its fibrous, osseous, and cartilaginous components. Ultrastruct. Pathol. **10:** 55–66.
119. RIMINUCCI, M. *et al.* 1997. Fibrous dysplasia of bone in the McCune-Albright syndrome: abnormalities in bone formation. Am. J. Pathol. **151:** 1587–1600.
120. RIMINUCCI, M. *et al.* 1999. The histopathology of fibrous dysplasia of bone in patients with activating mutations of the Gsα gene: site-specific patterns and recurrent histological hallmarks. J. Pathol. **187:** 249–258.
121. ISHIDA, T. & H.D. DORFMAN. 1993. Massive chondroid differentiation in fibrous dysplasia of bone (fibrocartilagenous dysplasia). Am. J. Surg. Pathol. **17:** 924–930.
122. LIENS, D., P.D. DELMAS & P.J. MEUNIER. 1994. Long-term effects of intravenous pamidronate in fibrous dysplasia of bone. Lancet **343:** 953–954.
123. YABUT, S.M. *et al.* 1988. Malignant transformation of fibrous dysplasia: a case report and review of the literature. Clin. Orthop. **228:** 281–289.
124. RUGGIERI, P. *et al.* 1994. Malignancies in fibrous dysplasia. Cancer **73:** 1411–1424.
125. BIANCO, P. & P.G. ROBEY. 1999. Diseases of bone and the stromal cell lineage. J. Bone Miner. Res. **14:** 336–341.
126. BIANCO, P. *et al.* 1998. Reproduction of human fibrous dysplasia of bone in immunocompromised mice by transplante d mosaics of normal and Gsα-mutated skeletal progenitor cells. J. Clin. Invest. **101:** 1737–1744.
127. BIANCO, P. *et al.* 2000. Mutations of the *GNAS1* gene, stromal cell dysfunction, and osteomalacic changes in non-McCune-Albright fibrous dysplasia of bone. J. Bone Miner. Res. **15:** 120–128.
128. MALCHOFF, C.D. *et al.* 1994. An unusual presentation of McCune-Albright syndrome confirmed by an activating mutation of the G_s α-subunit from a bone lesion. J. Clin. Endocrinol. Metab. **78:** 803–806.
129. SHENKER, A. *et al.* 1995. Osteoblastic cells derived from isolated lesions of fibrous dysplasia contain activating somatic mutations of the $G_s\alpha$ gene. Hum. Mol. Genet. **4:** 1675–1676.
130. ALMAN, B.A., H.J. WOLFE & D.A. GREEL. 1996. Activating mutations of Gs protein in monostotic fibrous lesions of bone. J. Orthop. Res. **14:** 311–315.
131. CANDELIERE, G.A., F.H. GLORIEUX & P.J. ROUGHLEY. 1997. Polymerase chain reaction-based technique for the selective enrichment and analysis of mosaic arg201 mutations in Gαs from patients with fibrous dysplasia of bone. Bone **21:** 201–206.

132. OKAMOTO, S. *et al.* 2000. Activating $Gs\alpha$ mutation in intramuscular myxomas with and without fibrous dysplasia of bone. Virchows Arch. **437:** 133–137.
133. TURKSEN, K. *et al.* 1990. Forskolin has biphasic effects on osteoprogenitor cell differentiation in vitro. J. Cell. Physiol. **142:** 61–69.
134. BELLOWS, C.G. *et al.* 1990. Parathyroid hormone reversibly suppresses the differentiation of osteoprogenitor cells into functional osteoblasts. Endocrinology **127:** 3111–3116.
135. TINTUT, Y. *et al.* 1999. Inhibition of osteoblast-specific transcription factor Cbfa1 by the cAMP pathway in osteoblasts. J. Biol. Chem. **274:** 28875–28879.
136. GAIDDON, C. *et al.* 1994. Genomic effects of the putative oncogene $G\alpha s$: chronic transcriptional activation of the *c-fos* proto-oncogene in endocrine cells. J. Biol. Chem. **269:** 22663–22671.
137. SASSONE-CORSI, P. 1995. Signaling pathways and *c-fos* transcriptional response-links to inherited diseases. N. Engl. J. Med. **332:** 1576–1577.
138. CANDELIERE, G.A. *et al.* 1995. Increased expression of the *c-fos* proto-oncogene in bone from patients with fibrous dysplasia. N. Engl. J. Med. **332:** 1546–1551.
139. RÜTHER, U. *et al.* 1987. Deregulated *c-fos* expression interferes with normal bone development in transgenic mice. Nature **325:** 412–416.
140. STEIN, G.S. & J.B. LIAN. 1993. Molecular mechanisms mediating proliferation/differentiation interrelationships during progressive development of the osteoblast phenotype. Endocr. Rev. **14:** 424–442.
141. MOTOMURA, T. *et al.* 1998. Increased interleukin-6 production in mouse osteoblastic MC3T3-E1 cells expressing activating mutant of the stimulatory G protein. J. Bone Miner. Res. **13:** 1084–1091.
142. FRASE, W.D. *et al.* 2000. Parathyroid hormone-related protein in the aetiology of fibrous dysplasia of bone in the McCune Albright syndrome. Clin. Endocrinol. **53:** 621–628.
143. ALMAN, B.A. *et al.* 1995. Platelet derived growth factor in fibrous musculoskeletal disorders: a study of pathologic tissue sections and *in vitro* primary cell cultures. J. Orthop. Res. **13:** 67–77.
144. WEINSTEIN, L.S. 1998. Albright hereditary osteodystrophy, pseudohypoparathyroidism and G_s deficiency. *In* G Proteins, Receptors, and Disease. A.M. Spiegel, Ed.: 23–56. Humana Press. Totowa, NJ.
145. SPIEGEL, A.M. & L.S. WEINSTEIN. 2001. Pseudohypoparathyroidism. *In* The Metabolic and Molecular Bases of Inherited Disease, Vol. 8. C.R. Scriver *et al.*, Eds.: 4205–4221. McGraw-Hill. New York.
146. STEINBACH, H.L. & D.A. YOUNG. 1966. The roentgen appearance of pseudohypoparathyroidism (PH) and pseudo-pseudohypoparathyroidism (PPH). Differentiation from other syndromes associated with short metacarpals, metatarsals, and phalanges. Am. J. Roentgenol. Radium. Ther. Nucl. Med. **97:** 49–66.
147. POZNANSKI, A.K. *et al.* 1977. The pattern of shortening of the bones of the hand in pseudohypoparathyroidism and pseudopseudohypoparathyroidism—a comparison with brachydactyly E, Turner syndrome, and acrodysostosis. Radiology **123:** 707–718.
148. GRAUDAL, N. *et al.* 1986. Coexistent pseudohypoparathyroidism and D brachydactyly in a family. Clin. Genet. **30:** 449–455.
149. KAARTINEN, J.M., M.-L. KÄÄR & J.J. OHISALO. 1994. Defective stimulation of adipocyte adenylate cyclase, blunted lipolysis, and obesity in pseudohypoparathyroidism 1a. Pediatr. Res. **35:** 594–597.
150. CAREL, J.C. *et al.* 1999. Resistance to the lipolytic action of epinephrine: a new feature of protein G_s deficiency. J. Clin. Endocrinol. Metab. **84:** 4127–4131.
151. WEINSTEIN, L.S. *et al.* 1990. Mutations of the G_s α-subunit gene in Albright hereditary osteodystrophy detected by denaturing gradient gel electrophoresis. Proc. Natl. Acad. Sci. USA **87:** 8287–8290.
152. MIRIC, A., J.D. VECHIO & M.A. LEVINE. 1993. Heterogeneous mutations in the gene encoding the alpha subunit of the stimulatory G protein of adenylyl cyclase in Albright hereditary osteodystrophy. J. Clin. Endocrinol. Metab. **76:** 1560–1568.
153. DAVIES, S.J. & H.E. HUGHES. 1993. Imprinting in Albright's hereditary osteodystrophy. J. Med. Genet. **30:** 101–103.

154. TSANG, R.C. *et al.* 1984. The development of pseudohypoparathyroidism. Involvement of progressively increasing serum parathyroid hormone concentrations, increased 1,25-dihydroxyvitamin D concentrations, and 'migratory' subcutaneous calcifications. Am. J. Dis. Child. **138:** 654–658.
155. WERDER, E.A. *et al.* 1978. Pseudohypoparathyroidism and idiopathic hypoparathyroidism: relationship between serum calcium and parathyroid hormone levels and urinary cyclic adenosine-3',5'-monophosphate response to parathyroid extract. J. Clin. Endocrinol. Metab. **46:** 872–879.
156. BARR, D.G.D., H.F. STIRLING & J.A.B. DARLING. 1994. Evolution of pseudohypoparathyroidism: an informative family study. Arch. Dis. Child. **70:** 337–338.
157. YU, D. *et al.* 1999. Identification of two novel deletion mutations within the $G_s\alpha$ gene (*GNAS1*) in Albright hereditary osteodystrophy. J. Clin. Endocrinol. Metab. **84:** 3254–3259.
158. DREZNER, M.K. & M.R. HAUSSLER. 1979. Normocalcemic pseudohypoparathyroidism. Association with normal vitamin D3 metabolism. Am. J. Med. **66:** 503–508.
159. BALACHANDAR, V. *et al.* 1975. Pseudohypoparathyroidism with normal serum calcium level. Am. J. Dis. Child. **129:** 1092–1095.
160. BRESLAU, N.A. *et al.* 1980. Studies on the attainment of normocalcemia in patients with pseudohypoparathyroidism. Am. J. Med. **68:** 856–860.
161. CHASE, L.R., G.L. MELSON & G.D. AURBACH. 1969. Pseudohypoparathyroidism: defective excretion of 3',5'-AMP in response to parathyroid hormone. J. Clin. Invest. **48:** 1832–1844.
162. LEVINE, M.A. *et al.* 1986. Activity of the stimulatory guanine nucleotide-binding protein is reduced in erythrocytes from patients with pseudohypoparathyroidism and pseudopseudohypoparathyroidism: biochemical, endocrine, and genetic analysis of Albright's hereditary osteodystrophy in six kindreds. J. Clin. Endocrinol. Metab. **62:** 497–502.
163. CAVERZASIO, J., R. RIZZOLI & J.P. BONJOUR. 1986. Sodium-dependent phosphate transport inhibited by parathyroid hormone and cyclic AMP stimulation in an opossum kidney cell line. J. Biol. Chem. **261:** 3233–3237.
164. BELL, N.H. *et al.* 1972. Effects of dibutyryl cyclic adenosine 3',5'-monophosphate and parathyroid extract on calcium and phosphorus metabolism in hypoparathyroidism and pseudohypoparathyroidism. J. Clin. Invest. **51:** 816–823.
165. MIURA, R. *et al.* 1990. Response of plasma 1,25-dihydroxyvitamin D in the human PTH(1-34) infusion test: an improved index for the diagnosis of idiopathic hypoparathyroidism and pseudohypoparathyroidism. Calcif. Tissue Int. **46:** 309–313.
166. LAMBERT, P.W. *et al.* 1980. Demonstration of a lack of change in serum $1\alpha,25$-dihydroxyvitamin D in response to parathyroid extract in pseudohypoparathyroidism. J. Clin. Invest. **66:** 782–791.
167. BRAUN, J.J. *et al.* 1981. Lack of response of 1,25-dihydroxycholecalciferol to exogenous parathyroid hormone in a patient with treated pseudohypoparathyroidism. Clin. Endocrinol. (Oxf.) **14:** 403–407.
168. YAMAOKA, K. *et al.* 1981. Effect of dibutyryl adenosine 3',5'-monophosphate administration on plasma concentrations of 1,25-dihydroxyvitamin D in pseudohypoparathyroidism type I. J. Clin. Endocrinol. Metab. **53:** 1096–1100.
169. BRESLAU, N.A. & R.S. WEINSTOCK. 1988. Regulation of 1,25 (OH)2D synthesis in hypoparathyroidism and pseudohypoparathyroidism. Am. J. Physiol. **255:** E730–E736.
170. DREZNER, M.K. *et al.* 1976. 1,25-dihydroxycholecalciferol deficiency: the probable cause of hypocalcemia and metabolic bone disease in pseudohypoparathyroidism. J. Clin. Endocrinol. Metab. **42:** 621–628.
171. EPSTEIN, S. *et al.* 1983. $1\alpha,25$-dihydroxyvitamin D3 corrects osteomalacia in hypoparathyroidism and pseudohypoparathyroidism. Acta Endocrinol. (Copenh.) **103:** 241–247.
172. LEVINE, M.A. *et al.* 1983. Resistance to multiple hormones in patients with pseudohypoparathyroidism. Association with deficient activity of guanine nucleotide regulatory protein. Am. J. Med. **74:** 545–556.
173. WERDER, E.A. *et al.* 1975. Excessive thyrotropin response to thyrotropin-releasing hormone in pseudohypoprathyroidism. Pediatric Res. **9:** 12–16.

174. LEVINE, M.A., T.S. JAP & W. HUNG. 1985. Infantile hypothyroidism in two sibs: an unusual presentation of pseudohypoparathyroidism type Ia. J. Pediatr. **107:** 919–922.
175. WEISMAN, Y. et al. 1985. Pseudohypoparathyroidism type Ia presenting as congenital hypothyroidism. J. Pediatr. **107:** 413–415.
176. YOKORO, S. et al. 1990. Hyperthyrotropinemia in a neonate with normal thyroid hormone levels: the earliest diagnostic clue for pseudohypoparathyroidism. Biol. Neonate **58:** 69–72.
177. MALLET, E. et al. 1982. Coupling defect of thyrotropin receptor and adenylate cyclase in a pseudohypoparathyroid patient. J. Clin. Endocrinol. Metab. **54:** 1028–1032.
178. WOLFSDORF, J.I. et al. 1978. Partial gonadotrophin-resistance in pseudohypoparathyroidism. Acta Endocrinol. (Copenh.) **88:** 321–328.
179. SHIMA, M. et al. 1988. Multiple associated endocrine abnormalities in a patient with pseudohypoparathyroidism type 1a. Eur. J. Pediatr. **147:** 536–538.
180. SHAPIRO, M.S. et al. 1980. Multiple abnormalities of anterior pituitary hormone secretion in association with pseudohypoparathyroidism. J. Clin. Endocrinol. Metab. **51:** 483–487.
181. NAMNOUM, A.B. et al. 1998. Reproductive dysfunction in women with Albright's hereditary osteodystrophy. J. Clin. Endocrinol. Metab. **83:** 824–829.
182. FAULL, C.M. et al. 1991. Pseudohypoparathyroidism: its phenotypic variability and associated disorders in a large family. Q. J. Med. **78:** 251–264.
183. SCHUSTER, V. et al. 1993. Endocrine and molecular biological studies in a German family with Albright hereditary osteodystrophy. Eur. J. Pediatr. **152:** 185–189.
184. CARLSON, H.E., A.S. BRICKMAN & G.F. BOTTAZZO. 1977. Prolactin deficiency in pseudohypoparathyroidism. N. Engl. J. Med. **296:** 140–144.
185. BRICKMAN, A.S., H.E. CARLSON & L.J. DEFTOS. 1981. Prolactin and calcitonin responses to parathyroid hormone infusion in hypoparathyroid, pseudohypoparathyroid, and normal subjects. J. Clin. Endocrinol. Metab. **53:** 661–664.
186. KRUSE, K. et al. 1981. Deficient prolactin response to parathyroid hormone in hypocalcemic and normocalcemic pseudohypoparathyroidism. J. Clin. Endocrinol. Metab. **52:** 1099–1105.
187. SCOTT, D.C. & W. HUNG. 1995. Pseudohypoparathyroidism type Ia and growth hormone deficiency in two siblings. J. Pediatr. Endocrinol. Metab. **8:** 205–207.
188. HENKIN, R.I. 1968. Impairment of olfaction and of the tastes of sour and bitter in pseudohypoparathyroidism. J. Clin. Endocrinol. Metab. **28:** 624–628.
189. WEINSTOCK, R.S. et al. 1986. Olfactory dysfunction in humans with deficient guanine nucleotide-binding protein. Nature **322:** 635–636.
190. IKEDA, K. et al. 1988. Clinical investigation of olfactory and auditory function in type I pseudohypoparathyroidism: participation of adenylate cyclase system. J. Laryngol. Otol. **102:** 1111–1114.
191. DOTY, R.L. et al. 1997. Olfactory dysfunction in type I pseudohypoparathyroidism: dissociation from $G_s\alpha$ deficiency. J. Clin. Endocrinol. Metab. **82:** 247–250.
192. MOSES, A.M. et al. 1986. Evidence for normal antidiuretic responses to endogenous and exogenous arginine vasopressin in patients with guanine nucleotide-binding stimulatory protein-deficient pseudohypoparathyroidism. J. Clin. Endocrinol. Metab. **62:** 221–224.
193. BRICKMAN, A.S., H.E. CARLSON & S.R. LEVIN. 1986. Responses to glucagon infusion in pseudohypoparathyroidism. J. Clin. Endocrinol. Metab. **63:** 1354–1360.
194. CARLSON, H.E. & A.S. BRICKMAN. 1983. Blunted plasma cyclic adenosine monophosphate response to isoproterenol in pseudohypoparathyroidism. J. Clin. Endocrinol. Metab. **56:** 1323–1326.
195. LEVINE, M.A. et al. 1980. Deficient activity of guanine nucleotide regulatory protein in erythrocytes from patients with pseudohypoparathyroidism. Biochem. Biophys. Res. Commun. **94:** 1319–1324.
196. FARFEL, Z. et al. 1980. Defect of receptor-cyclase coupling protein in pseudohypoparathyroidism. N. Engl. J. Med. **303:** 237–242.
197. FARFEL, Z. & H.R. BOURNE. 1980. Deficient activity of receptor-cyclase coupling protein in platelets of patients with pseudohypoparathyroidism. J. Clin. Endocrinol. Metab. **51:** 1202–1204.

198. LEVINE, M.A. et al. 1988. Genetic deficiency of the α subunit of the guanine nucleotide-binding protein G_s as the molecular basis for Albright hereditary osteodystrophy. Proc. Natl. Acad. Sci. USA **85:** 617–621.
199. CARTER, A. et al. 1987. Reduced expression of multiple forms of the α subunit of the stimulatory GTP-binding protein in pseudohypoparathyroidism type Ia. Proc. Natl. Acad. Sci. USA **84:** 7266–7269.
200. PATTEN, J.L. & M.A. LEVINE. 1990. Immunochemical analysis of the α-subunit of the stimulatory G-protein of adenylyl cyclase in patients with Albright's hereditary osteodystrophy. J. Clin. Endocrinol. Metab. **71:** 1208–1214.
201. ALDRED, M.A. & R.C. TREMBATH. 2000. Activating and inactivating mutations in the human *GNAS1* gene. Hum. Mutat. **16:** 183–189.
202. WEINSTEIN, L.S. et al. 1992. A heterozygous 4-bp deletion mutation in the $G_s\alpha$ gene (*GNAS1*) in a patient with Albright hereditary osteodystrophy. Genomics **13:** 1319–1321.
203. YU, S. et al. 1995. A deletion hot-spot in exon 7 of the $G_s\alpha$ gene (*GNAS1*) in patients with Albright hereditary osteodystrophy. Hum. Mol. Genet. **4:** 2001–2002.
204. WALDEN, U. et al. 1999. Stimulatory guanine nucleotide binding protein subunit 1 mutation in two siblings with pseudohypoparathyroidism type Ia and mother with pseudopseudohypoparathyroidism. Eur. J. Pediatr. **158:** 200–203.
205. NAKAMOTO, J.M. et al. 1998. Pseudohypoparathyroidism type Ia from maternal but not paternal transmission of a $G_s\alpha$ gene mutation. Am. J Med. Genet. **77:** 61–67.
206. YOKOYAMA, M. et al. 1996. A 4-base pair deletion mutation of Gs alpha gene in a Japanese patient with pseudohypoparathyroidism. J. Endocrinol. Invest. **19:** 236–241.
207. AHMED, S.F. et al. 1998. *GNAS1* mutational analysis in pseudohypoparathyroidism. Clin. Endocrinol. **49:** 525–531.
208. MANTOVANI, G. et al. 2000. Mutational analysis of *GNAS1* in patients with pseudohypoparathyroidism: identification of two novel mutations. J. Clin. Endocrinol. Metab. **85:** 4243–4248.
209. DE SANCTIS, L. et al. 2000. Albright hereditary osteodystrophy (AHO) and pseudohypoparathyroidism: three new mutations and a common deletion in *GNAS1* [abstract]. Am. J. Hum. Genet. **67** (Suppl. 2): 295.
210. PATTEN, J.L. et al. 1990. Mutation in the gene encoding the stimulatory G protein of adenylate cyclase in Albright's hereditary osteodystrophy. N. Engl. J. Med. **322:** 1412–1419.
211. IIRI, T. et al. 1994. Rapid GDP release from Gs-alpha in patients with gain and loss of endocrine function. Nature **371:** 164–167.
212. WARNER, D.R. et al. 1997. A novel mutation adjacent to the switch III domain of $G_{s\alpha}$ in a patient with pseudohypoparathyroidism. Mol. Endocrinol. **11:** 1718–1727.
213. SCHWINDINGER, W.F. et al. 1994. A novel $G_s\alpha$ mutant in a patient with Albright hereditary osteodystrophy uncouples cell surface receptors from adenylyl cyclase. J. Biol. Chem. **269:** 25387–25391.
214. WU, W.I. et al. 2001. Selective resistance to parathyroid hormone caused by a novel uncoupling mutation in the carboxyl terminus of $G_{\alpha s}$. A cause of pseudohypoparathyroidism type Ib. J. Biol. Chem. **276:** 165–171.
215. FARFEL, Z. et al. 1996. Pseudohypoparathyroidism: a novel mutation in the βγ-contact region of $G_s\alpha$ impairs receptor stimulation. J. Biol. Chem. **271:** 19653–19655.
216. WARNER, D.R. & L.S. WEINSTEIN. 1999. A mutation in the heterotrimeric stimulatory guanine nucleotide binding protein α-subunit with impaired receptor-mediated activation because of elevated GTPase activity. Proc. Natl. Acad. Sci. USA **96:** 4268–4272.
217. FISCHER, J.A. et al. 1998. An inherited mutation associated with functional deficiency of the α-subunit of the guanine nucleotide-binding protein G_s in pseudo- and pseudopseudohypoparathyroidism. J. Clin. Endocrinol. Metab. **83:** 935–938.
218. WILSON, L.C. et al. 1994. Parental origin of Gsα gene mutations in Albright's hereditary osteodystrophy. J. Med. Genet. **31:** 835–839.
219. YU, S. et al. 1998. Variable and tissue-specific hormone resistance in heterotrimeric G_s protein α-subunit ($G_s\alpha$) knockout mice is due to tissue-specific imprinting of the $G_s\alpha$ gene. Proc. Natl. Acad. Sci. USA **95:** 8715–8720.

220. CAMPBELL, R., C.M. GOSDEN & D.T. BONTHRON. 1994. Parental origin of transcription from the human GNAS1 gene. J. Med. Genet. **31:** 607–614.
221. WEINSTEIN, L.S., S. YU & C.A. ECELBARGER. 2000. Variable imprinting of the heterotrimeric G protein G_s α-subunit within different segments of the nephron. Am. J. Physiol. **278:** F507–F514.
222. CARLSON, H.E. *et al.* 1985. Normal free fatty acid response to isoproterenol in pseudohypoparathyroidism. J. Clin. Endocrinol. Metab. **61:** 382–384.
223. ECELBARGER, C.A. *et al.* 1999. Decreased Na-K-2Cl cotransporter expression in mice with heterozygous disruption of the Gsα gene. Am. J. Physiol. **277:** F235–F244.
224. STONE, M.D. *et al.* 1993. The renal response to exogenous parathyroid hormone in treated pseudohypoparathyroidism. Bone **14:** 727–735.
225. VORTKAMP, A. *et al.* 1996. Regulation of rate of cartilage differentiation by Indian hedgehog and PTH-related protein. Science **273:** 613–622.
226. KARAPLIS, A.C. *et al.* 1994. Lethal skeletal dysplasia from targeted disruption of the parathyroid hormone-related peptide gene. Genes Dev. **8:** 277–289.
227. LANSKE, B. *et al.* 1996. PTH/PTHrP receptor in early development and Indian hedgehog-regulated bone growth. Science **273:** 663–666.
228. JOBERT, A.S. *et al.* 1998. Absence of functional receptors for parathyroid hormone and parathyroid hormone-related peptide in Blomstrand chondrodysplasia. J. Clin. Invest. **102:** 34–40.
229. CHUNG, U. *et al.* 2000. *In vivo* function of stimulatory G protein (Gs) in the growth plate [abstract]. J. Bone Miner. Res. **15** (Suppl. 1): S175.
230. CUMMINGS, D.E. *et al.* 1996. Genetically lean mice result from targeted disruption of the RIIβ subunit of protein kinase A. Nature **382:** 622–626.
231. VALET, P. *et al.* 2000. Expression of human α2-adrenergic receptors in adipose tissue of β3-adrenergic receptor deficient mice promotes diet-induced obesity. J. Biol. Chem. **275:** 34797–34802.
232. YU, S. *et al.* 2000. Paternal versus maternal transmission of a stimulatory G protein α subunit knockout produces opposite effects on energy metabolism. J. Clin. Invest. **105:** 615–623.
233. WANG, H., D.C. WATKINS & C.C. MALBON. 1992. Antisense oligodeoxynucleotides to G_s α-subunit sequence accelerate differentiation of fibroblasts to adipocytes. Nature **358:** 334–337.
234. LEVIN, L.R. *et al.* 1992. The Drosophila learning and memory gene *rutabaga* encodes a Ca^{2+}/calmodulin-responsive adenylyl cyclase. Cell **68:** 479–489.
235. WU, Z.L. *et al.* 1995. Altered behavior and long-term potentiation in type I adenylyl cyclase mutant mice. Proc. Natl. Acad. Sci. USA **92:** 220–224.
236. KAPLAN, F.S. & E.I. SHORE. 2000. Progressive osseous heteroplasia. J. Bone Miner. Res. **15:** 2084–2094.
237. YEH, G.L. *et al.* 2000. GNAS1 mutation and Cbfa1 misexpression in a child with severe congenital platelike osteoma cutis. J. Bone Miner. Res. **15:** 2063–2073.
238. EDDY, M.C. *et al.* 2000. Deficiency of the α-subunit of the stimulatory G protein and severe extraskeletal ossification. J. Bone Miner. Res. **15:** 2074–2083.
239. JUPPNER, H. *et al.* 1998. The gene responsible for pseudohypoparathyroidism type Ib is paternally imprinted and maps in four unrelated kindreds to chromosome 20q13.3. Proc. Natl. Acad. Sci. USA **95:** 11798–11803.
240. SILVE, C. *et al.* 1986. Selective resistance to parathyroid hormone in cultured skin fibroblasts from patients with pseudohypoparathyroidism type Ib. J. Clin. Endocrinol. Metab. **62:** 640–644.
241. BASTEPE, M. *et al.* 2001. Positional dissociation between the genetic mutation responsible for pseudohypoparathyroidism type Ib and the associated methylation defect at exon A/B: evidence for a long-range regulatory element within the imprinted *GNAS1* locus. Hum. Mol. Genet. **10:** 1231–1241.
242. JAN DE BEUR, S.M. *et al.* 2001. Loss of imprinting on the maternal *GNAS1* allele in pseudohypoparathyroidism type Ib [abstract]. Endocrine Society 83rd Annual Meeting Program and Abstracts.

243. BASTEPE, M., A.H. LANE & H. JÜPPNER. 2001. Paternal uniparental disomy of chromosome 20q—and the resulting changes in *GNAS1* methylation—as a plausible cause of pseudohypoparathyroidism. Am. J. Hum. Genet. **68:** 1283–1289.
244. BELL, A.C. & G. FELSENFELD. 2000. Methylation of a CTCF-dependent boundary controls imprinted expression of the *Igf2* gene. Nature **405:** 482–485.
245. Hark, A. T. *et al.* 2000. CTCF mediates methylation-sensitive enhancer-blocking activity at the *H19/Igf2* locus. Nature **405:** 486–489.
246. CATTANACH, B.M. & M. KIRK. 1985. Differential activity of maternally and paternally derived chromosome regions in mice. Nature **315:** 496–498.
247. YU, S. *et al.* 2001. Increased insulin sensitivity in $G_s\alpha$ knockout mice. J. Biol. Chem. **276:** 19994–19998.
248. CATTANACH, B.M. *et al.* 2000. Two imprinted gene mutations: three phenotypes. Hum. Mol. Genet. **9:** 2263–2273.

Regulation of Phospholipase D and Secretion in Mast Cells by Protein Kinase A and Other Protein Kinases

WAHN SOO CHOI, AHMED CHAHDI, YOUNG MI KIM, PAUL F. FRAUNDORFER, AND MICHAEL A. BEAVEN

Laboratory of Molecular Immunology, National Heart, Lung, and Blood Institute, National Institutes of Health, Bethesda, Maryland 20892-1760, USA

ABSTRACT: Functions attributed to phospholipase (PL) D include the regulation of intracellular trafficking of Golgi-derived vesicles and secretion of granules from mast cells. We have reported that activation of PLD and secretion in a rat mast cell (RBL-2H3) line is substantially enhanced by cholera toxin, a known activator of protein kinase (PK) A. Here we review the evidence that (1) the synergistic interactions of cholera toxin and other pharmacological agents on mast cell secretion are attributable to the synergistic activation of PLD via PKA, CaM kinase II, and PKC and (2) both PLD1 and PLD2 participate in this process. For example, treatment with cholera toxin, thapsigargin, and phorbol 12-myristate 13-acetate (which activate PKA, CaM kinase II, and PKC, respectively) exhibit synergy in the stimulation of both PLD and secretion. These kinases and PLD are likely confined to membrane components, as similar synergistic interactions could be demonstrated in permeabilized cells. The regulation of PLD and secretion by these kinases is also apparent from studies of inhibitors of PKA and other kinases. Also, by overexpression of either PLD1 or PLD2 it is apparent that both isoforms respond to the same stimuli as endogenous PLD, although PLD1 is largely associated with secretory granules and PLD2 with plasma membrane. The studies reveal interesting differences in the regulation of the translocation of granules (regulated by PKA) and the fusion of these granules with the plasma membrane (regulated by Ca^{2+} and PKC). The pathological/physiological implications of the regulation of PLD by PKA require further evaluation in other cell systems.

KEYWORDS: phospholipase D; protein kinases; mast cells; exocytosis

Address for correspondence: Dr. Michael A. Beaven, Laboratory of Immunology, NHLBI, Room 8N109/Building 10, National Institutes of Health, Bethesda, MD 20896-1760. Voice: 301-496-6188; fax: 301-402-0171.
beavenm@nhlbi.nih.gov

ABBREVIATIONS: ARF, ADP-ribosylation factor; BAD, Bcl-2/Bcl-X_L antagonist causing death; CaM, Ca^{2+}/calmodulin regulated; DNP-BSA, dintrophenylated bovine serum albumin—the antigen used in these studies along with DNP-specific IgE for sensitizing RBL-2H3 cells; FceRI, the high-affinity receptor for IgE; PK, protein kinase; PKI, peptide inhibitor of protein kinase A; PL, phospholipase; PMA, phorbol 12-myristate 13-acetate.

Mast cells, because they express large numbers of receptors with high affinity for IgE (FcεRI), are responsible for acute allergic reactions to IgE-directed antigens. Antigen stimulation of these cells leads to rapid secretion of granules with release of preformed inflammatory mediators, the generation of arachidonate-derived inflammatory lipids, and the production of inflammatory cytokines through gene transcription. Mast cells are intrinsically interesting models for study because different signaling pathways are used for each of the responses described above. These pathways have been elucidated mainly from studies of the RBL-2H3 mast cell line and its genetically modified counterparts, which include the RBL-2H3(m1) cell line made to express the G protein–coupled muscarinic m1 receptors.[1]

Stimulation of RBL-2H3 cells with antigen via FcεRI leads to recruitment of cytosolic tyrosine kinases and the tyrosine phosphorylation of proteins including phospholipase (PL) Cγ^2 or, in the case of G protein–coupled receptors, to activation of PLCβ by the α subunit of the G protein, $G_{q/11}$. PLC-mediated increase in cytosolic calcium (otherwise referred to as the calcium signal) and the activation of protein kinase (PK) C isozymes provide essential signals for secretion of granules.[3] Arachidonate metabolism and production of cytokines, in contrast, are dependent on mitogen-activated protein (MAP) kinases and a calcium signal.[4] However, recent studies with RBL-2H3 cells has renewed interest in the role of PLD in secretion in mast cells. Tissue mast cells[5] and RBL-2H3 cells[6–8] contain PLD activity, which is stimulated by receptor ligands,[5,9] calcium ionophores,[6] thapsigargin,[8,10] and phorbol 12-myristate 13-acetate (PMA)[6] to cause sustained increases in phosphatidic acid and its metabolite, diacylglycerol.[5–7] Both these products stimulate PKC, and PLD could indirectly promote secretion through activation of PKC. However, PLD might be directly involved in the secretory process, as will be discussed later.

PHOSPHOLIPASE D AND ITS ISOFORMS

PLD catalyzes the hydrolysis of phosphatidylcholine to form phosphatidic acid, which can be rapidly converted to diacylglycerol by the action of phosphatidate hydrolase.[11] Both products are active signaling molecules and may regulate a variety of cellular enzymes, including the activation of PKC (as noted above) and type 1 phosphatidylinositol phosphate 5-kinases, in addition to various intracellular activities such as vesicular trafficking.[12]

A unique feature of PLD is its preference for primary alcohols over water such that phosphatidylalcohols are produced instead of phosphatidic acid in the presence of low concentrations of primary alcohols by a process known as transphosphatidylation.[13] This reaction is used experimentally to assay PLD activity *in vivo* by diverting production of phosphatidic acid to phosphatidylalcohol, which is metabolically inert. Typically, and as used in the studies described herein, intracellular phospholipids are labeled by incubating cells with radioactively labeled myristic acid and then stimulating them in the presence of 0.5% ethanol (or other primary alcohols). Labeled phosphatidylethanol is then isolated for assay by extraction and thin-layer chromatography. The transphosphatidylation reaction is also used to unmask physiological roles of the PLD product phosphatidic acid. Secondary or tertiary alcohols are weak substrates for transphosphatidylation and are useful controls for assessing nonspecific affects of the corresponding primary alcohol. For example, comparison of the effects of 1-butanol with *tert*-butanol may reveal phosphatidate-dependent responses of intact cells.

Information on the mechanisms of activation of PLD is incomplete. Two isoforms of PLD have been cloned, PLD1 and PLD2, both of which exist as spliced variants (see refs. 11, 14). *In vitro*, PLD1 is activated in the presence of phosphatidylinositol 4,5-bisphosphate by various small GTPases, which include ARF, RhoA, Rac1, and Cdc42, as well as by PKCα in an ATP-independent manner apparently by direct interaction of these molecules with PLD1.[14] PLD2, in contrast, is constitutively active in the presence of phosphatidylinositol 4,5-bisphosphate, and this activity is not affected by the small GTPases or PKCα, either alone or in combination.[15] The only described mechanism for activation of PLD2 is through the generation of phosphatidylinositol 4,5-bisphosphate by type 1 phosphatidylinositol 5-kinase.[16] As the latter enzyme is stimulated by phosphatidic acid (as noted above), it is theoretically possible that PLD1 activation might result in PLD2 activation through further stimulation of type I phosphatidylinositol 5-kinase by PLD1-generated phosphatidic acid. This scenario would require proximity of both isoforms of PLD, but such proximity may occur during fusion of granules with the plasma membrane in the final steps of exocytosis, as will be discussed later.

In addition to the above pathways, PLD1 is probably regulated *in vivo* through phosphorylation by several serine/threonine kinases. Early studies with intact cell systems showed that PLD can be activated by pharmacological stimulants of PKC and calcium mobilization.[11] More recent studies implicate PKC,[17,18] Ca^{2+}/calmodulin-dependent (CAM) kinase II,[19] and Rho Kinase.[20] The phosphorylation-dependent activation of PLD1 by PKC [17] probably involves phosphorylation at multiple sites. Activation of PKC by PMA *in vivo* results in phosphorylation of the PLD1 at three sites the mutation of which attenuates, but does not abolish, the subsequent activation of PLD1.[18,21] Activation of PLD2 by PKC is not apparent *in vitro*[15] nor reported *in vivo*. CaM kinase II [19,22] and rho kinase [20] have been implicated on the basis of the effects of inhibitors and coexpression of mutated forms of rho kinase. However, there are no reports of phosphorylation of PLD1 or 2 by these two kinases. Our finding that PKA may be an additional regulatory kinase for PLD[23] is the topic of this paper and is discussed in the context of our studies of secretory mechanisms in mast cells.

EVIDENCE FOR PLD AS A REGULATOR OF SECRETION IN MAST CELLS

The RBL-2H3 cell has become the preeminent model for studies of PLD and secretion, and several lines of evidence indicate a close relationship between secretion and PLD activation in these cells. For example, activation of PLD and secretion are highly correlated under a wide variety of circumstances.[8,23] Treatment of these cells with cholera toxin (see next section) enhances PLD activation and secretion to the same extent without affecting the activation of other phospholipases.[8] Also, inhibitors of protein kinases that regulate PLD activity suppress PLD activation and secretion equally.[23] A second line of evidence is illustrated in FIGURE 1. Primary alcohols suppress secretion in RBL-2H3 cells by diverting production of phosphatidic acid by PLD to phosphatidylalcohol through the transphosphatidylation reaction.[8,23,24] Se-

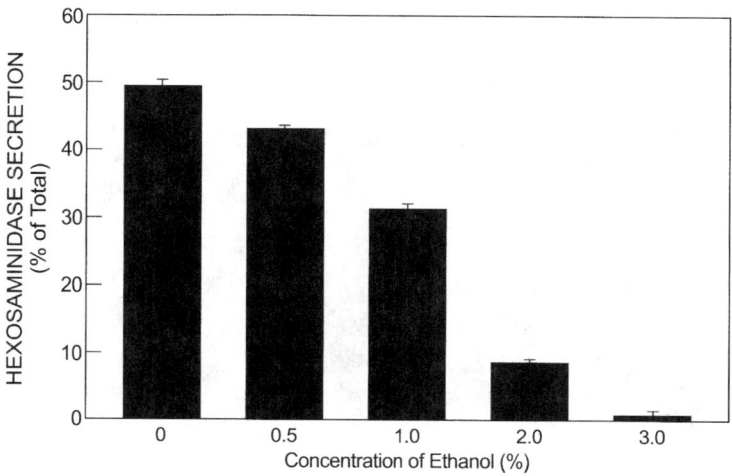

FIGURE 1. Inhibition of secretion by ethanol. RBL-2H3(m1) cells[8] were incubated in medium containing the indicated concentration of ethanol and stimulated with antigen (10 ng/mL DNP-BSA) for 15 min for measurement of secretion of the granule marker, hexosaminidase, according to previously described protocols.[1,44] The data are the mean ± standard error of the mean (MEM) from three separate experiments. Secretion is similarly inhibited by 1-butanol but not by *tertiary*-butanol, as described elsewhere.[23]

cretion is not suppressed (or minimally so) by secondary or tertiary alcohols, which are weak substrates for transphosphatidylation. Finally, the secretory response to antigen can be reconstituted in permeabilized RBL-2H3 cells by provision of ARF1 or phosphatidylinositol transfer protein, either of which increase levels of phosphatidylinositol 4,5-bisphosphate and restore PLD activity.[24]

It seems likely from the above lines of evidence that production of phosphatidic acid by PLD is required for secretion in RBL-2H3 cells, and this is our presumption for the remainder of this paper. The pertinent questions then become: how is PLD regulated, how does it in turn regulate secretion, and are both isoforms of PLD involved? With respect to the latter question, RBL-2H3 cells express message for PLD1b and PLD2 (P. Holbrook *et al.*, unpublished data).

PLD AND SECRETION ARE REGULATED BY A CHOLERA TOXIN–ACTIVATED PATHWAY IN ADDITION TO PKC AND CAM KINASE II

It became apparent in our initial work that an additional pathway (or pathways) to calcium and PKC participated in the activation of PLD. A significant component of the PLD response is resistant to the actions of inhibitors of the calcium signal and PKC. As shown in FIGURE 2, stimulation of PLD by antigen or carbachol in intact RBL-2H3(m1) cells was only partially inhibited by substituting K^+ for Na^+ ions in the medium to suppress calcium influx[25,26] in combination with the PKC inhibitor Ro31-7549[3,27] (FIG. 2A). In contrast, K^+ blocked completely activation of PLD by

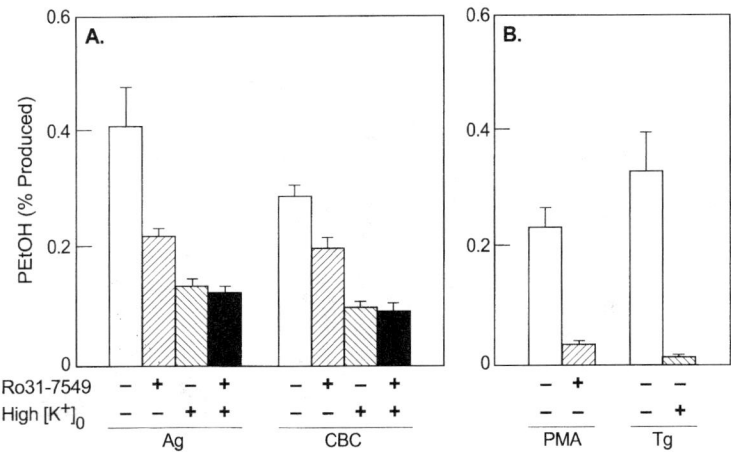

FIGURE 2. Effects of inhibitors of PKC and Ca^{2+}-influx on the activation of PLD; evidence for Ca^{2+}/PKC-independent pathway(s) of activation. [^{3}H]Myristate-labeled RBL-2H3(m1) cells[8] were incubated in medium containing 10 μM Ro31-7549 (an inhibitor of PKC) or high $[K^+]_o$ (to inhibit calcium influx) or both as indicated for 10 min and then stimulated for 15 min with antigen (Ag; DNP-BSA, 10 ng/mL), carbachol (CBC; 1 mM) **(Panel A)**, phorbol 12-myristate 13-acetate (PMA; 50 nM) or thapsigargin (Tg; 300 nM) **(Panel B)** before measurement of [^{3}H]phosphatidylethanol production (PEtOH). Data are corrected for values in the absence of stimulants and inhibitors (<0.07% conversion of intracellular [^{3}H]lipids to [^{3}H]phosphatidylethanol) and are mean ± SEM of values from three separate experiments.

thapsigargin, which stimulates calcium influx in RBL-2H3 cells;[26] and Ro31-7549 blocked PLD activation by PMA, which activates PKC (FIG. 2B). Approximately one-third of the PLD response to the receptor ligands appeared to be resistant to the inhibitors (see FIG. 2A). Of note, the same inhibitors at the concentrations used (140 mM K^+ and 10 μM Ro31-7549) totally blocked (>95%) secretion in response to antigen, carbachol, and thapsigargin (data not shown), which is consistent with previous studies.[3,8,26] Clearly, signals generated through calcium and PKC have critical roles in secretion, independently of PLD and through PLD.

We next examined the effects of cholera toxin in detail with the thought that the additional cryptic pathway(s) might be activated by this toxin. Stimulation of PLD by antigen, carbachol, thapsigargin, and PMA is increased two- to threefold in cholera toxin–treated cells.[23] Secretion is similarly enhanced by cholera toxin in response to these stimulants with the exception of PMA, which by itself does not generate a calcium signal nor does it induce secretion in these cells.[26] These effects are not observed with pertussis toxin or the B subunit of cholera toxin (which actually prevents binding and the actions of the holotoxin); these effects thus appear to be specific for cholera holotoxin.[23]

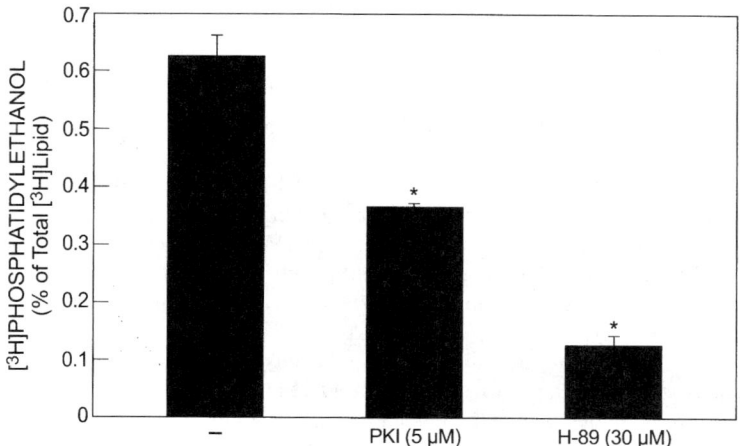

FIGURE 3. Suppression of PLD activation by PKA inhibitors. [^{3}H]Myristate-labeled RBL-2H3(m1) cells were incubated in medium containing the indicated concentrations of PKI, an inhibitory peptide of PKA and H-89 for 10 min and then stimulated for 15 min with antigen (Ag; DNP-BSA, 10 ng/mL) before measurement of [^{3}H]phosphatidylethanol production. As in the previous figure, data are corrected for values in the absence of stimulants and inhibitors and are mean ± SEM of values from three separate experiments.

EVIDENCE THAT CHOLERA TOXIN ACTS VIA ADENYLYL CYCLASE AND PKA

Cholera toxin is known to ADP-ribosylate and activate the α subunit of the G protein, G_s and, in turn, adenylyl cyclase to form cAMP.[28] A primary target for cAMP is PKA (see other articles in this volume). It is known that cholera toxin induces substantial ADP-ribosylation of $G_{\alpha s}$ in RBL-2H3 cells,[29] although the activation of PKA and its possible connection to PLD have not been previously investigated in mast cells. Therefore, studies were undertaken to fill this gap.

Our studies[23] with inhibitors of the adenylate cyclase/PKA pathway show that stimulation of PLD activity by antigen, carbachol, and thapsigargin were partially suppressed by the adenylate cyclase inhibitor, 2′,3′-dideoxyadenosine[30] and substantially inhibited by the PKA inhibitor H-89.[31] As shown in FIGURE 3, the PKA inhibitor peptide (PKI) also inhibits antigen-induced activation of PLD. Both PLD activation and secretion were suppressed equally by these inhibitors in agreement with the idea that PLD activation and secretion were linked.[23] Collectively, the effects of cholera toxin and inhibitors suggest that PLD and secretion are regulated by PKA.

Adenylyl cyclase and PKA are activated in a manner consistent with the above results.[23] The product of adenylyl cyclase activity, cAMP, is present at relatively high levels in unstimulated RBL-2H3 cells. These levels are increased two- to threefold in cholera toxin–treated cells and minimally so by antigen and carbachol. These increases are blocked by 2′,3′-dideoxyadenosine. It would appear that adenylyl cyclase is constitutively active in RBL-2H3 cells and that substantial increases in cellular levels of cAMP could be achieved only by treatment with cholera toxin. PKA also is constitutively active in RBL-2H3 cells, as indicated by direct measurement of PKA activity or of the PKA-specific phosphorylation of BAD at serine-155.[32] Cholera toxin and the activation of receptors by antigen or carbachol further enhances PKA activity. Basal and stimulated PKA activities are markedly inhibited by H-89, as is PLD activation and secretion.[23]

The constitutively elevated levels of cAMP and PKA activity in RBL-2H3 cells explained the paradoxical observation that had long puzzled us. Our early studies showed that cAMP analogues such as dibutyryl cAMP and forskolin had variable effects on PLD activity and secretion (unpublished data), but more recent studies showed that under the same conditions these agents had minimal effects on PKA activity in these cells.[23]

The reason for the high basal activities of adenylate cyclase and PKA in RBL-2H3 cells is unknown, and studies of other cell lines without this defect are desirable. Information on the mechanism by which PKA regulates PLD activity awaits ongoing studies of the effects of cholera toxin and inhibitors on the phosphorylation of the PLD. Nevertheless, in the absence of this information, the effects of cholera toxin and inhibitors provide reasonable evidence for such regulation by PKA. On this basis, and for simplicity of presentation, we shall assume here that the effects of cholera toxin and H-89 are mediated via PKA.

ACTIVATION OF PLD VIA PKA AND OTHER KINASES IS PRESERVED IN PERMEABILIZED CELLS AND PLASMA MEMBRANE VESICLES

The potentiating effects of cholera toxin on PLD activity were still evident when cholera toxin–treated cells were permeabilized with streptolysin O and then stimulated with antigen at low (FIG. 4A) or high (FIG. 4B) concentrations of free calcium. In these experiments, free calcium was buffered at 0.1 µM and 1.0 µM or concentrations that mimic those found in unstimulated and stimulated cells, respectively.[33] Comparison of the two panels reveals that both basal and stimulated enzyme activities as well as the potentiating effects of cholera toxin are further enhanced by merely increasing the concentration of free calcium. An identical pattern was obtained when PMA was used as the stimulant (data not shown). In these experiments cells were minimally permeabilized so that ≥95% of the cells were permeabilized without detectable loss of ADP-ribosylated proteins including $G_{\alpha s}$,[29] and with minimal losses of lactate dehydrogenase (≤5%) and PKCβ and δ (<20%).[3] These simple experiments clearly demonstrate that PLD activity is modulated by calcium (presumably via CaM kinase II), PKC, and cholera toxin (presumably via PKA) and that the mechanisms for receptor-mediated activation of PLD remained intact. The results imply that integrity of the various kinase pathways for PLD activation is retained in these permeabilized cells. The magnitude of the responses shown in FIGURE 4 also suggest synergy among these pathways in the stimulation of PLD.

The effects of cholera toxin and calcium were retained in plasma membrane vesicles from [³H]myristate-labeled RBL-2H3 cells (FIG. 5). PLD activity was increased by raising levels of free calcium from zero to 0.1 µM and 1 µM, and prior treatment of cells with cholera toxin further enhanced these activities. These prepa-

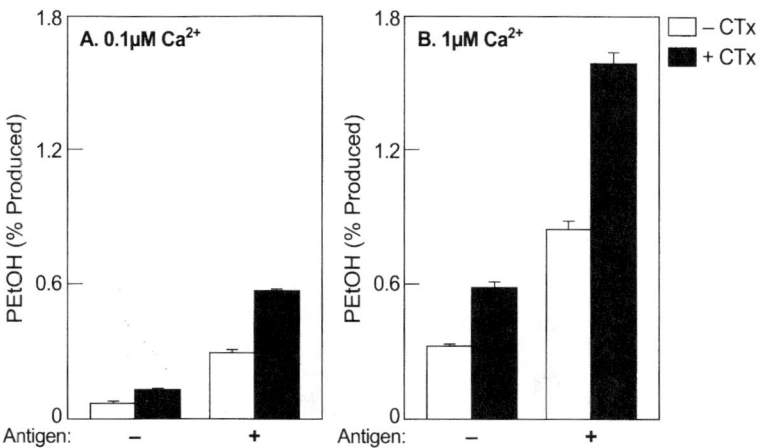

FIGURE 4. Sensitivity of PLD to calcium and cholera toxin is preserved in permeabilized cells. Untreated (*open columns*) and cholera toxin–treated (1 μg/ml for 4 h, *solid columns*) [^{3}H]myristate-labeled RBL-2H3(m1) cells[8] were permeabilized with streptolysin-O in a potassium/calcium-buffered medium.[3] Permeabilized cells were left unstimulated or stimulated with antigen (100 ng/mL) for 15 min in the presence of the indicated concentration of free calcium before measurement of [^{3}H]phosphatidylethanol (PEtOH). Values are uncorrected and are the mean ± SEM for three cultures from one of three similar experiments.

rations contained PLD2 (FIG. 5, inset), which accounted for a substantial proportion (<70%) of the PLD2 in whole-cell lysates as calculated by densitometric scanning of the immunoblots and determining recoveries of FcεRI as determined by IgE binding. As the levels of PLD2 in untreated and cholera toxin–treated cells were similar (data not shown), differences in the amounts of PLD2 did not account for the effects of cholera toxin.

SYNERGISTIC ACTIVATION OF PLD AND SECRETION BY PKA, CALCIUM, AND PKC

The indication from the studies with permeabilized cells that PKA, increased calcium, and PKC activated PLD synergistically was reinforced from studies with intact cells. As reported elsewhere,[23] concentrations of PMA that minimally stimulated PLD markedly potentiated the PLD response to thapsigargin. This potentiation was further enhanced by treatment with cholera toxin, which by itself had minimal or no effect on PLD activity. The synergistic components of each response could be selectively blocked by the appropriate inhibitor. Overall, the data indicated that the PLD response to the combination of all three stimulants was three- to fourfold greater than the sum of the responses to the individual stimulants. The data suggested that PKA by itself is not a strong promotor of PLD activity but that it markedly potentiates the effects of CaM kinase II and PKC.

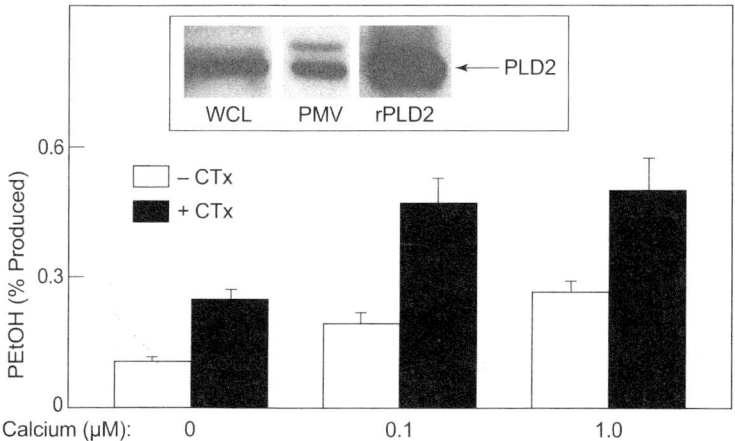

FIGURE 5. Effects of calcium and cholera toxin on PLD activity in PLD2-containing plasma membrane preparations from RBL-2H3(m1) cells. Plasma membrane vesicles[45] from untreated (*open columns*) and cholera toxin–treated (*solid columns*) [^{3}H]myristate-labeled RBL-2H3(m1) were incubated at the indicated concentration of free calcium cells for 15 min for measurement of production of [^{3}H]phosphatidylethanol (PEtOH). Values are uncorrected and are mean ± SEM from six separate preparations. The **inset** shows immunoblots for PLD2 in whole-cell lysates (WCL) of RBL-2H3(m1) cells and plasma membrane vesicles (PMV) prepared from these lysates (see ref. 45). The primary antibody was raised against residues 28–42 of PLD2. The band for PLD2 (*arrow*) was identified by comigration with recombinant and histidine-tagged PLD2. PLD1 was undetectable in these preparations (see text).

PLD1 AND PLD2 HAVE DIFFERENT LOCATIONS BUT ARE COORDINATELY REGULATED: STUDIES WITH EXPRESSED PROTEINS

As previously mentioned, RBL-2H3 cells contain message for PLD1b and PLD2, and PLD2 protein appeared to be located predominantly in the plasma membrane. However, the location of PLD1 was unclear, as this isoform was not readily detected in subcellular fractions with the antibodies available to us. Nevertheless, useful information was obtained by expression of cDNA (by electroporation) for wild-type and mutated forms of PLD (fused with enhanced-green fluorescent protein) in RBL-2H3 cells. In studies to be reported elsewhere almost all of the fluorescent-tagged PLD1 colocalized with rat mast cell protease II, a granule marker, in granule-like structures; whereas most of the tagged PLD2 colocalized with FcεRI in the plasma membrane. This was the first indication that the two PLDs may be situated at different but strategic locations within the mast cell, although the association of PLD1 with secretory granules in RBL-2H3 cells had been reported.[34] Interestingly, one-third to half of the granules were devoid of the tagged PLD1. On stimulating the cells with antigen or thapsigargin, most if not all PLD1-labeled granules migrated to the cell periphery and fused with the plasma membrane, whereupon the PLD1 label redistributed onto the plasma membrane. Yet these stimulants caused release of only about 40% of another granule marker, hexosaminidase, into the medium. This result raised the intriguing possibility that only those granules associated with PLD1 are capable of exocytosis.

The expression studies also revealed that both isoforms, even though present in different locations of the cell, were similarly stimulated by antigen and pharmacological stimulants. This stimulation was apparent from the marked enhancement in PLD activity upon overexpression of either form of PLD. The potentiation of responses by cholera toxin was evident also with both isoforms of PLD, although PLD2 was less sensitive to the effects of the toxin than PLD1. Catalytically inactive mutants of the PLDs, in contrast, showed no response or blunted the response of endogenous PLD to these stimulants. The mutant forms of PLD were expressed at the same location as their wild-type counterpart. The mutations were in motifs (the HKD motif) that are critical for catalytic activity but not for intracellular localization (see ref. 35).

REGULATION OF MIGRATION OF PLD1-LABELED GRANULES BY PKA

Migration of granules to the cell periphery and secretion were blocked by expression of a catalytically inactive mutant of PLD1 and 1-butanol but not by *tert*-butanol. Therefore, the catalytic activity of PLD1 and production of phosphatidic acid appeared to be essential for migration as well as secretion. Interestingly, treatment with cholera toxin alone induced migration of PLD1-labeled granules to the cell periphery, but the granules did not fuse with the plasma membrane. Also, unlike antigen stimulation, there was no release of hexosaminidase or redistribution of PLD1 into the plasma membrane. However, subsequent stimulation with a minimal stimulatory dose of antigen or thapsigargin resulted in rapid fusion and secretion of hexosaminidase into the medium. Secretion was complete within minutes and was much more rapid than that induced with high doses of antigen. The accelerated secretion in cholera toxin–treated cells was markedly attenuated by the presence of 1-butanol but not of *tert*-butanol. The results of these experiments will be reported in detail elsewhere,[36] but our impression was that treatment with cholera toxin left the granules poised along the plasma membrane for rapid release. Furthermore, granule migration was promoted by PKA and possibly PKC in a PLD-dependent manner. Secretion of granules was dependent on PKC, calcium, and PLD.

BOTH PLD1 AND PLD2 PARTICIPATE IN SECRETION

PLD2 in the plasma membrane probably participates in the secretory process, possibly in the fusion of granules with the plasma membrane. Overexpression of PLD2 enhanced secretion and expression of the catalytically inactive mutants of PLD2-suppressed secretion in thapsigargin-stimulated cells. Nonetheless, PLD2 remains associated with the plasma membrane during cell stimulation. This location suggests that PLD2 is appropriately positioned for promoting fusion of granules with the plasma membrane. If so, the PLDs probably have distinct but complementary roles in secretion, PLD1 for migration of granules to the cell periphery and PLD2 for subsequent fusion of granules with the plasma membrane.

The question arises as to how the regulation of the two PLDs is coordinated for the orchestration of translocation and fusion of granules with the plasma membrane. Both isoforms appear to be similarly regulated in a synergistic manner by several protein kinases; yet the only known regulator of PLD2 is phosphatidylinositol 4,5-bisphosphate. An explanation for this paradox may lie in the ability of phosphatidate to stimulate phosphatidylinosital 5-kinase and production of phosphatidylinositol 4,5-bisphosphate, as described earlier. Once granules begin to fuse with the plasma membrane, phosphatidic acid generated by granule-associated PLD1 would then lead to activation of membrane-bound PLD2 via the 5-kinase pathway. This scenario predicts a model whereby serine/threonine kinases such as PKA primarily regulate PLD1, and PLD1 in turn regulates PLD2 activity by virtue of the mechanics of secretion. The validity of this model is now being examined in our current studies.

In addition to the provision of activators for PKC, PLD may facilitate secretion by other mechanisms. Phosphatidylinositiol 4,5-bisphosphate is thought to recruit proteins such as the SNAREs, synaptotagmin, and the calcium-dependent activator protein for secretion (CAPS) that promote docking and fusion of granules with the plasma membrane. PKC, CaM kinase II, and PKA have been proposed as regulators of these processes (reviewed in ref. 37) although not in the context of PLD activation. The regulation of PLD by the same protein kinases and the stimulation of phosphatidylinositol 4,5-bisphosphate synthesis via PLD point to the possible involvement of PLD in the same processes. In addition, PLD could have a more direct role through recruitment and activation of essential components to membranes by virtue of the charged nature of phosphatidic acid[38] and the promotion of membrane fusion by phosphatidic acid itself.[39] However, these remain speculative conjectures at this time.

IMPLICATIONS FOR THE MAST CELL

Although PKA itself has not been previously investigated in mast cells, its activator, cAMP, has been described as either a mediator or an inhibitor of mast cell activation (reviewed in ref. 40); reinvestigation of these systems in different subtypes of mast cells is clearly desirable. Mast cells from different tissues vary in their responses to agents known to alter cAMP levels. It is now recognized that mast cells exhibit phenotypic plasticity in regard to the types of receptors expressed (other than FcεRI, which are expressed in varying numbers in all types of mast cells), cytokines produced, and granule constituents. The particular phenotype expressed in a given tissue is probably governed by the cytokine environment of that tissue, and this environment can influence the magnitude of response to antigen and other stimulants (T.R. Hundley & M.A. Beaven, unpublished data). An additional consideration is that PKA negatively regulates PLC[41,42] and positively regulates PLD (this paper), and this may result in different outcomes for different stimulants. Therefore, the therapeutic implications of our studies on PKA must await additional studies with other mast cell populations and full evaluation of the effects of mast cell growth-sustaining cytokines such as IL-3 and stem cell factor (c-kit ligand).

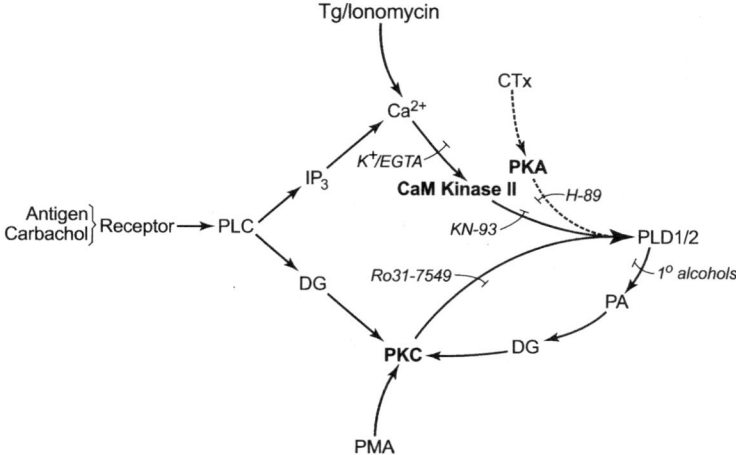

FIGURE 6. Activation of PLD by PKA and other protein kinases in RBL-2H3(m1) cells. The scheme depicts the pathways for activation by receptor ligands, thapsigargin (Tg) or ionomycin, cholera toxin, and phorbol 12-myristae 13-acetate (PMA) and the sites of action of inhibitors used in our studies. As discussed in the text, the scheme depicts a separate input for cholera toxin (CTx) and a reinforcement of the PLD response via PKC through formation of the PLD product, phosphatidic acid (PA) and its metabolite, diacylglycerol (DG). Activation of PLD1 may result in activation of PLD2 because of stimulation of type 1 phosphatidylinositol 5-kinase by PLD1-generated phosphatidic acid, as discussed in the text.

CONCLUDING COMMENTS

The pathways leading to PLD activation are presented schematically in FIGURE 6. The scheme depicts the sites of action of all stimulants and inhibitors that were used in this study. In addition, it shows the predicted feedback reinforcement of signals via phosphatidic acid and PKC. It is uncertain whether PLD is activated directly by the various kinase (i.e., by phosphorylation) or through recruitment of molecules such as ARF and Rho GTPases for the activation of PLD1. If the latter mechanism operates, then additional mechanisms must come into play for the activation of PLD2, which is not activated by these molecules. The possibility that PLD1 activation leads to activation of PLD2 upon fusion of granules with plasma membrane via activation of phosphatidylinositol 5-kinase has been discussed. We should note that the apparent activation of PLD by PKA in our studies is consistent with two previous reports of enhancement of PLD activity by cholera toxin, which was attributed to cAMP in one study[43] and to an undefined cAMP-independent mechanism in another.[44] Finally, we note that the actions of cholera toxin on PLD and secretion reported here have pathological as well as mechanistic implications if PLD activation underlies some of the clinical manifestations of the toxin.

REFERENCES

1. JONES, S.V.P., O.H. CHOI & M.A. BEAVEN. 1991. Carbachol induces secretion in a mast cell line (RBL-2H3) transfected with the m1 muscarinic receptor gene. FEBS Lett. **289:** 47–50.
2. BEAVEN, M.A. & R.A. BAUMGARTNER. 1996. Downstream signals initiated in mast cells by FcεRI and other receptors. Curr. Opin. Immunol. **8:** 766–772.
3. OZAWA, K., Z. SZALLASI, M.G. KAZANIETZ, et al. 1993. Ca^{2+}-dependent and Ca^{2+}-independent isozymes of protein kinase C mediate exocytosis in antigen-stimulated rat basophilic RBL-2H3 cells: reconstitution of secretory responses with Ca^{2+} and purified isozymes in washed permeabilized cells. J. Biol. Chem. **268:** 1749–1756.
4. ZHANG, C. & M.A. BEAVEN. 1998. The MAP kinases and their role in mast cells and basophils. In Signal Transduction in Mast Cells and Basophils. E. Razin & J. Rivera, Eds.: 247–273. Springer-Verlag. New York.
5. DINH, T.T. & D.A. KENNERLY. 1991. Assessment of receptor-dependent activation of phosphatidylcholine hydrolysis by both phospholipase D and phospholipase C. Cell Regul. **2:** 299–309.
6. LIN, P. & A.M. GILFILLAN. 1992. The role of calcium and protein kinase C in the IgE-dependent activation of phosphatidylcholine-specific phospholipase D in a rat mast (RBL-2H3) cell line. Eur. J. Biochem. **207:** 163–168.
7. KUMADA, T., S. NAKASHIMA, H. MIYATA & Y. NOZAWA. 1994. Potent activation of phospholipase D by phenylarsine oxide in rat basophilic leukemia (RBL-2H3) cells. Biochem. Biophys. Res. Commun. **199:** 792–798.
8. CISSEL, D.S., P.F. FRAUNDORFER & M.A. BEAVEN. 1998. Thapsigargin-induced secretion is dependent on activation of a cholera toxin-sensitive and a phosphatidylinositol-3-kinase-regulated phospholipase D in a mast cell line. J. Pharmacol. Exp. Ther. **285:** 110–118.
9. ALI, H., O.H. CHOI, P.F. FRAUNDORFER, et al. 1996. Sustained activation of phospholipase D via adenosine A_3 receptors is associated with enhancement of antigen- and Ca^{2+}-ionophore-induced secretion in a rat mast cell line. J. Pharm. Exp. Ther. **276:** 837–845.
10. SMITH, B.S., G.G. DEANIN & J.M. OLIVER. 1991. Regulation of IgE receptor-mediated secretion from RBL-2H3 cells by GTP binding-proteins and calcium. Biochem. Biophys. Res. Commun. **174:** 1064–1069.
11. EXTON, J.H. 1997. Phospholipase D: enzymology, mechanisms of regulation, and function. Physiol. Rev. **77:** 303–320.
12. JONES, D., C. MORGAN & S. COCKCROFT. 1999. Phospholipase D and membrane traffic. Potential roles in regulated exocytosis, membrane delivery and vesicle budding. Biochim. Biophys. Acta **1439:** 229–244.
13. MORRIS, A.J., M.A. FROHMAN & J. ENGEBRECHT. 1997. Measurement of phospholipase D activity. Anal. Biochem. **252:** 1–9.
14. HAMMOND, S.M., J.M. JENCO, S. NAKASHIMA, et al. 1997. Characterization of two alternately spliced forms of phospholipase D1. Activation of the purified enzymes by phosphatidylinositol 4,5-bisphosphate, ADP-ribosylation factor, and Rho family monomeric GTP-binding proteins and protein kinase C-α. J. Biol. Chem. **272:** 3860–3868.
15. COLLEY, W.C., T.-C. SUNG, R. ROLL, et al. 1997. Phospholipase D2, a distinct phospholipase D isoform with novel regulatory properties that provokes cytoskeletal reorganization. Curr. Biol. **7:** 191–201.
16. DIVECHA, N., M. ROEFS, J.R. HALSTEAD, et al. 2000. Interaction of the type 1α PIPkinase with phospholipase D: a role for the local generation of phosphatidylinositol 4,5-bisphosphate in the regulation of PLD2 activity. EMBO J. **19:** 5440–5449.
17. ZHANG, Y., Y.M. ALTSHULLER, S.M. HAMMOND & M.A. FROHMAN. 1999. Loss of receptor regulation by a phospholipase D1 mutant unresponsive to protein kinase C. EMBO J. **18:** 6339–6348.
18. KIM, Y., J.M. HAN, J.B. PARK, et al. 1999. Phosphorylation and activation of phospholipase D1 by protein kinase C in vivo: determination of multiple phosphorylation sites. Biochemistry **38:** 10344–10351.

19. MIN, D.S., N.J. CHO, S.H. YOON, et al. 2000. Phospholipase C, protein kinase C, Ca^{2+}/calmodulin-dependent protein kinase II, and tyrosine phosphorylation are involved in carbachol-induced phospholipase D activation in Chinese hamster ovary cells expressing muscarinic acetylcholine receptor of *Caenorhabditis elegans*. J. Neurochem. **75:** 274–281.
20. SCHMIDT, M., M. VOB, P.A. OUDE WEERNINK, et al. 1999. A role for Rho-kinase in Rho-controlled phospholipase D stimulation by the m3 muscarinic acetylcholine receptor. J. Biol. Chem. **274:** 14648–14654.
21. KIM, Y., M.H. JUNG, B.R. HAN, et al. 2000. Phospholipase D1 is phosphorylated and activated by protein kinase C in caveolin-enriched microdomains within the plasma membrane. J. Biol. Chem. **275:** 13621–13627.
22. SARRI, E., I. BOCKMANN, U. KEMPTER, et al. 1998. Regulation of phospholipase D activity in synaptosomes permeabilized with *Staphylocccus aureus* α-toxin. FEBS Lett. **440:** 287–290.
23. CHAHDI, A., W.S. CHOI, Y.M. KIM, et al. 2002. Serine/threonine kinases synergistically regulate phospholipase D 1 and 2 and secretion in RBL-2H3 mast cells. Mol. Immunol. In press.
24. WAY, G., N. O'LUANAIGH & S. COCKCROFT. 2000. Activation of exocytosis by cross-linking of the IgE receptor is dependent on ADP-ribosylation factor 1-regulated phospholipase D in RBL-2H3 mast cells: evidence that the mechanism of activation is via regulation of phosphatidylinositol 4,5-bisphosphate synthesis. Biochem. J. **346:** 63–70.
25. MOHR, F.C. & C. FEWTRELL. 1987. Depolarization of rat basophilic leukemia cells inhibits calcium uptake and exocytosis. J. Cell Biol. **104:** 783–792.
26. ALI, H., K. MAEYAMA, R. SAGI-EISENBERG & M.A. BEAVEN. 1994. Antigen and thapsigargin promote influx of Ca^{2+} in rat basophilic RBL-2H3 cells by ostensibly similar mechanisms that allow filling of inositol 1,4,5-trisphosphate-sensitive and mitochondrial Ca^{2+} stores. Biochem. J. **304:** 431–440.
27. WILKINSON, S.E., P.J. PARKER & J.S. NIXON. 1993. Isoenzyme specificity of bisindolylmaleimides, selective inhibitors of protein kinase C. Biochem. J. **294:** 335–337.
28. LEVIS, M.J. & H.R. BOURNE. 1992. Activation of the alpha subunit of Gs in intact cells alters its abundance, rate of degradation, and membrane avidity. J. Cell. Biol. **119:** 1297–1307.
29. HIDE, M., H. ALI, S.R. PRICE, et al. 1991. The GTP-binding protein, Gαz: its down regulation by dexamethasone and its credentials as a mediator of antigen-induced responses in RBL-2H3 cells. Mol. Pharm. **40:** 473–479.
30. DESSAUER, C.W., J.J.G. TESMER, S.R. SPRANG & A.G. GILMAN. 1999. The interactions of adenylate cyclases with P-site inhibitors. Trends Pharmacol. Sci. **20:** 205–210.
31. HIDAKA, H. & T. ISHIKAWA. 1992. Molecular pharmacology of calmodulin pathways in the cell functions. Cell Calcium. **13:** 465–472.
32. LIZCANO, J.M., N. MORRICE & P. COHEN. 2000. Regulation of BAD by cAMP-dependent protein kinase is mediated via phosphorylation of a novel site, Ser^{155}. Biochem. J. **349:** 547–557.
33. BEAVEN, M.A., J. ROGERS, J.P. MOORE, et al. 1984. The mechanism of the calcium signal and correlation with histamine release in 2H3 cells. J. Biol. Chem. **259:** 7129–7136.
34. BROWN, F.D., N. THOMPSON, K.M. SAQID, et al. 1998. Phospholipase D1 localises to secretory granules and lysosomes and is plasma-membrane translocated on cellular stimulation. Curr. Biol. **8:** 835–838.
35. FROHMAN, M.A., T.C. SUNG & A.J. MORRIS. 1999. Mammalian phospholipase D structure and regulation. Biochim. Biophys. Acta **1439:** 175–186.
36. CHOI, W.S., Y.M. KIM, C. COMBS, et al. 2002. Phospholipase D1 and D2 regulate different phases of exocytosis in mast cells. J. Immunol. In press.
37. EASON, R.A. 2000. β-granule transport and exocytosis. Semin. Cell Develop. Biol. **11:** 53–266.
38. LISCOVITCH, M., M. CZARNY, G. FIUCCI & X. TANG. 2000. Phospholipase D: molecular and cell biology of a novel gene family. Biochem. J. **345:** 401–415.
39. HARSH, D.M. & R.A. BLACKWOOD. 2001. Phospholipase A_2-mediated fusion of neutrophil-derived membranes is augmented by phosphatidide acid. Biochem. Biophys. Res. Commun. **282:** 480–486.

40. WESTON, M.C. & P.T. PEACHELL. 1998. Regulation of human mast cells and basophil function by cAMP. Gen. Pharmacol. **31:** 715–719.
41. ALI, H., S. SOZZANI, I. FISHER, *et al.* 1998. Differential regulation of formyl peptide and platelet-activating factor receptors. Role of phospholipase Cβ_3 phosphorylation by protein kinase A. J. Biol. Chem. **273:** 11012–11016.
42. YUE, C., K.L. DODGE, G. WEBER & B.M. SANBORN. 1998. Phosphorylation of serine 1105 by protein kinase A inhibits phospholipase Cβ_3 stimulation by Gα_q. J. Biol. Chem. **273:** 18023–18027.
43. GARCIA, J.G.N., J.W. FENTON & V. NATARAJAN. 1992. Thrombin stimulation of human endothelial cell phospholipase D activity. Regulation by phospholipase C, protein kinase C, and cyclic adenosine 3'5'-monophosphate. Blood. **79:** 2056–2067.
44. QUIAN, Z. & L.R. DREWES. 1989. Muscarinic acetylcholine receptor regulates phosphatidylcholine phospholipase D in canine brain. J. Biol. Chem. **264:** 21720–21724.
45. ALI, H., J.R. CUNHA-MELO, W.F. SAUL & M.A. BEAVEN. 1990. The activation of phospholipase C via adenosine receptors provides synergistic signals for secretion in antigen stimulated RBL-2H3 cells: evidence for a novel adenosine receptor. J. Biol. Chem. **265:** 745–753.
46. ISERSKY, C., J. RIVERA, T.J. TRICHE & H. METZGER. 1982. Characterization of the receptors for IgE on membranes isolated from rats basophilic leukemia cells. Mol. Immunol. **19:** 925–941.

Role of PTEN, a Lipid Phosphatase Upstream Effector of Protein Kinase B, in Epithelial Thyroid Carcinogenesis

CHARIS ENG

Clinical Cancer Genetics and Human Cancer Genetics Programs, Comprehensive Cancer Center, and Division of Human Genetics, Department of Internal Medicine, The Ohio State University, Columbus, Ohio 43210, USA

Cancer Research Campaign Human Cancer Genetics Research Group, University of Cambridge, Cambridge, UK

ABSTRACT: Both benign and malignant thyroid disease are well-established components of Cowden syndrome (CS), an autosomal dominant disorder characterized by multiple hamartomas and breast cancer that may be considered a phakomatosis. The susceptibility gene for CS is *PTEN*, a tumor suppressor gene on 10q23.3 that encodes a lipid phosphatase that lies upstream of protein kinase B (Akt). Interestingly, Carney complex is also a phakomatosis where multiple endocrine neoplasias are prominent and thyroid cancer might be a rare component. One of its susceptibility genes is the regulatory subunit of protein kinase A. Over the course of the last four years, investigators have found the increasing clinical spectrum of syndromes characterized by germline loss-of-function *PTEN* mutation. In addition to CS, subsets of such disparate syndromes as Bannayan-Riley-Ruvalcaba syndrome, Proteus syndrome, and possibly VATER with hydrocephalus and megencephaly with autistic features have been found to have germline *PTEN* mutations. Paradoxically, somatic intragenic *PTEN* mutations were rare in uncultured primary epithelial thyroid tumors, although hemizygous deletion occurred in 10–20% of thyroid adenomas and carcinomas. However, with subsequent study, it was discovered that epigenetic silencing of *PTEN* and perhaps inappropriate subcellular compartmentalization were two novel mechanisms of PTEN inactivation pertinent in thyroid carcinogenesis. Ectopic expression studies *in vitro* have borne out the importance of PTEN in the pathogenesis of epithelial thyroid neoplasias.

KEYWORDS: Cowden syndrome; Bannayan-Riley-Ruvalcaba syndrome; follicular thyroid neoplasm; chromosome 10; PTEN; MMAC1; TEP1; protein kinase B; Akt; PI3K

Address for correspondence: Charis Eng, M.D., Ph.D., Human Cancer Genetics Program, The Ohio State University, 420 W. 12th Avenue, Suite 690 TMRF, Columbus, OH 43210. Voice: 614-292-2347; fax: 614-688-3582 or 688-3205.
eng.25@osu.edu

INTRODUCTION

Carney complex is an autosomal dominant disorder that can be viewed as a multiple endocrine neoplasia and a phakomatosis. In that manner, Carney complex may be considered to belong to a family of syndromes of multiple endocrine neoplasias and phakomatoses that also includes Cowden syndrome (CS). CS is an autosomal dominant multiple hamartoma syndrome characterized by an increased risk of benign and malignant breast, epithelial thyroid, and endometrial tumors.[1,2] One of the susceptibility genes for Carney complex was recently identified as the gene encoding the type 1a regulatory subunit of protein kinase A.[3] Relatedly, the susceptibility gene for CS is *PTEN*,[4,5] encoding a lipid phosphatase that lies upstream of protein kinase B, also known as Akt, a proapoptotic factor.[6–8]

COWDEN SYNDROME

While follicular thyroid adenomas and even invasive ductal carcinomas of the breast can be rarely found in individuals with Carney complex, breast and thyroid neoplasias are major component tumors of CS. The clinical diagnosis of CS is extremely difficult, so the precise incidence in the population is unknown. Prior to the identification of the susceptibility gene, the incidence was thought to be one in a million.[9] However, after identification of *PTEN* as the CS gene,[4] a molecularly based study revealed the incidence to be one in 200,000,[10,11] although this is likely still to be an underestimate. Because CS is underdiagnosed, a true count of the fraction of isolated cases (defined as no obvious family history) and familial cases (defined as two or more related affected individuals) cannot be performed. From the literature and the experience of both major CS centers in the US, the majority of CS cases are isolated. As a broad estimate, perhaps 10–50% of CS cases are familial.[12]

It is believed that >90% of individuals affected by CS manifest a phenotype by their 20s.[2,9] By the third decade, it has been noted that 99% of affected individuals would have developed the mucocutaneous sitgmata, although any of the features could be present already (TABLES 1 and 2). The most commonly reported manifestations are mucocutaneous lesions, thyroid abnormalities, fibrocystic disease, carcinoma of the breast, multiple, early-onset uterine leiomyoma, and macrocephaly (specifically, megencephaly) (TABLE 1).[2,13–16] Pathognomonic mucocutaneous lesions are trichilemmomas and papillomatous papules (TABLE 2). Because of the lack of uniform diagnostic criteria for CS prior to 1995, a group of individuals, the International Cowden Consortium, interested in systematically studying this syndrome to localize the susceptibility gene, arrived at a set of consensus operational diagnostic criteria.[9,17] These criteria have been revised recently in the context of new data and are reflected in the practice guidelines of the US-based National Comprehensive Cancer Network(NCCN) Genetics/High Risk Panel (TABLE 2).[2,18]

The two documented component cancers in CS are carcinoma of the breast and thyroid.[13] By contrast, in the general population, lifetime risks for breast and thyroid cancers are approximately 11% (in women) and 1%, respectively. In women with CS, lifetime risk estimates for the development of breast cancer range from 25 to 50%.[13,14,16,19] The mean age at diagnosis is likely to be 10 years earlier than that for breast cancer occurring in the general population.[13,16] Although Rachel Cowden

TABLE 1. Common manifestations of Cowden syndrome

Mucocutaneous lesions (90–100%)
    trichilemmomas
    acral keratoses
    verucoid or papillomatous papules

Thyroid abnormalities
    goiter
    adenoma
    cancer (3–10%)

Breast lesions
    fibroadenomas/fibrocystic disease (76% of affected females)
    adenocarcinoma (25–50% of affected females)

Gastrointestinal lesions (40%)
    hamartomatous polyps

Macrocephaly (38%)

Genitourinary abnormalities (44% of females)
    uterine leiomyoma (multiple, early onset)

---

died of breast cancer at the age of 31,[20,21] and the earliest recorded age of diagnosis of breast cancer is 14,[13] the majority of CS breast cancers are diagnosed after the age of 30–35 years (range 14–65).[16] Until genotype-phenotype analyses were performed with the discovery of the susceptibility gene, it was thought that male breast cancer was not a component of CS. However, male breast cancer does occur in *PTEN*–mutation positive CS but with unknown frequency.[5,22]

The lifetime risk for epithelial thyroid cancer can be as high as 10% in males and females with CS. It is unclear if the age at onset is truly earlier than that of the general population. Histologically, the thyroid cancer is predominantly follicular carcinoma, although papillary histology has also rarely been observed (Eng, unpublished observations).[13,14,16] After identification of *PTEN* as the susceptibility gene, preliminary data suggest that endometrial carcinoma is a component cancer of CS.[2,23,24] What its frequency is in mutation carriers is as yet unknown.

Benign tumors predominate in CS. Apart from those of the skin, benign tumors or disorders of breast and thyroid are the most frequently noted and likely represent true component features of this syndrome (TABLE 1). Fibroadenomas and fibrocystic disease of the breast are common signs in CS, as are follicular adenomas and multinodular goiter of the thyroid. An unusual central nervous system tumor, cerebellar dysplastic gangliocytoma or Lhermitte-Duclos disease, has only recently been associated with CS.[25,26]

TABLE 2. International Cowden Syndrome Consortium Operational Criteria for the Diagnosis of Cowden Syndrome (ver. 2000)[a]

Pathognonomic criteria

Mucocutaneous lesions
    trichilemmomas, facial
    acral keratoses
    papillomatous papules
    mucosal lesions

Major criteria

Breast carcinoma
    Thyroid carcinoma (nonmedullary), especially follicular thyroid carcinoma
    Macrocephaly (megalencephaly) (say, ≥97th percentile)
    Lhermitte-Duclos disease (LDD)
    Endometial carcinoma

Minor criteria

    Other thyroid lesions (e.g., adenoma or multinodular goiter)
    Mental retardation (say, IQ ≤75)
    GI hamartomas
    Fibrocystic disease of the breast
    Lipomas
    Fibromas
    GU tumors (e.g., renal cell carcinoma, uterine fibroids) or malformation
    GU tumors (e.g., renal cell carcinoma, uterine fibroids) or malformation

Operational diagnosis in an individual

1. Mucocutaneous lesions alone if:
    a. there are 6 or more facial papules, of which 3 or more must be trichilemmoma, or
    b. cutaneous facial papules and oral mucosal papillomatosis, or
    c. oral mucosal papillomatosis, or
    d. palmoplantar keratoses, 6 or more
2. Two major criteria, but one must include macrocephaly or LDD
3. One major and three minor criteria
4. Four minor criteria

Operational diagnosis in a family where one individual is diagnostic for Cowden

1. The pathognonomic criterion/criteria
2. Any one major criterion with or without minor criteria
3. Two minor criteria

[a]Operational diagnostic criteria are reviewed and revised on a continuous basis as new clinical and genetic information becomes available. The 1995 version and 2000 version have been accepted by the US-based National Comprehensive Cancer Network High Risk/Genetics Panel.

GENETICS

CS is an autosomal dominant disorder with age-related penetrance. It is believed that the penetrance is 90% after the age of 20 years.[9] The *CS* gene was mapped to 10q22-q24 without apparent genetic heterogeneity,[9] although one study suggests that rare locus heterogeneity might exist.[27] Subsequently, germline mutations of *PTEN*, which encodes a dual specificity phosphatase, were found in CS.[4] That *PTEN* is the CS gene has been confirmed by other groups.[10,27,28] When CS is strictly defined by the operational diagnostic criteria of the International Cowden Consortium, 80% have been found to harbor germline *PTEN* mutations.[5] Approximately two-thirds of these mutations were found in exons 5, 7, and 8, and 40% of all CS germline mutations are located in exon 5. Genotype-phenotype analyses revealed an association of the presence of germline mutation and malignant breast disease.[5] In addition, missense mutations and those within the phosphatase core motif and 5′ of it appeared to be associated with involvement of five or more organs, a surrogate phenotype for severity of disease.[5] Another group examined families for germline *PTEN* mutations and found mutations in only 13 probands.[11] They could not find any clear genotype-phenotype associations, most likely due to their small sample size.

Germline *PTEN* mutations in families with Bannayan-Riley-Ruvalcaba syndrome (BRR), characterized by macrocephaly, lipomatosis, hemangiomatosis, and speckled penis,[29] have also been found.[30] Thus, at least a subset of BRR and CS may be considered allelic. In contrast to CS, 60% of BRR were found to have germline *PTEN* mutations.[12] In addition, two of these mutations included one with a cytogenetically detectable deletion of 10q23, encompassing *PTEN*, and another with a translocation involving 10q23.[12,31,32] The mutational spectra of BRR and CS seemed to overlap, thus lending formal proof that CS and BRR are allelic.[12]

Interestingly, the clinical spectrum of *PTEN*-related disorders has recently been extended. The clue came when an individual with an unclassified Proteus-like syndrome, who did not meet the criteria for diagnosis of either BRR or Proteus syndrome, was found to harbor a germline *PTEN* mutation as well as a germline mosaic *PTEN* mutation.[33] Extended studies, albeit with small sample sizes, revealed that approximately 50% of Proteus-like syndromes and up to 20% of classic Proteus syndrome ascertained by published operational diagnostic criteria[34] carry germline *PTEN* mutations.[35] Single cases of a child with hydrocephalus and VATER association and a child with macrocephaly and autism, who do not meet diagnostic criteria for CS, BRR, or Proteus syndrome, have been described to have germline *PTEN* mutations.[36,37] What the extent of *PTEN*-defining disorders is remains the subject of intense investigation.

SOMATIC GENETIC AND EPIGENETIC *PTEN* ALTERATIONS IN SPORADIC EPITHELIAL THRYOID NEOPLASMS

Somatic *PTEN* mutations and/or deletions occur with variable frequency depending on neoplasia type. When *PTEN* was initially isolated, a wide variety of neoplastic cell line models were reported to have high frequencies of homozygous *PTEN* mutations and deletions,[38,39] and served as somewhat misleading indicators for predicting the frequency of "two-hit" genetic alterations in noncultured primary malignancies.

Deletions represented by LOH of anonymous polymorphic markers residing on chromosome 10 have been well documented in both benign and malignant epithelial thyroid neoplasia over the course of the last 10 years.[40] Three series, based mainly on epithelial thyroid tumors of European origin, have demonstrated that hemizygous deletion of *PTEN* occurs with a higher frequency in follicular adenomas (20–25%) compared to follicular carcinomas (5–10%).[41–43] The apparent higher LOH frequency in adenomas compared to carcinomas was not observed in a single US-based series.[44] The only intragenic point mutation amongst all series to date was a somatic frameshift mutation in a single papillary thyroid carcinoma.[42] These observations suggest that the pathogenesis of adenomas and carcinomas may proceed along two different pathways and that the adenoma-carcinoma sequence is not the rule in epithelial thyroid neoplasia.[43] The data were initially surprising in that epithelial thyroid malignancies occur in 3–10% of CS patients,[13,19] and one would expect that a larger proportion of sporadic thyroid carcinomas would be associated with somatic *PTEN* alteration. It was rationalized that benign thyroid disease occurs in 50–67% of CS individuals, far outnumbering the frequency of thyroid carcinomas. However, a recent expression and genetic analysis of 139 benign and malignant nonmedullary (epithelial) thyroid tumors yielded some interesting data that may begin to address this apparent paradox.[45] In this series, follicular adenomas, follicular carcinomas, and papillary thyroid carcinomas all had a 20–30% frequency of hemizygous deletion, while almost 60% of undifferentiated (anaplastic) carcinomas had hemizygous *PTEN* deletion. Of note, hemizygous deletion and decreased PTEN expression were associated. Decreasing PTEN expression was observed with a declining degree of differentiation. Decreasing nuclear PTEN expression seemed to precede that in the cytoplasm. The thyroid data suggest that in addition to structural deletion, epigenetic silencing and inappropriate subcellular compartmentalization might also contribute to PTEN inactivation.[45]

MECHANISM OF PTEN TUMOR SUPPRESSION IN THE THYROID

PTEN is a phosphatase tumor suppressor. PTEN is the major 3-phosphatase of phosphoinositol-(3,4,5)-triphosphate,[6,7] and signals down the phosphoinositide 3-kinase (PI3K) and the Akt/PKB pathways.[7,8,46] Proper PTEN function leads to decreased phosphorylated Akt (P-Akt), which is proapoptotic. Dysfunctional/absent PTEN leads to high levels of P-Akt, which is antiapoptotic. Consistent with the role of PTEN in PI3K/Akt signaling, PTEN negatively regulates cell survival by causing apoptosis and/or G1 cell cycle arrest.[7,47–49] In a breast cancer line stably transfected with wild-type PTEN, growth suppression was noted to be due to G1 arrest in the first 48 hours of PTEN induction followed by both G1 arrest and apoptosis after 72-h induction.[49] From this work, it is believed that G1 arrest is the primary event and apoptosis follows. Subsequently, it was found that the cells that were arrested at G1 are not the ones that undergo apoptosis.[50]

PI3K also plays an important role in cell proliferation. Several proteins including Akt, PAK and PKC, which phosphorylate and activate p70s6k, as well as MEK/MAPK could potentially transduce mitogenic signals mediated by PI3K; but thus far, Akt is the only one that has been found to be affected by PTEN. The antiapoptotic role of Akt is well documented, but the role of Akt in PI3K-mediated cell cycle regulation is not clear. Cheney et al. (1999) reported that PTEN inhibits S phase entry by recruiting p27 into the cyclin E/CDK2 complex and blocking CDK2 kinase activity without affecting cyclin E or p27.[51] In contrast, Li et al. (1998) demonstrated that PTEN-transfected glioma lines resulted in an increase in the G1 population, with a concomitant increase in p27.[47] Further, LY294002, a PI3K inhibitor, can mimic the effect of PTEN. The decrease in CDK2 kinase activity or p27 increase was accompanied by the reduction of P-Akt, which suggests that the inhibition of cell cycle progression by PTEN is mediated through its lipid phosphatase function that blocks the PI3K/Akt signaling pathway. However, direct evidence is lacking. Further, it remains unclear whether the alteration of p27 levels or p27-associated CDK2 activity is the direct cause or consequence of G1 arrest. Nonetheless, at least in the MCF-7 model, PTEN-mediated apoptosis and cell cycle arrest seem to result from both PI3K/Akt-dependent and -independent pathways.[50]

Few functional studies have been performed with epithelial thyroid cancer models. In a rare study, 10 epithelial thyroid cancer lines, originating from follicular thyroid carcinomas, papillary thyroid carcinomas, and undifferentiated (anaplastic) thyroid carcinomas, were first examined for structural alterations of *PTEN*. Interestingly, mirroring noncultured primary epithelial thyroid malignancies, only one, FTC-133, a follicular thyroid carcinoma line, was found to harbor an IVS4-19G>A mutation and hemizygous deletion of the remaining allele, resulting in decreased transcript and no PTEN protein production.[52] Of the 10 assorted lines, four, including FTC-133, showed decreased PTEN transcript, and in general, decreased or no PTEN protein levels and correspondingly increased phosphorylation of Akt. Transient expression of wild-type PTEN by adenoviral infection of seven of these lines resulted in G1 cell cycle arrest in two well-differentiated papillary thyroid carcinoma lines, and G1 arrest followed by apoptosis in the remaining lines, which comprised three follicular carcinoma lines, one poorly differentiated papillary thyroid carcinoma line, and one undifferentiated thyroid carcinoma line.[52] These observations suggested that down-regulation of PTEN transcript—and, rarely, structural abnormalities—play a role in PTEN inactivation in thyroid carcinogenesis and that PTEN exerts its tumor-suppressive activities on epithelial thyroid cancer by inhibiting cell cycle progression at G1 and cell death.

REFERENCES

1. ENG, C. & R. PARSONS. 2001. Cowden syndrome. *In* The Metabolic and Molecular Bases of Inherited Disease, 8th edit. C. Scriver, A.L. Beaudet, W.S. Sly & D. Valle, Eds.: 979–988.
2. ENG, C. 2000. Will the real Cowden syndrome please stand up: revised diagnostic criteria. J. Med. Genet. **37:** 828–830.
3. KIRSCHNER, L.S. *et al.* 2000. Mutations of the gene encoding the protein kinase A type 1-alpha regulatory subunit in patients with Carney complex. Nat. Genet. **26:** 89–92.
4. LIAW, D. *et al.* 1997. Germline mutations of the *PTEN* gene in Cowden disease, an inherited breast and thyroid cancer syndrome. Nat. Genet. **16:** 64–67.
5. MARSH, D.J. *et al.* 1998. Mutation spectrum and genotype-phenotype analyses in Cowden disease and Bannayan-Zonana syndrome, two hamartoma syndromes with germline *PTEN* mutation. Hum. Mol. Genet. **7:** 507–515.

6. MAEHAMA, T. & J.E. DIXON. 1998. The tumor suppressor, PTEN/MMAC1, dephosphorylates the lipid second messenger phosphoinositol 3,4,5-triphosphate. J. Biol. Chem. **273:** 13375–13378.
7. STAMBOLIC, V. *et al.* 1998. Negative regulation of PKB/Akt-dependent cell survival by the tumor suppressor PTEN. Cell **95:** 1–20.
8. DAHIA, P.L.M. *et al.* 1999. PTEN is inversely correlated with the cell survival factor PKB/Akt and is inactivated by diverse mechanisms in haematologic malignancies. Hum. Mol. Genet. **8:** 185–193.
9. NELEN, M.R. *et al.* 1996. Localization of the gene for Cowden disease to 10q22-23. Nat. Genet. **13:** 114–116.
10. NELEN, M.R. *et al.* 1997. Germline mutations in the *PTEN/MMAC1* gene in patients with Cowden disease. Hum. Mol. Genet. **6:** 1383–1387.
11. NELEN, M.R. *et al.* 1999. Novel *PTEN* mutations in patients with Cowden disease: absence of clear genotype-phenotype correlations. Eur. J. Hum. Genet. **7:** 267–273.
12. MARSH, D.J. *et al.* 1999. *PTEN* mutation spectrum and genotype-phenotype correlations in Bannayan-Riley-Ruvalcaba syndrome suggest a single entity with Cowden syndrome. Hum. Mol. Genet. **8:** 1461–1472.
13. STARINK, T.M. *et al.* 1986. The Cowden syndrome: a clinical and genetic study in 21 patients. Clin. Genet. **29:** 222–233.
14. HANSSEN, A.M.N. & J.P. FRYNS. 1995. Cowden syndrome. J. Med. Genet. **32:** 117–119.
15. MALLORY, S.B. 1995. Cowden syndrome (multiple hamartoma syndrome). Dermatol. Clin. **13:** 27–31.
16. LONGY, M. & D. LACOMBE. 1996. Cowden disease. Report of a family and review. Ann. Genet. **39:** 35–42.
17. ENG, C. 1998. Genetics of Cowden syndrome—through the looking glass of oncology. Int. J. Oncol. **12:** 701–710.
18. NCCN. 1999. NCCN Practice Guidelines: Genetics/Familial High Risk Cancer. Oncology **13** (11A): 161–186.
19. ENG, C. 1997. Cowden syndrome. J. Genet. Counsel. **6:** 181–191.
20. LLOYD, K.M. & M. DENIS. 1963. Cowden's disease: a possible new symptom complex with multiple system involvement. Ann. Intern. Med. **58:** 136–142.
21. BROWNSTEIN, M.H. *et al.* 1978. Cowden's disease. Cancer **41:** 2393–2398.
22. FACKENTHAL, J. *et al.* 2001. Male breast cancer in Cowden syndrome patients with germline PTEN mutations. J. Med. Genet. **38:** 159–164.
23. MARSH, D.J. *et al.* 1998. Germline *PTEN* mutations in Cowden syndrome-like families. J. Med. Genet. **35:** 881–885.
24. DE VIVO, I. *et al.* 2000. Novel germline mutations in the *PTEN* tumour suppressor gene found in women with multiple cancers. J. Med. Genet. **37:** 336–341.
25. PADBERG, G.W. *et al.* 1991. Lhermitte-Duclos disease and Cowden syndrome: a single phakomatosis. Ann. Neurol. **29:** 517–523.
26. ENG, C. *et al.* 1994. Cowden syndrome and Lhermitte-Duclos disease in a family: a single genetic syndrome with pleiotropy? J. Med. Genet. **31:** 458–461.
27. TSOU, H.C. *et al.* 1997. Role of *MMAC1* mutations in early onset breast cancer: causative in association with Cowden's syndrome and excluded in *BRCA1*-negative cases. Am. J. Hum. Genet. **61:** 1036–1043.
28. LYNCH, E.D. *et al.* 1997. Inherited mutations in *PTEN* that are associated with breast cancer, Cowden syndrome and juvenile polyposis. Am. J. Hum. Genet. **61:** 1254–1260.
29. GORLIN, R.J. *et al.* 1992. Bannayan-Riley-Ruvalcaba syndrome. Am. J. Med. Genet. **44:** 307–314.
30. MARSH, D.J. *et al.* 1997. Germline mutations in *PTEN* are present in Bannayan-Zonana syndrome. Nat. Genet. **16:** 333–334.
31. ARCH, E.M. *et al.* 1997. Deletion of *PTEN* in a patient with Bannayan-Riley-Ruvalcaba syndrome suggests allelism with Cowden disease. Am. J. Med. Genet. **71:** 489–493.
32. AHMED, S.F. *et al.* 1999. Balanced translocation of 10q and 13q, including the PTEN gene, in a boy with an HCG-secreting tumor and the Bannayan-Riley-Ruvalcaba syndrome. J. Clin. Endocrinol. Metab. **84:** 4665–4670.

33. ZHOU, X.P. *et al.* 2000. Germline and germline mosaic mutations associated with a Proteus-like syndrome of hemihypertrophy, lower limb asymmetry, arterio-venous malformations and lipomatosis. Hum. Mol. Genet. **9:** 765–768.
34. BIESECKER, L.G. *et al.* 1999. Proteus syndrome: diagnostic criteria, differential diagnosis and patient evaluation. Am. J. Med. Genet. **84:** 389–395.
35. ZHOU, X.P. *et al.* 2001. Association of germline mutation in the PTEN tumour suppressor gene and a subset of Proteus and Proteus-like syndromes. Lancet **358:** 210–211.
36. REARDON, W. *et al.* 2001. A novel germline mutation of the PTEN gene in a patient with macrocephaly, ventricular dilatation and features of VATER association. J. Med. Genet. In press.
37. DASOUKI, M.J. *et al.* 2001. Macrocephaly, macrosomia and autistic behavior due to a *de novo PTEN* germline mutation. Am. J. Hum. Genet. **69S:** 280 (Abstract 564).
38. LI, J. *et al.* 1997. *PTEN*, a putative protein tyrosine phosphatase gene mutated in human brain, breast and prostate cancer. Science **275:** 1943–1947.
39. TENG, D.H.-F. *et al.* 1997. *MMAC1/PTEN* mutations in primary tumor specimens and tumor cell lines. Cancer Res. **57:** 5221–5225.
40. ZEDENIUS, J. *et al.* 1995. Allelotyping of follicular thyroid tumors. Hum. Genet. **96:** 27–32.
41. MARSH, D.J. *et al.* 1997. Differential loss of heterozygosity in the region of the Cowden locus within 10q22-23 in follicular thyroid adenomas and carcinomas. Cancer Res. **57:** 500–503.
42. DAHIA, P.L.M. *et al.* 1997. Somatic deletions and mutations in the Cowden disease gene, *PTEN*, in sporadic thyroid tumors. Cancer Res. **57:** 4710–4713.
43. YEH, J.J. *et al.* 1999. Fine structure deletion analysis of 10q22-24 demonstrates novel regions of loss and suggests that sporadic follicular thyroid adenomas and follicular thyroid carcinomas develop along distinct parallel neoplastic pathways. Genes Chromosomes Cancer **26:** 322–328.
44. HALACHMI, N. *et al.* 1998. Somatic mutations of the *PTEN* tumor suppressor gene in sporadic follicular thyroid tumors. Genes Chromosomes Cancer **23:** 239–243.
45. GIMM, O. *et al.* 2000. Differential nuclear and cytoplasmic expression of PTEN in normal thyroid tissue, and benign and malignant epithelial thyroid tumors. Am. J. Pathol. **156:** 1693–1700.
46. MYERS, M.P. *et al.* 1998. The lipid phosphatase activity of PTEN is critical for its tumor suppressor function. Proc. Natl. Acad. Sci. USA **95:** 13513–13518.
47. LI, J. *et al.* 1998. The *PTEN/MMAC1* tumor suppressor induces cell death that is rescued by the AKT/protein kinase B oncogene. Cancer Res. **58:** 5667–5672.
48. DAVIES, M.A. *et al.* 1998. Adenoviral transgene expression of MMAC/PTEN in human glioma cells inhibits Akt activation and induces anoikis. Cancer Res. **58:** 5285–5290.
49. WENG, L.-P. *et al.* 1999. PTEN suppresses breast cancer cell growth by phosphatase function-dependent G1 arrest followed by apoptosis. Cancer Res. **59:** 5808–5814.
50. WENG, L.P. *et al.* 2001. PTEN induces apoptosis and cell cycle arrest through phosphoinositol-3-kinase/Akt-dependent and independent pathways. Hum. Mol. Genet. **10:** 237–242.
51. CHENEY, I.W. *et al.* 1999. Adenovirus-mediated gene entry of *MMAC1/PTEN* to glioblastoma inhibits S phase entry by the recruitment of p27Kip2 and cyclin E/CDK2 complexes. Cancer Res. **59:** 2318–2323.
52. WENG, L.P. *et al.* 2001. Transient ectopic expression of PTEN in thyroid cancer cell lines induces cell cycle arrest and cell type-dependent cell death. Hum. Mol. Genet. **10:** 251–258.

Signaling Pathways in Adrenocortical Cancer

LAWRENCE S. KIRSCHNER

Unit on Genetics and Endocrinology, DEB, NICHD, National Instutes of Health, Bethesda, Maryland 20892-1862, USA

ABSTRACT: Adrenocortical carcinoma is a rare tumor that carries a very poor prognosis. Despite efforts to develop new therapeutic regimens to treat this disease, surgery remains the mainstay of treatment. Laboratory studies of adrenocortical cancers have revealed a wide variety of signaling pathways that can be altered in these neoplasms. Although ACTH signaling through adenylyl cyclase and protein kinase A is important for normal adrenal cellular physiology, there is evidence to suggest that this pathway may inhibit the growth of adrenocortical tumors, and that inactivation of the ACTH receptor may promote tumor formation. Although multiple signal transduction pathways are essential for normal adrenal growth and hormone secretion, efforts to identify events required for neoplastic transformation have met with limited success. Alterations that have frequently been observed in adrenocortical carcinoma include up-regulation of the IGF-II system, as well as mutations in *TP53* and *RAS*. Current studies aim to elucidate the mechanisms of tumor growth by studying proproliferative signaling pathways, such as those involving Akt/PKB and the mitogen-activated protein kinases (MAPKs). Although studies of single pathways have been helpful in guiding investigations, new tools to study the integration and multiplicity of signaling pathways hold the hope of improved understanding of the signaling pathway alterations in adrenocortical cancer.

KEYWORDS: adrenal cancer; adrenocortical carcinoma; growth factors; signal transduction; G proteins; protein kinase A (PKA); mitogen-activated protein kinase (MAPK); ACTH receptor; p53; ras; insulin-like growth factor;EGF

INTRODUCTION

Cancers of the endocrine glands are rare entities, with only thyroid cancer found among the top 50 causes of cancer deaths. The reasons for this observation may lie in the highly differentiated nature of these tissues and the fact that endocrine cells reach the end of their proliferative potential early in life. Again, with the exception of thyroid carcinoma, endocrine cancers tend to be resistant to therapy other than surgical excision. For this reason, the prognosis for a patient diagnosed with an endocrine cancer is generally poor.

Because of its low incidence and resultant low emphasis as a health care issue, experimental investigations of endocrine carcinomas have not been common. Most

---

Address for correspondence: Lawrence S. Kirschner, M.D., Ph.D., Division of Endocrinology, Ohio State University, 491D McCampbell Hall, 1581 Dodd Drive, Columbus, OH 43210. Voice: 614-292-2995.

studies to date have studied small numbers of tumors, investigating a single molecule or pathway to look for changes associated with malignancy.

Adrenocortical carcinoma (ACC) is a good example of this paradigm. There are good clinical studies describing its behavior in patients, but relatively few laboratory investigations aimed at examining its molecular etiology. In this review, I will attempt to summarize the data from these published investigations, focusing on this information as a means to describe current understanding of the alterations in signal transduction that may lead to the development of adrenocortical malignancy.

Because studies of ACC are limited by the rarity of tumor specimens, many investigators have turned to model systems in an attempt to address questions of adrenocortical cell proliferation and carcinogenesis. Although these systems can provide vital insights into adrenal cell function, each also has certain limitations, which must be kept in mind when the data are analyzed.

Although the majority of currently published studies have analyzed only single pathways of signal transduction, it is now becoming clear that most cellular responses involve multiple pathways. For this reason, studies examining the integration of signals and the interplay of downstream pathways are becoming more important than understanding the effects of any individual molecular alteration. As we usher in the new age of analysis, including the tools of genomics and proteomics, better techniques for addressing these questions are becoming available, with the expectation that the answers will also become clearer. Like all good science, though, better understanding may pave the way to new, as yet unappreciated, questions.

CLINICAL ASPECTS OF ADRENAL CANCER

In terms of tumor epidemiology, adrenocortical cancer (ACC) is a rare cancer, accounting in large studies for 0.04 to 0.2% of all cancer deaths.[1,2] The incidence of the disease appears highest in the fifth decade,[1–3] although there is also another peak of incidence in children under 5 years of age.[3,4] Most series agree that adrenal cancer is more common in women than in men, with a ratio of approximately 1.5:1.[1,3,5]

The clinical presentation of ACC can be variable, and patients can present with symptoms of a specific hormonal syndrome or can present with nonspecific symptoms resulting from an abdominal mass. In larger case series, 45–75% of adult patients present with a hormonal oversecretion syndrome,[1,3,6] of which Cushing syndrome is the most common.[1,4] Other hormonal hypersecretion syndromes associated with ACC include virilization (from androgen-producing tumors), feminization (estrogen-producing tumors), and hyperaldosteronism. Multiple hormones can be produced by a single tumor, leading to a mixed clinical picture (e.g., Cushing syndrome plus virilization).[2,6] In children, there is a greater tendency for tumors to present with a hormonal syndrome, and virilization is more common in this age group.[4]

Of the patients presenting without a hormonal syndrome, nonspecific abdominal or dorsal pain is the most common presentation, being the presenting symptom in up to 30% of cases.[1,5] In a recent study of ACC from Poland, this diagnosis was found in 6.8% of patients found to have an adrenal mass,[7] in line with previous estimates that 0–25% of incidentalomas will be ACCs.[8,9] Of all adrenal cancers, incidentally detected lesions form approximately 30% of the group.[10]

Although ACC is a rare tumor, it tends to be quite aggressive and carries a very poor prognosis, which appears to depend little on the initial presentation. Approximately 50% of adults found to have ACC will not survive beyond two years of the diagnosis, and the 5-year mortality rate hovers around 80%.[1] Other than improved surgical management,[11] the prognosis for an individual diagnosed with ACC has not changed significantly over the past 40 years.[2,3] Adrenalectomy for benign lesions is now most commonly performed via a laparoscopic approach, and a low-level suspicion for cancer does not preclude this type of operation.[12] However, in cases where adrenal cancer is expected, an open approach provides a better opportunity for margin-free resection and resection of isolated metastases, and thus is the preferred surgical approach.[13–16]

Medical therapy of ACC consists mainly of mitotane (o, p`-DDD, 1,1 dichlorodiphenyldichloroethane), which has therapeutic effects in approximately 30% of patients.[3,6] Data to support that mitotane significantly improves survival are lacking, however, despite the fact that it appears to decrease hormone secretion in steroid-producing tumors. Efforts to use other chemotherapeutic agents, either alone or in combination, have generally not been more effective than mitotane alone,[17,18] although a recent study suggested that combination chemotherapy plus mitotane induced at least partial response in up to 54% of patients.[19]

ALTERATIONS IN SIGNAL TRANSDUCTION PATHWAYS

G Protein–Coupled Receptors

The ACTH-cAMP/PKA Pathway

Corticotropin (ACTH) signaling has long been one of the classic paradigms in endocrinology, and this pathway has been well worked out at the molecular level.[20] Treatment of adrenal cells with ACTH leads to the binding of this hormone to its receptor, termed MC2R, which belongs to the seven-transmembrane G protein–coupled receptor (GPCR) family. Activation of the receptor causes dissociation of the heterotrimeric stimulatory G protein (Gs), leading to the release of the alpha-subunit (Gs_α) and stimulation of adenylyl cyclase. This enzyme in turn causes the production of cyclic AMP (cAMP) from ATP. Cyclic AMP binds to the regulatory subunits of PKA, causing release of the catalytic subunits with subsequent transduction of the signal via phosphorylation of proteins both in the cytoplasm and in the nucleus.

There was ample reason to suspect that this pathway might be involved in adrenal tumorigenesis. First, adrenal cells under the constant stimulation of ACTH in patients (such as in those with Cushing disease) have enlarged adrenals.[21] Second, GPCR gain-of-function mutations have been well characterized in other endocrine tumors,[22] most notably activation of the TSH receptor in toxic thyroid adenomas.[23] Last, in the inherited syndrome Carney complex (CNC), the causative mutations were recently found to be inactivating mutations of the *PRKAR1A* gene, leading presumably to an increase in PKA activity.[24] Patients with this disease get multiple, hyperfunctioning adrenal nodules, suggesting that the pathway may play a role in proliferation and/or hyperfunction in the adrenal cortex.[25,26] Of note, adrenocortical carcinomas have not been shown to be associated with the complex, despite the well-described (benign) adrenal pathology.

To investigate the role of ACTH signaling in ACC, adrenal tumors were screened for mutations in *MC2R*. This receptor was initially cloned from humans in 1992,[27] and the number of investigations published is small. To date, there are only 38 adrenal tumors that have been studied (25 adenomas and 13 carcinomas), and no activating mutations of the ACTH receptor have been detected.[28,29] Failure to detect such mutations is likely not a cause for these findings, given that inactivating mutations of the ACTH receptor have been described in patients with the syndrome of ACTH insensitivity.[30]

The role of Gs_α is less clear. Although mutations have been sought in adrenal tumors in this gene, there is only a single report of an activating mutation in one aldosteronoma from Japan.[31] Efforts to detect mutations in other series of both adrenal adenomas and adrenal carcinomas have not yielded any positive findings.[32,33]

Mutations in the subunits of PKA itself have not been described other than the mutations seen in CNC, and, as mentioned above, it is not clear if these alterations play a role in adrenal cancer. Studies of the cAMP/PKA pathway *in vitro* suggest a reason why activation of the ACTH receptor pathway may not be involved in carcinogenesis. It has long been known that ACTH is the most potent stimulus for steroid secretion in adrenal cells isolated either from rodent or bovine species.[20] The same observation is made in the Y1 cell line, an adrenocortical cancer cell line initially derived from a mouse adrenocortical tumor (www.atcc.org, cat. #CCL-79). In each of these systems, treatment of cells with ACTH leads to a rapid and dose-dependent increase in steroidogenesis. However, analysis of the proliferative effects of ACTH has not been so clear. Studies of the proliferative effects of ACTH suggest that this hormone leads to decreased proliferation, or at least growth arrest.[34,35] This growth arrest is thought to be the result of ACTH's action as a factor that promotes differentiation, rather than proliferation. At the molecular level, treatment of cells with ACTH appears to have a biphasic effect on quiescent cells. Treatment for short periods of time (less than 2 hours) leads to cells exiting G0 and entering the G1 phase of the cell cycle.[35] However, prolonged treatment (18–24 hours) leads to an arrest of cells before S phase, so that the overall effect is an inhibition of cell cycle progression.[34] At the molecular level, ACTH or cAMP treatment leads to a rise in *c-fos* and *c-jun* mRNA, but appears to inhibit the accumulation of *c-myc*.[34,36,37] This blockade of *c-myc* appears sufficient to prevent cells from entering the cell cycle.[37]

Furthermore, there is ample evidence in adrenal cells that ACTH, acting through the PKA pathway, is able to inhibit proliferative signals initiated through other signaling pathways.[38] One proposed mechanism of this effect was demonstrated in Y1 cells, where ACTH treatment led to a dephosphorylation of the Akt/PKB kinase, a key modulator of cell cycle progression.[39,40] In Y1 adrenal cells, ACTH has also been suggested to play a role in suppressing signal transduction via the Mek kinase, a key player in activation of the MAP kinases.[41] These observations support the notion that ACTH is a differentiation-maintaining factor and is therefore anticarcinogenic. As this concept has become more appreciated, studies have been performed to address the potential role that loss of the ACTH response (via loss of MC2R) may play as a permissive factor for adrenal carcinogenesis. In a recent small study, loss-of-heterozygosity (LOH) at the *MC2R* locus was rarely detected in benign adrenal lesions, but was present in two of four adrenal cancers. This allelic loss of the ACTH receptor was correlated with reduced levels of the receptor in malignant tumors.[42]

Other G Protein–Coupled Systems

In addition to ACTH, many other hormones signal in the adrenal cortex via GPCRs. For example, the recently described family of hormones known as orexins also stimulates the adrenal gland through the OR1 receptor.[43] These signals appear to be mediated by stimulation of adenylyl cyclase and the PKA system.

In addition to the hormones described above, there are several other hormones that bind to GPCRs in the adrenal glands, but whose actions are modulated through signaling systems other than AC-PKA system.

The best characterized of these hormones is angiotensin-II (A-II), which is though to act primarily in the adrenal glomerulosa to stimulate release of aldosterone from the normal adrenal.[44,45] It has also been shown that A-II increases adrenal cell proliferation in tissue culture systems from human, rat, and bovine cells.[44,46,47] At the receptor level, A-II has two molecules that mediate its signal, AT-1R and AT-2R. In the adrenals, it is the former of these that is present and modulates signaling through this pathway.[44] In tissue culture studies, the AT-1R has been demonstrated to couple through multiple G proteins, including both the inhibitory G subunit G_i2 and the phospholipase C coupled subunit G_q.[45] In bovine adrenal cells, the proliferative effect of A-II was shown to be pertussis toxin (PTX) insensitive, implying that the majority of the signaling occurred through G_q.[48] However, similar experiments carried out in a slightly different culture system suggest that there were both PTX-sensitive and -insensitive components to the observations, implicating G_q- and G_i-mediated signaling.[49] G_q is known to transduce signals through protein kinase C (PKC), the diacylglycerol- and calcium-sensitive kinase that is also stimulated by phorbol esters (e.g., TPA). Effects mediated by this kinase can be blocked by using the specific PKC inhibitors Ro-8339 or calphostin. When adrenal cells were treated with these agents, most of the proliferative effect of A-II was blocked, although there appeared to be some persistence of signaling.[45] The signaling mechanism for G_i2 is less clear, although there is mounting evidence that this G protein complex signals through its βγ subunits through the ras pathway to activate the MAP kinase signaling pathway (ref. 50, and see below).

Other investigators have examined the role that the endothelins (ETs) play in adrenal hormone secretion. Because ETs are involved in vascular tone and blood pressure control, there has been significant interest in studying the role of these hormones in the regulation of aldosterone secretion. It has clearly been demonstrated that ET-1 stimulates aldosterone secretion both in normal adrenals[51,52] and in aldosterone producing adenomas (APAs).[53] The mechanism of these effects is quite complex, but appears to involve a multiplicity of pathways, including those of protein kinase C (PKC), cyclooxygenase, PI3-kinase, and phospholipase C (PLC).[51,52]

There has also been significant interest in the role of adrenomedullin, a peptide hormone initially identified from pheochromocytomas and thought to play a role in interaction between the cortex and the medulla.[54] This protein, which signals in the adrenal cortex through the CGRP-1 receptor, also induced proliferation via the MAPK pathway.[55] As with A-II, there appear to be multiple signaling mechanisms involved, as the proliferative effect is sensitive not only to modulators of the MAPK pathway, but also to molecules which modulate the PKC/PLC and PKA pathways.[56]

Despite the evidence for their involvement in adrenocortical cell proliferation provided from these studies, no mutations of the receptors for any of these hormones have been described in adrenal tumors. Interestingly, mutations in the $G\alpha_i 2$ subunit have been described in a small number of tumors, although these results were found only in a single study.[32] In that investigation, 11 adrenocortical tumors were studied, and activating mutations of $G\alpha_i 2$ were detected in three tumors. In two adrenal adenomas, an activating mutation of $G\alpha_i 2$ was present in the heterozygous state. In the third tumor, an adrenal cancer, only the mutated allele was detected, indicating loss of the normal allele (either by LOH or gene conversion). In order to verify these findings, the mutated form of $G\alpha_i 2$ was introduced into tissue culture cells. In NIH3T3 fibroblast cells, the mutant allele led to a marked reduction of cAMP in the cells, although no growth effect was noted.[57] However, in a different fibroblast cell line, Rat-1, the same construct not only decreased cAMP levels, but also led to enhanced proliferation, including the gain of the ability to grow in soft agar.[58] The authors interpreted these findings to suggest a tissue specificity of the transforming effect of this oncogene, dubbed the *gip2* oncogene; unfortunately, no studies introducing the construct into adrenal cells have been carried out. Subsequent investigations of the $G\alpha_i 2$ gene from other groups have not detected mutations in over 60 additional tumors.[33,59,60]

The Role of Ectopic Receptors

Interestingly, there are a number of reports suggesting that ectopic GPCRs may play a role in adrenal tumorigenesis. The pioneering work of Lacroix and colleagues has clearly demonstrated that ectopic expression of a small number of Gs-coupled GPCRs can lead to the phenomenon of massive macronodular adrenocortical disease (MMAD, also known as ACTH-independent macronodular adrenal hyperplasia, AIMAH).[61–67] In this condition, there is marked bilateral enlargement of the adrenal glands associated with hypersecretion of cortisol under the control of an aberrant hormone receptor. In all of the cases described to date, the receptors involved are thought to recapitulate the hormone stimulation by ACTH, but there is obvious disruption of the negative feedback loop. As in states associated with ACTH excess (e.g., Cushing disease or ectopic ACTH production), there is massive hypertrophy of the adrenals, although there have been no cases of malignant transformation reported.

Similarly, ectopic expression of a specific receptor can lead to the development of a single adenoma, as has been shown in a small number of cases. In older studies, the coupling of ectopic receptors to the adenylate cyclase system has clearly been demonstrated in *in vitro* studies using extracts from adrenal adenomas.[68,69] In a single study, similar observations were made about the ectopic coupling of the β-adrenergic receptor to PKA,[70] although large studies of this phenomenon have not been performed to date. MMAD and the concept of signaling through ectopic GPCRs are discussed elsewhere in this volume.

Growth Factor Signaling Pathways

Insulin-like Growth Factors

The insulin-like growth factor (IGF) system is perhaps the best characterized in terms of being involved in adrenocortical tumorigenesis,[71] although it has many roles in normal cell growth and development.[72–74] There are two distinct forms of IGFs, known as IGF-I and IGF-II. IGF-I, coded for by a gene at 12q22-q24.1, is thought to be the major signal elucidated by growth hormone (GH) in postnatal life. IGF-II, located at 11p15.5, is felt to play a much more important role during fetal life, and its importance as a growth mediator is minimal after birth. The two proteins signal through two receptors, which bind IGF-I and IGF-II with similar affinities. They also bind insulin, albeit at much lower affinity. The IGF1R has a structure analogous to that of the insulin receptor, being composed of two heterodimeric chains that possess an intrinsic tyrosine kinase activity. The IGF2R is a single polypeptide chain that also functions as the receptor for mannose-6-phosphate, a key sugar moiety involved in intracellular trafficking of lysosomal enzymes. The mechanism by which the IGF2R signals is not as clear, although mutations of the gene have been shown to be involved in hepatocellular carcinoma.[75]

Studies of the IGF system were initially suggested by correlation with the Beckwith-Wiedemann syndrome (BWS) (www.ncbi.nlm.nih.gov/omim OMIM #130650). This syndrome is characterized by generalized overgrowth, macroglossia, and exomphalos. Adrenal carcinoma is also a feature of the syndrome. Genetically, the disease had been mapped to the 11p15.5 region harboring not only the IGF-II gene, but the genes coding for insulin (INS), p57Kip2 and H19, a transcript of unknown function that is not translated into protein.

With this background, IGF-I and -II levels were investigated in a variety of adrenal lesions. In early studies, no changes in IGF-I were detected, although marked elevation of IGF-II was detected.[76–80] This change was observed both by analysis of mRNA levels and at the protein level by immunohistochemistry.[76] These findings were infrequently seen in benign lesions, but were prominent in over 60% of adrenocortical carcinomas. Studies searching for potential receptors for IGF-II in ACC have looked at levels of both the IGF1R and IGF2R, as IGF-II can signal through both of these. There have been consistent findings of elevation in the IGF1R, without marked changes in IGF2R levels. Again, this has been observed at both the level of mRNA and protein, but also at the functional level by receptor binding studies.[79] In more recent studies, transgenic mice expressing IGF-II postnatally were generated and were demonstrated to have adrenocortical hyperplasia, although frank malignancy was not observed.[81] This observation suggests that IGF-II is important for the abnormal proliferation of adrenal cells, but that additional steps are required for transformation to neoplasia.

At the genetic level, the IGF-II gene and its surrounding genomic region (including H19 and p57Kip2) are known to be subject to regulation by genomic imprinting. Imprinting is a specific form of methylation that occurs early in life (often in embryonic stages) leading to stable patterns of activation or inactivation of genes.[82] This locus, like most that are subject to imprinting, shows specific parent-of-origin effects of imprinting on gene expression. Specifically, the IGF-II gene is maternally imprinted, such that expression in the adult comes almost exclusively from the paternal allele. In contrast, the other two genes in the region, H19 and p57Kip2, are paternally imprinted, such that expression derives solely from the maternal allele.

In genetic studies to address the mechanism of IGF-II overexpression in adrenal lesions, it has been observed that there are genetic changes at the IGF-II locus. Most commonly, LOH is observed, leading to specific deletion of the maternal allele.[83]

This is often accompanied (although the exact mechanism of the genetic changes is not clear) by gain of a second copy of the paternal allele, a process called uniparental disomy (UPD), such that there are two copies of the father's allele and none of the mother's.[77] Not only is the genetic sequence of the mother lost or changed to the father's, but also the imprinting status reflects the chromosome of origin. In this manner, the replacement of the maternal allele with the paternal allele leads to the presence of a double dose of paternal allele, with the effect that both copies of the gene are now expressed. Whether the level of expression is similar to what would be predicted solely on the basis of two copies of the paternal allele is not clear, but this disruption of the normal genomic structure of the locus clearly plays a role.

Conversely, as mentioned above, the H19 gene and p57Kip2 are normally expressed solely from the maternal allele. When this region of the genome is lost and/or replaced by the paternal region, the expression of these two genes is also lost, or at least markedly reduced. The role of H19 is not clear, although it has been posited to function as a tumor suppressor, so its loss is of unknown functional consequence. However, p57Kip2 is known to function in the suppression of cell cycle progression. Indeed, it is loss of this function that is thought to explain why mutations of this gene lead to BWS. Although it has been studied in only a small number of tumors to date, down-regulation and/or loss of both H19 and p57Kip2 may contribute to the phenotype of adrenal cancers that are found to have alterations in the 11p15.5 locus.[78,84]

Other Tyrosine Kinase Family Growth Factors

Growth factors that signal through tyrosine kinase mechanisms are widely varied, and the signals can be mediated either directly through a receptor with tyrosine kinase activity (such as the IGF1R) or via accessory signaling molecules, which mediate the transduction of the signaling through accessory kinases (such as the IGF2R).

One of the most influential general growth factors is epithelial growth factor (EGF), which was one of the first peptide growth factors to be identified, and its cognate receptor, the EGF receptor (EGFR). This latter protein was initially described as a growth-promoting oncogene known as the v-*erb-B* oncogene, which causes avian erythroblastic leukemia.[85] The EGF system has been studied in a small number of adrenal tumors, and expression of EGF receptor was found to be present in over 90% of adrenocortical carcinomas.[86–88] However, the levels were not markedly different from those observed in both cortisol- and aldosterone-producing adrenal adenomas, suggesting that the EGFR system may be a nonspecific growth factor that plays a permissive but nonspecific role in adrenocortical transformation.[87,88] Interestingly, EGF itself does not appear to be expressed in adrenal tumors, suggesting either that EGF functions as an endocrine hormone (i.e., as opposed to a locally acting factor) or that the EGFR is stimulated by another compound. A good candidate ligand for this receptor is the growth factor TGF-α, which is expressed in adrenocortical lesions; again, there was not a marked difference between benign and malignant tumors, suggesting that the growth-promoting effect may not be specific for tumors.[88]

Last, the role of cytokines has also been assessed in the adrenal, although studies specific for adrenocortical cancer are quite rare. It was described a number of years ago that a number of interleukins could affect steroid hormone production and proliferation of adrenocortical cells. Specific studies have examined the role of IL-3 and IL-6 in this process. Cytokines signal through a heterodimeric receptor complex consisting of an α-chain that is specific for the cytokine (but that lacks kinase activity) and a common cytokine transducing component known as gp130. IL-3 is able to mediate steroid production through its receptor, and the effect has been shown to be dependent on an intact lipoxygenase pathway, as treatment of adrenal cells with indomethacin is able to block the action of this cytokine. Similarly, IL-6 has similar effects in adrenocortical cells, but the effect is mediated instead through the cyclooxygenase pathway.[89,90] Neither of these cytokines has been shown to have a specific role in adrenal tumorigenesis.

In contrast to these *in vitro* studies, two recent studies have suggested a similar action may be occurring *in vivo*. In one case, a benign adrenal adenoma was found to express receptors for IL-1, which were thought to mediate both the growth and cortisol hypersecretion found in the patient.[91] Within the last few months, a similar report has appeared for an adrenal cancer[92] in which a tumor was found to secrete a variety of cytokines of the CXC family, including Gro-α and -β, as well as NAC70 and others. When archived ACCs were examined for the same cytokines, expression levels were also found at appreciable levels in six of seven samples tested. When the index tumor was explanted into nude mice, it grew aggressively, recapitulating its behavior in the patient from whom it was isolated. In contrast, blockade of the cytokines using specific antisera led to a marked reduction in tumor growth, implying a direct connection between cytokine signaling and tumor aggressiveness. Although this last manuscript clearly proved that these pathways could play a significant role in adrenocortical cancer, more studies will need to be done to see how common these pathways are.

Intracellular Pathways of Signaling in Adrenal Cancer

Intracellular mediators of signaling pathways are multiple, and sorting out the role of any particular molecule is difficult. Before attempting to integrate the signaling pathways into a larger scheme, it is worthwhile to review those molecules in which mutations have been shown to be associated with adrenal cancer. These are few, and so I will discuss each briefly.

TP53 *Mutations*

The first intracellular molecule clearly shown to be associated with adrenal cancer is the tumor suppressor gene *TP53*. *TP53* gene mutations have been known for a long time to cause cancer, either through inactivation or, more commonly, through a dominant-negative effect. The direct connection between *TP53* and adrenal cancer was made when it was discovered that patients with the Li-Fraumeni cancer syndrome harbored mutations in this gene.[93] Li-Fraumeni syndrome (LFS) is an inherited cancer syndrome consisting primarily of soft-tissue sarcomas, tumors of the brain and breast, as well as a wide variety of other malignant tumors, including adrenocortical cancer (www.ncbi.nlm.nih.gov/omim OMIM #151623). The *TP53* gene, which mediates an important cell cycle checkpoint control, is inactivated in these tumors. It is largely a target for many proliferative systems, so it is not possible to place this gene in the setting of one particular signaling cascade. However, it is certainly worth mentioning that large percentages of patients with ACC have mutations in *TP53*.[94] Indeed, it has recently been demonstrated that in southern Brazil, where the incidence of pediatric ACC is markedly higher than anywhere else in the world, the majority of these tumors possess mutations in *TP53*. In fact, 35/36 unrelated patients with ACC contained an identical point mutation in codon 337 (R337H), suggesting a potential mutation hot spot, as well as the possible involvement of a specific environmental factor in adrenal tumors of that region.[95]

RAS *Mutations*

The *ras* gene family is another important mediator of cancers, and mutations in this family of genes have been found in a wide variety of malignant tumors. The ras family is composed of three family members, termed v-Harvey-ras (Ha-ras, or *HRAS*), Kirsten-ras (Ki-ras, or *KRAS*), and *NRAS*. These genes are localized in humans to 11p15.5, 12p12.1, and 1p13.2, respectively. The first two were identified from murine sarcoma virus, and the third was identified as an oncogene from human sarcomas. They are the founding members of a large family of small GTP binding proteins and their cohorts.

The *RAS* proteins themselves can be oncogenic, and they have been analyzed in large numbers of cancers in the body. In tumors containing *RAS* mutations, there is frequent alteration in codons 12, 13, or 61, leading to a decrease in the intrinsic GTPase activity of the protein and increased proliferative signals. Initial studies of adrenal tumors did not identify *RAS* mutations, although only the mutations in the three common codons were evaluated.[96] When a more general screening approach was used, mutations were detected in each of the *RAS* genes. To date, mutations of *KRAS*[97] or *HRAS*[98] have been detected in small numbers of ACCs, whereas mutations in *NRAS* have been seen in both adenomas and carcinomas.[60] To study the effects of the *KRAS* mutations, tumor-derived mutant forms of the protein were tested for intrinsic GTPase activity. Like other oncogenic mutations of *RAS*, the adrenal mutants were found to have a decrease in intrinsic GTPase activity.[99] Interestingly, transfection of normal adrenocortical cells with *KRAS* mutants led to a marked increase in expression of steroidogenic enzymes, accompanied by a 20–30-fold increase in hormone secretion.[100]

Transcription Factors

Clearly, mitogenic signals have their ultimate end point in the nucleus, at the level of altered gene transcription. Genes that are turned on (or off) in cancers represent a stable mechanism to perpetuate the signal to continue to proliferate. To generate these changes, there must also be changes in the presence or activity of the factors governing mRNA transcription. Small numbers of studies have examined alterations in transcription associated with adrenal cancers.

The GATA family of transcription factors is ubiquitous and has been shown to be involved in a variety of cellular processes, including both proliferation and differentiation. The two most important GATA family members for the endocrine glands are GATA-4 and GATA-6. It was reported in 1999[101] that GATA-6 is present in fetal and postnatal adrenal glands of the mouse, whereas GATA-4 levels are low. In a mouse tumor model, adrenal tumorigenesis is accompanied by the loss of GATA-6 and marked up-regulation of GATA-4. In human tissues, GATA-4 was found to be absent from the normal adrenal, whereas it was easily detected in a series of benign and malignant adrenal tumors. The downstream effects of these alterations are unknown in the adrenal gland.

More recently, and perhaps more interestingly, the transcription factors associated with cAMP signaling were evaluated in a small series of adrenal tumors.[102] Specifically, levels of the cAMP response element (CRE) binding protein (CREB), the CRE modulator (CREM), and the inducible cAMP early repressor (ICER) were evaluated in adrenal adenomas and carcinomas. CREB is an interesting protein, as it has long been known to be subject to phosphorylation, which causes it to become active. This phosphorylation was initially attributed to PKA, but later studies demonstrated that CREB can be phosphorylated by many kinases, including MAPK. Less is known about the function of the other proteins in this pathway. In the study of Peri *et al.*[102] CREM levels were found not to vary significantly between benign and malignant adrenal tumors. However, whereas levels of CREB and ICER were steady in the adenomas, they were markedly reduced in about half of the carcinomas.

Antiapoptotic Signals

Study of bcl-2 in tumors suggests that alterations in this antiapoptotic factor do not play a significant role in tumors, nor did related family members mcl-1 or bax. The role of other family members (e.g., Bcl-x) was not studied.[103] Recently, there was a report describing alterations in the human homologue of the *C. Elegans* Diminuto protein. This protein, which is thought to be involved in steroid biosynthesis in plants, has been suggested to play an antiapoptotic role in human neurons. This study demonstrated that it is up-regulated in adrenal tumors, suggesting that it may play a role in preventing apoptotic death of steroid secreting cells undergoing unscheduled cell cycling.[104] However, these data are very preliminary at present and will need to be verified in more systematic studies.

INTEGRATION OF SIGNALING PATHWAYS IN ADRENAL TUMORS

Although the large majority of studies aimed at examining changes in the signaling pathways in adrenal cancer have studied single pathways, it is now becoming clear that this approach provides only a limited view of the picture. In other experimental systems, recent work has turned towards elucidating the interaction of signaling pathways for growth and/or differentiation, and these interactions have been shown to be quite complex.[105,106] Clearly, a better understanding of adrenal cancer will require a similar approach.

As of now, what can be said about alterations in signaling events responsible for adrenal cancer? The change with the strongest supporting evidence is for overexpression of IGF-II and signal transduction through a tyrosine kinase–mediated IGF1R pathway. There may be a similar role for EGFR signaling, as well. Slightly downstream from these molecules is *ras*, where mutations have been found in significant numbers of ACC samples. The importance of cytokine overexpression for tumor proliferation has been demonstrated in only a single tumor, although the detailed characterization of that tumor's behavior both *in vitro* and *in vivo* clearly showed that this pathway is sufficient to drive tumor growth.[92]

Activating mutations of *MC2R* are clearly *not* observed in ACC, although loss of this gene (by LOH or other means) may play a role in overcoming antiproliferative signals. The role of activating mutations in the GPCR-associated G proteins Gs_α and $G_i\alpha 2$ is unclear, since there have been single reports of mutations, but they have not been confirmed.

How can these disparate lines of evidence be assembled into a coherent pathway that may explain the signaling pathways that are important in adrenal cancer? As in all cancers, carcinogenesis is a combination of two factors: hyperproliferation and dedifferentiation. Current thinking using Knudsen's "two-hit" hypothesis states that hyperproliferation increases the chance of obtaining a secondary mutation, which then leads to cancer. This is especially true in terms of the necessary alterations that need to occur in tumor cells—namely, reactivation of telomerase and deregulation of cell cycle checkpoints.

In highly differentiated cells such as are found in the endocrine system, the most important ultimate regulator of proliferation is likely the mitogen-activated protein kinase (MAPK), also known as extracellular signal-regulated kinase (ERKs). There are two proteins in the family, ERK-1 and ERK-2, also known as p44 and p42, respectively. In the adrenal, as in most other tissues, activation of the MAPK system appears to be a prerequisite to enhanced proliferation.

Regulation of the MAPKs is extremely complex, and their ultimate activation can occur via many pathways. In adrenal cell culture systems (either cell lines -Y1 mouse adrenocortical cancer cells or H295R human adrenocortical cells) or in primary cell cultures from human, mouse, or cow, each of the pathways described above has been implicated in the activation of the MAPK pathway.

The role of ACTH signaling in adrenal cancers appears quite complex, and our current understanding is far from clear. It is well established from patient data that stimulation of the adrenal by ACTH eventually leads to proliferation (manifested as hyperplastic adrenals), although neoplasia does not appear to occur. It has been demonstrated that ACTH can lead, via a PKA-independent pathway, to the activation of MAPKs after only a very short (<5-min) pulse of ACTH.[41] This observation is congruent with the findings that transient ACTH treatment stimulates adrenal cells to leave G0 and enter G1,[34,39] although the role of the hormone in long-term proliferation seems doubtful.

In fact, most data suggest that, on balance, ACTH is antiproliferative. Again, the data are not clear, but there appear to be two mechanisms for this action. First is the reduction in c-myc, most likely caused by increased protein degradation as well as blockade of synthesis.[34] More recent studies have also demonstrated a second mode of action of ACTH. It appears to markedly stimulate dephosphorylation of the Akt/PKB kinase.[39,40] This kinase, which is also activated primarily by a PI3-K–dependent pathway, is a major component of mitogenic signaling. Dephosphorylation reduces its activity and diminishes the signals for cell proliferation.

In the case of MMAD, another state with hyperplastic but nonmalignant adrenal growth, it has been demonstrated that the growth is due to the presence of ectopic GPCRs.[61] It is also possible that, like the presence of PKA-coupled receptor expression itself, these receptors may acquire the facility to couple with the growth-promoting effects of the MAPK pathway. Their retention of coupling to cAMP pathways likely leads to a check on growth, and hence these tumors are not malignant.

For kinases that fall in the tyrosine kinase family (IGF2R, EGFR, and so on), their direct coupling to the MAPK provides adequate explanation for their role in the abnormal proliferation seen in adrenal cancer. The same paradigm would be expected to apply to cytokine signaling pathways, such as in the CXC cytokine–producing tumor described above.[92] Experimentally, when cytokine signaling was blocked, so was proliferation.

Rather than gain of activity, dedifferentiation likely occurs by loss of molecules that keep proliferation in check. Chief among them may be p53, but loss of the MC2R (ACTH receptor) also may play a permissive role in the transformation of adrenal cells from benign hyperplasia to malignancy.

SUMMARY

Adrenal cancer remains a rare and, unless caught in its early stages, almost uniformly fatal disease. The large majority of studies of this disease have dealt with studying the presence or absence of a given mRNA or protein thought to be involved in the tumorigenic process. As the tools for molecular medicine improve, it will likely become possible to analyze large numbers of transcripts, which, in turn, will lead to the identification of new steps in the malignant pathways. In addition to this better understanding, these new approaches hold open the hope that new therapeutic strategies may be developed that will render this disease less burdensome.

REFERENCES

1. CRUCITTI, F. *et al.* 1996. The Italian Registry for Adrenal Cortical Carcinoma: analysis of a multiinstitutional series of 129 patients. The ACC Italian Registry Study Group. Surgery **119:** 161–170.
2. LIPSETT, M.B., R. HERTZ & G.T. ROSS. 1963. Clinical and pathophysiologic aspects of adrenocortical carcinoma. Am. J. Med. **35:** 374–383.
3. WOOTEN, M.D. & D.K. KING. 1993. Adrenal cortical carcinoma. Epidemiology and treatment with mitotane and a review of the literature. Cancer **72:** 3145–3155.
4. MENDONCA, B.B. *et al.* 1995. Clinical, hormonal and pathological findings in a comparative study of adrenocortical neoplasms in childhood and adulthood. J. Urol. **154:** 2004–2009.
5. DIDOLKAR, M.S., R.A. BESCHER, E.G. ELIAS & R.H. MOORE. 1981. Natural history of adrenal cortical carcinoma: a clinicopathologic study of 42 patients. Cancer **47:** 2153–2161.
6. KASPERLIK-ZALUSKA, A.A., B.M. MIGDALSKA, S. ZGLICZYNSKI & A.M. MAKOWSKA. 1995. Adrenocortical carcinoma. A clinical study and treatment results of 52 patients. Cancer **75:** 2587–2591.
7. KASPERLIK-ZALUSKA, A.A., B.M. MIGDALSKA & A.M. MAKOWSKA. 1998. Incidentally found adrenocortical carcinoma. A study of 21 patients. Eur. J. Cancer **34:** 1721–1724.
8. KLOOS, R.T. *et al.* 1995. Incidentally discovered adrenal masses. Endocr. Rev. **16:** 460–484.
9. KLOOS, R.T. *et al.* 1997. Incidentally discovered adrenal masses. Cancer Treat. Res. **89:** 263–292.
10. KASPERLIK-ZALUSKA, A.A. *et al.* 1997. Incidentally discovered adrenal mass (incidentaloma): investigation and management of 208 patients. Clin. Endocrinol. **46:** 29–37.
11. KENDRICK, M.L. *et al.* 2001. Adrenocortical carcinoma: surgical progress or status quo? Arch. Surg. **136:** 543–549.
12. TERACHI, T. *et al.* 1997. Transperitoneal laparoscopic adrenalectomy: experience in 100 patients. J. Endourol. **11:** 361–365.
13. IINO, K., Y. OKI & H. SASANO. 2000. A case of adrenocortical carcinoma associated with recurrence after laparoscopic surgery. Clin. Endocrinol. **53:** 243–248.

14. HENIFORD, B.T., M.J. ARCA, R.M. WALSH & I.S. GILL. 1999. Laparoscopic adrenalectomy for cancer. Semin. Surg. Oncol. **16:** 293–306.
15. GILL, I.S. 2001. The case for laparoscopic adrenalectomy. J. Urol. **166:** 429–436.
16. HAMOIR, E., M. MEURISSE & T. DEFECHEREUX. 1998. Is laparoscopic resection of a malignant corticoadrenaloma feasible? Case report of early, diffuse and massive peritoneal recurrence after attempted laparoscopic resection. Ann. Chir. **52:** 364–368.
17. AHLMAN, H. et al. 2001. Cytotoxic treatment of adrenocortical carcinoma. World J. Surg. **25:** 927–933.
18. DOGLIOTTI, L. et al. 1995. Cytotoxic chemotherapy for adrenocortical carcinoma. Minerva Endocrinol. **20:** 105–109.
19. BERRUTI, A. et al. 1998. Mitotane associated with etoposide, doxorubicin, and cisplatin in the treatment of advanced adrenocortical carcinoma. Italian Group for the Study of Adrenal Cancer. Cancer **83:** 2194–2200.
20. ABDEL–MALEK, Z.A. 2001. Melanocortic receptors: their functions and regulation by physiological agonists and antagonists. Cell. Mol. Life Sci. **58:** 434–441.
21. PARKER, K.L. & B.P. SCHIMMER. 2001. Genetics of the development and function of the adrenal cortex. Rev. Endocrinol. Metab. Disorders **2:** 245–252.
22. ARVANITAKIS, L., E. GERAS-RAAKA & M.C. GERSHENGORN. 1998. Constitutively signaling G-protein-coupled receptors and human disease. Trends Endocrinol. Metab. **9:** 27–31.
23. TRULZSCH, B. et al. 2001. Detection of thyroid-stimulating hormone receptor and Gs alpha mutations in 75 toxic thyroid nodules by denaturing gradient gel electrophoresis. J. Mol. Med. **78:** 684–691.
24. KIRSCHNER, L.S. et al. 2000. Mutations of the gene encoding the protein kinase A type I-alpha regulatory subunit in patients with the Carney complex. Nat. Genet. **26:** 89–92.
25. CARNEY, J.A. et al. 1985. The complex of myxomas, spotty pigmentation, and endocrine overactivity. Medicine (Baltimore) **64:** 270–283.
26. STRATAKIS, C.A. et al. 1996. Carney complex, a familial multiple neoplasia and lentiginosis syndrome. Analysis of 11 kindreds and linkage to the short arm of chromosome 2. J. Clin. Invest. **97:** 699–705.
27. MOUNTJOY, K.G., L.S. ROBBINS, M.T. MORTRUD & R.D. CONE. 1992. The cloning of a family of genes that encode the melanocortin receptors. Science **257:** 1248–1251.
28. LIGHT, K. et al. 1995. Are activating mutations of the adrenocorticotropin receptor involved in adrenal cortical neoplasia? Life Sci. **56:** 1523–1527.
29. LATRONICO, A.C. et al. 1995. No evidence for oncogenic mutations in the adrenocorticotropin receptor gene in human adrenocortical neoplasms. J. Clin. Endocrinol. Metab. **80:** 875–877.
30. CLARK, A.J., L. METHERELL, F.M. SWORDS & L.L. ELIAS. 2001. The molecular pathogenesis of ACTH insensitivity syndromes. Ann. Endocrinol. **62:** 207–211.
31. YOSHIMOTO, K. et al. 1993. Rare mutations of the Gs alpha subunit gene in human endocrine tumors. Mutation detection by polymerase chain reaction-primer-introduced restriction analysis. Cancer **72:** 1386–1393.
32. LYONS, J. et al. 1990. Two G protein oncogenes in human endocrine tumors. Science **249:** 655–659.
33. REINCKE, M., M. KARL, W. TRAVIS & G.P. CHROUSOS. 1993. No evidence for oncogenic mutations in guanine nucleotide-binding proteins of human adrenocortical neoplasms. J. Clin. Endocrinol. Metab. **77:** 1419–1422.
34. ARMELIN, H.A., C.F. LOTFI & A.P. LEPIQUE. 1996. Regulation of growth by ACTH in the Y-1 line of mouse adrenocortical cells. Endocrin. Res. **22:** 373–383.
35. LOTFI, C.F., Z. TODOROVIC, H.A. ARMELIN & B.P. SCHIMMER. 1997. Unmasking a growth-promoting effect of the adrenocorticotropic hormone in Y1 mouse adrenocortical tumor cells. J. Biol. Chem. **272:** 29886–29891.
36. MIYAMOTO, N. et al. 1992. A 3′,5′-cyclic adenosine monophosphate-dependent pathway is responsible for a rapid increase in c-fos messenger ribonucleic acid by adrenocorticotropin. Endocrinology **130:** 3231–3236.
37. LOTFI, C.F. et al. 2000. Proliferative signaling initiated in ACTH receptors. Braz. J. Med. Biol. Res. **33:** 1133–1140.

38. KAPAS, S. & J.P. HINSON. 1996. Inhibition of endothelin- and phorbol ester-stimulated tyrosine kinase activity by corticotrophin in the rat adrenal zona glomerulosa. Biochem. J. **313:** 867–872.
39. LEPIQUE, A.P., F.L. FORTI, M.S. MORAES & H.A. ARMELIN. 2000. Signal transduction in G0/G1-arrested mouse Y1 adrenocortical cells stimulated by ACTH and FGF2. Endoc. Res. **26:** 825–832.
40. FORTI, F.L. & A. ARMELIN. 2000. ACTH inhibits A Ras-dependent anti-apoptotic and mitogenic pathway in mouse Y1 adrenocortical cells. Endocr. Res. **26:** 911–914.
41. LE, T. & B.P. SCHIMMER. 2001. The regulation of MAPKs in Y1 mouse adrenocortical tumor cells. Endocrinology **142:** 4282–4287.
42. REINCKE, M. *et al.* 1997. Deletion of the adrenocorticotropin receptor gene in human adrenocortical tumors: implications for tumorigenesis. J. Clin. Endocrinol. Metab. **82:** 3054–3058.
43. MAZZOCCHI, G. *et al.* 2001. Orexin A stimulates cortisol secretion from human adrenocortical cells through activation of the adenylate cyclase-dependent signaling cascade. J. Clin. Endocrinol. Metab. **86:** 778–782.
44. TIAN, Y., T. BALLA, A.J. BAUKAL & K.J. CATT. 1995. Growth responses to angiotensin II in bovine adrenal glomerulosa cells. Am. J. Physiol. **268:** E135–144.
45. KAPAS, S., A. PURBRICK & J.P. HINSON. 1995. Role of tyrosine kinase and protein kinase C in the steroidogenic actions of angiotensin II, alpha-melanocyte-stimulating hormone and corticotropin in the rat adrenal cortex. Biochem. J. **305:** 433–438.
46. COTE, M., J. MUYLDERMANS, L. CHOUINARD & N. GALLO-PAYET. 1998. Involvement of tyrosine phosphorylation and MAPK activation in the mechanism of action of ACTH, angiotensin II and vasopressin. Endocrin. Res. **24:** 415–419.
47. WATANABE, G. *et al.* 1996. Angiotensin II activation of cyclin D1-dependent kinase activity. J. Biol. Chem. **271:** 22570–22577.
48. CHABRE, O. *et al.* 1995. Hormonal regulation of mitogen-activated protein kinase activity in bovine adrenocortical cells: cross-talk between phosphoinositides, adenosine 3′,5′-monophosphate, and tyrosine kinase receptor pathways. Endocrinology **136:** 956–964.
49. SMITH, R.D., A.J. BAUKAL, P. DENT & K.J. CATT. 1999. Raf-1 kinase activation by angiotensin II in adrenal glomerulosa cells: roles of Gi, phosphatidylinositol 3-kinase, and Ca2+ influx. Endocrinology **140:** 1385–1391.
50. TIAN, Y., R.D. SMITH, T. BALLA & K.J. CATT. 1998. Angiotensin II activates mitogen-activated protein kinase via protein kinase C and Ras/Raf-1 kinase in bovine adrenal glomerulosa cells. Endocrinology **139:** 1801–1809.
51. NUSSDORFER, G.G., G.P. ROSSI, L.K. MALENDOWICZ & G. MAZZOCCHI. 1999. Autocrine-paracrine endothelin system in the physiology and pathology of steroid-secreting tissues. Pharmacol. Rev. **51:** 403–438.
52. REBUFFAT, P. *et al.* 2001. Signaling pathways involved in the A and B receptor-mediated cortisol secretagogue effect of endothelins in the human adrenal cortex. Int. J. Mol. Med. **7:** 301–305.
53. ROSSI, G.P. *et al.* 2000. Endothelin-1 stimulates aldosterone synthesis in Conn's adenomas via both A and B receptors coupled with the protein kinase C- and cyclooxygenase-dependent signaling pathways. J. Invest. Med. **48:** 343–350.
54. BORNSTEIN, S.R., C.A. STRATAKIS & G.P. CHROUSOS. 1999. Adrenocortical tumors: recent advances in basic concepts and clinical management. Ann. Intern. Med. **130:** 759–771.
55. SEMPLICINI, A. *et al.* 2001. Adrenomedullin stimulates DNA synthesis of rat adrenal zona glomerulosa cells through activation of the mitogen-activated protein kinase-dependent cascade. J. Hypertens. **19:** 599–602.
56. ANDREIS, P.G. *et al.* 2000. Adrenomedullin enhances cell proliferation and deoxyribonucleic acid synthesis in rat adrenal zona glomerulosa: receptor subtype involved and signaling mechanism. Endocrinology **141:** 2098–2104.
57. WONG, Y. H. *et al.* 1991. Mutant alpha subunits of Gi2 inhibit cyclic AMP accumulation. Nature **351:** 63–65.
58. PACE, A.M., Y.H. WONG. & H.R. BOURNE. 1991. A mutant alpha subunit of Gi2 induces neoplastic transformation of Rat-1 cells. Proc. Natl. Acad. Sci. USA **88:** 7031–7035.

59. GICQUEL, C. et al. 1995. Oncogenic mutations of alpha-Gi2 protein are not determinant for human adrenocortical tumourigenesis. Eur. J. Endocrinol. **133:** 166–172.
60. YASHIRO, T. et al. 1994. Point mutations of ras genes in human adrenal cortical tumors: absence in adrenocortical hyperplasia. World J. Surg. **18:** 455–460; discussion 460–461.
61. LACROIX, A., H. MIRCESCU & P. HAMET. 1999. Clinical evaluation of the presence of abnormal hormone receptors in adrenal Cushing's syndrome. Endocrinologist **9:** 9–15.
62. LACROIX, A. et al. 1992. Gastric inhibitory polypeptide-dependent cortisol hypersecretion—a new cause of Cushing's syndrome. N. Engl. J. Med. **327:** 974–980.
63. LACROIX, A. et al. 1997. Propranolol therapy for ectopic beta-adrenergic receptors in adrenal Cushing's syndrome. N. Engl. J. Med. **337:** 1429–1434.
64. LACROIX, A. et al. 1997. Abnormal adrenal and vascular responses to vasopressin mediated by a V1- vasopressin receptor in a patient with adrenocorticotropin-independent macronodular adrenal hyperplasia, Cushing's syndrome, and orthostatic hypotension. J. Clin. Endocrinol. Metab. **82:** 2414–2422.
65. LACROIX, A. et al. 1998. Abnormal expression and function of hormone receptors in adrenal Cushing's syndrome. Endocr. Res. **24:** 835–843.
66. LACROIX, A., P. HAMET & J.M. BOUTIN. 1999. Leuprolide acetate therapy in luteinizing hormone-dependent Cushing's syndrome. N. Engl. J. Med. **341:** 1577–1581.
67. BOURDEAU, I. et al. 2001. Aberrant membrane hormone receptors in incidentally discovered bilateral macronodular adrenal hyperplasia with subclinical Cushing's syndrome. J. Clin. Endocrinol. Metab. **86:** 5534–5540.
68. HIRATA, Y. et al. 1981. Presence of ectopic beta-adrenergic receptors on human adrenocortical cortisol-producing adenomas. J. Clin. Endocrinol. Metab. **53:** 953–957.
69. MATSUKURA, S. et al. 1980. Multiple hormone receptors in the adenylate cyclase of human adrenocortical tumors. Cancer Res. **40:** 3768–3771.
70. KATZ, M.S. et al. 1985. Ectopic beta-adrenergic receptors coupled to adenylate cyclase in human adrenocortical carcinomas. J. Clin. Endocrinol. Metab. **60:** 900–909.
71. WEBER, M.M., C. FOTTNER & E. WOLF. 2000. The role of the insulin-like growth factor system in adrenocortical tumourigenesis. Eur. J. Clin. Invest. **30** (Suppl. 3): 69–75.
72. BASERGA, R. 2000. The contradictions of the insulin-like growth factor 1 receptor. Oncogene **19:** 5574–5581.
73. LEROITH, D., R. BASERGA, L. HELMAN & C.T. ROBERTS, JR. 1995. Insulin-like growth factors and cancer. Ann. Intern. Med. **122:** 54–59.
74. NAKAE, J., Y. KIDO & D. ACCILI. 2001. Distinct and overlapping functions of insulin and IGF-I receptors. Endocr. Rev. **22:** 818–835.
75. DE SOUZA, A.T. et al. 1995. M6P/IGF2R gene is mutated in human hepatocellular carcinomas with loss of heterozygosity. Nat. Genet. **11:** 447–449.
76. ILVESMAKI, V., A.I. KAHRI, P.J. MIETTINEN & R. VOUTILAINEN. 1993. Insulin-like growth factors (IGFs) and their receptors in adrenal tumors: high IGF-II expression in functional adrenocortical carcinomas. J. Clin. Endocrinol. Metab. **77:** 852–858.
77. GICQUEL, C. et al. 1994. Rearrangements at the 11p15 locus and overexpression of insulin-like growth factor-II gene in sporadic adrenocortical tumors. J. Clin. Endocrinol. Metab. **78:** 1444–1453.
78. LIU, J. et al. 1995. H19 and insulin-like growth factor-II gene expression in adrenal tumors and cultured adrenal cells. J. Clin. Endocrinol. Metab. **80:** 492–496.
79. WEBER, M.M., C.J. AUERNHAMMER, W. KIESS & D. ENGELHARDT. 1997. Insulin-like growth factor receptors in normal and tumorous adult human adrenocortical glands. Eur. J. Endocrinol. **136:** 296–303.
80. BOULLE, N. et al. 1998. Increased levels of insulin-like growth factor II (IGF-II) and IGF-binding protein-2 are associated with malignancy in sporadic adrenocortical tumors. J. Clin. Endocrinol. Metab. **83:** 1713–1720.
81. WEBER, M.M. et al. 1999. Postnatal overexpression of insulin-like growth factor II in transgenic mice is associated with adrenocortical hyperplasia and enhanced steroidogenesis. Endocrinology **140:** 1537–1543.

82. WEINSTEIN, L.S., S. YU, D.R. WARNER & J. LIU. 2001. Endocrine manifestations of stimulatory G protein alpha-subunit mutations and the role of genomic imprinting. Endocr. Rev. **22:** 675–705.
83. WILKIN, F. *et al.* 2000. Pediatric adrenocortical tumors: molecular events leading to insulin-like growth factor II gene overexpression. J. Clin. Endocrinol. Metab. **85:** 2048–2056.
84. LIU, J., A.I. KAHRI, P. HEIKKILA & R. VOUTILAINEN. 1997. Ribonucleic acid expression of the clustered imprinted genes, p57KIP2, insulin-like growth factor II, and H19, in adrenal tumors and cultured adrenal cells. J. Clin. Endocrinol. Metab. **82:** 1766–1771.
85. DOWNWARD, J. *et al.* 1984. Close similarity of epidermal growth factor receptor and v-erb-B oncogene protein sequences. Nature **307:** 521–527.
86. EDGREN, M. *et al.* 1997. Biological characteristics of adrenocortical carcinoma: a study of p53, IGF, EGF-r, Ki-67 and PCNA in 17 adrenocortical carcinomas. Anticancer Res. **17:** 1303–1309.
87. KAMIO, T. *et al.* 1990. Immunohistochemical expression of epidermal growth factor receptors in human adrenocortical carcinoma. Hum. Pathol. **21:** 277–282.
88. SASANO, H. *et al.* 1994. Transforming growth factor alpha, epidermal growth factor, and epidermal growth factor receptor expression in normal and diseased human adrenal cortex by immunohistochemistry and in situ hybridization. Mod. Pathol. **7:** 741–746.
89. MICHL, P., T. BEIKLER, D. ENGELHARDT & M.M. WEBER. 2000. Interleukin-3 and interleukin-6 stimulate bovine adrenal cortisol secretion through different pathways. J. Neuroendocrinol. **12:** 23–28.
90. WEBER, M.M., P. MICHL, C.J. AUERNHAMMER & D. ENGELHARDT. 1997. Interleukin-3 and interleukin-6 stimulate cortisol secretion from adult human adrenocortical cells. Endocrinology **138:** 2207–2210.
91. WILLENBERG, H.S. *et al.* 1998. Aberrant interleukin-1 receptors in a cortisol-secreting adrenal adenoma causing Cushing's syndrome. N. Engl. J. Med. **339:** 27–31.
92. SCHTEINGART, D.E. *et al.* 2001. Overexpression of CXC chemokines by an adrenocortical carcinoma: a novel clinical syndrome. J. Clin. Endocrinol. Metab. **86:** 3968–3974.
93. MALKIN, D. *et al.* 1990. Germ line p53 mutations in a familial syndrome of breast cancer, sarcomas, and other neoplasms. Science **250:** 1233–1238.
94. BARZON, L. *et al.* 2001. Molecular analysis of CDKN1C and TP53 in sporadic adrenal tumors. Eur. J. Endocrinol. **145:** 207–212.
95. RIBEIRO, R.C. *et al.* 2001. An inherited p53 mutation that contributes in a tissue-specific manner to pediatric adrenal cortical carcinoma. Proc. Natl. Acad. Sci. USA **98:** 9330–9335.
96. MOUL, J.W., J.T. BISHOFF, S.M. THEUNE & E.H. CHANG. 1993. Absent ras gene mutations in human adrenal cortical neoplasms and pheochromocytomas. J. Urol. **149:** 1389–1394.
97. LIN, S.R., TSAI, J. H., Y.C. YANG & S.C. LEE. 1998. Mutations of K-ras oncogene in human adrenal tumours in Taiwan. Br. J. Cancer **77:** 1060–1065.
98. YANO, T. *et al.* 1989. Genetic changes in human adrenocortical carcinomas. J. Natl. Cancer Inst. **81:** 518–523.
99. LIN, S.R. *et al.* 2000. Decreased GTPase activity of K-ras mutants deriving from human functional adrenocortical tumours. Br. J. Cancer **82:** 1035–1040.
100. HSU, C. H. *et al.* 2001. Significantly increased cortisol secretion in normal adrenocortical cells transfected with K-ras mutants derived from human functional adrenocortical tumors. DNA Cell. Biol. **20:** 231–238.
101. KIIVERI, S. *et al.* 1999. Reciprocal changes in the expression of transcription factors GATA-4 and GATA-6 accompany adrenocortical tumorigenesis in mice and humans. Mol. Med. **5:** 490–501.
102. PERI, A. *et al.* 2001. Variable expression of the transcription factors cAMP response element-binding protein and inducible cAMP early repressor in the normal adrenal cortex and in adrenocortical adenomas and carcinomas. J. Clin. Endocrinol. Metab. **86:** 5443–5449.

103. ANDO, T. et al. 2000. Expression and regulation of BCL-2 family genes in human adrenocortical adenomas in comparison with adrenal hyperplasia of Cushing's disease. Endocr. Res. **26:** 853–859.
104. SARKAR, D. et al. 2001. The human homolog of Diminuto/Dwarf1 gene (hDiminuto): a novel ACTH-responsive gene overexpressed in benign cortisol-producing adrenocortical adenomas. J. Clin. Endocrinol. Metab. **86:** 5130–5137.
105. GUTKIND, J.S. 1998. The pathways connecting G protein-coupled receptors to the nucleus through divergent mitogen-activated protein kinase cascades. J. Biol. Chem. **273:** 1839–1842.
106. DUMONT, J.E., F. PECASSE & C. MAENHAUT. 2001. Crosstalk and specificity in signalling. Are we crosstalking ourselves into general confusion? Cell. Signalling **13:** 457–463.

Cyclic AMP–Dependent Signaling Aberrations in Macronodular Adrenal Disease

ISABELLE BOURDEAU AND CONSTANTINE A. STRATAKIS

Unit on Genetics and Endocrinology (UGEN), Developmental Endocrinology Branch, National Institute of Child Health and Human Development, Bethesda, Maryland 20892-1862, USA

ABSTRACT: The adrenal glands are a major source of steroid hormone biosynthesis. In normal physiology, the pituitary hormone corticotropin (ACTH) regulates the secretion of glucocorticoids via its G protein–coupled receptor (ACTHR), the product of the *MC2R* gene. Aldosterone is another major product of the adrenal gland; its regulation is controlled mainly by the renin-angiotensin system, although ACTH plays a role, too, especially under certain pathological conditions. The adrenal gland also secretes lesser amounts of androgens and intermediate metabolites of all these steroids. Unregulated secretion of any of these hormones can be caused by tumors, adrenocortical adenomas or carcinomas, and/or bilateral (or, rarely, unilateral) hyperplasia. Cortisol-producing hyperplasia of the adrenal glands is caused by two distinct syndromes, both of which have been directly or indirectly associated with protein kinase A signaling: (i) primary pigmented nodular adrenocortical disease (PPNAD) (a micronodular form of bilateral adrenal hyperplasia), either isolated (rarely) or in the context of Carney complex, is caused (in most cases) by mutations of the *PRKAR1A* gene; and (ii) ACTH-independent macronodular adrenal hyperplasia (AIMAH), or massive macronodular adrenal disease (MMAD), has been associated with aberrant (ectopic) expression, and presumably regulation, of various G protein–coupled receptors. AIMAH is a rare, sporadic condition affecting predominantly middle-aged men and women with an almost equal ratio (the latter in contrast to other forms of endogenous Cushing's syndrome). Some familial cases of AIMAH have also been described, and it appears that the pathophysiological phenomena underlying AIMAH may be present in the far more common, sporadic adrenocortical tumors and, perhaps, in the nodular growth detected in the adrenal glands of the elderly in the general population. Thus, the study of ectopic receptor expression and cAMP-dependent PKA activity in AIMAH may have wider implications for adrenal and, indeed, endocrine tumorigenesis.

KEYWORDS: cyclic AMP; signaling; benign adrenal tumors; macronodular hyperplasia; GIP; GIPR

Address for correspondence: Constantine A. Stratakis, M.D., D.Sc., Chief, Unit on Genetics and Endocrinology, DEB, NICHD, NIH, Building 10, Room 10N262, 10 Center Dr. MSC1862, Bethesda, Maryland 20892-1862.Voice: 301-496-4686/402-1998; fax: 301-435-4358.
stratakc@cc1.nichd.nih.gov

INTRODUCTION

Adrenocortical tumors (ACT) may produce cortisol and lead to corticotropin (ACTH)-independent Cushing's syndrome (CS). Approximately, 90% of these tumors are benign unilateral adenomas. Adrenocortical carcinoma is rare, accounting for less than 1% of all ACTs. The remaining 10% of all ACTH-independent ACTs causing CS are due to two forms of bilateral hyperplasia: primary pigmented nodular adrenocortical disease (PPNAD) and ACTH-independent macronodular adrenal hyperplasia (AIMAH) or massive macronodular adrenocortical disease (MMAD).[1] Macronodular adrenal hyperplasia had not been well recognized as a distinct primary adrenal disease until the beginning of the 1990s.[2] This lack of recognition was, most likely, caused by the fact that nodular adrenocortical hyperplasia may be present in ACTH-dependent Cushing's disease. Indeed, chronic adrenal stimulation by ACTH leads to bilateral adrenocortical hyperplasia and, if long-standing, nodular transformation.[3–5] However, the adrenocortical changes seen in AIMAH are distinct from ACTH-induced hyperplasia.[6] These differences may be related to the activation of different signaling pathways in AIMAH.[7] Along these lines, over the last 15 years, *in vivo* and *in vitro* studies have shown the activation of ectopic hormonal signaling in adrenocortical tissue affected by AIMAH (reviewed in ref. 7 by Lacroix *et al.*). In the present report, we review these clinical and laboratory findings and speculate that alternate PKA signaling and interactions with other factors eutopically and/or ectopically expressed in adrenocortical tissue may account for the distinct features of AIMAH/MMAD.

IDENTIFICATION AND CLINICAL FEATURES OF AIMAH

In 1964, Kirschner *et al.*[8] described a 40-year-old woman with long-standing CS. Although her disease was not ACTH dependent, her testing showed hyperresponsiveness to ACTH. Her glucocorticoid production was not suppressed by the administration of dexamethasone, and retroperitoneal air insufflation revealed bilateral suprarenal masses. The patient underwent bilateral adrenalectomy; her adrenal glands had multiple nodules and a combined weight of 94 g (the normal weight of adrenal glands is usually 8–12 g).

Since 1964, this patient's disease has been seen in several patients and described under different names including AIMAH, MMAD, autonomous macronodular adrenal hyperplasia, ACTH-independent massive bilateral adrenal disease, and "giant" or "huge" macronodular adrenal disease.[6] By 1994, 24 patients with AIMAH had been described in the literature.[9] Several other cases have been described since.[10–12] In addition, bilateral adrenal nodules or tumors have been reported to be present in as many as 10–15% of the patients with adrenal masses that are discovered incidentally ("incidentalomas").[13–17]

AIMAH seems to have a bimodal age distribution: a minority of the patients present during the first year of the life, and this form of the disease may be associated with McCune-Albright syndrome;[18] whereas the majority present in the fifth decade of life.[9] The male:female ratio in the largest recent series of 24 cases was almost 1:1, which contrasts with the usual female predominance in other etiologies of endogenous CS.[9]

The most frequent clinical presentation of AIMAH is CS. Cases of AIMAH with excess of glucomineralocorticoid[19] and cortisol-estrone secretion[20] have also been reported. As mentioned above, patients with bilateral incidentalomas may also be detected in the process of investigating another disease or trauma cases; most of these patients have some evidence of cortisol or other hormone-secretion abnormality upon dynamic testing; or, if they are followed long term, they develop evidence of some form of CS (atypical, periodic, subclinical, or preclinical CS).[17,21,22]

In macronodular adrenal hyperplasia, steroid hormone secretion is ACTH independent; most of the time, plasma ACTH levels are undetectable, and high-dose dexamethasone administration fails to suppress cortisol secretion. In a recent series of nine patients with AIMAH who were followed for up to 8.5 years after bilateral adrenalectomy, no patient developed Nelson's syndrome, another indication that AIMAH is not caused by ACTH excess.[10] Experimental evidence supports this notion: Cheitlin et al. studied in vitro cells from a patient with AIMAH; the cells were cultured on an extracellular matrix and demonstrated rapid growth and a high rate of cortisol secretion in the absence of ACTH.[23] It is noteworthy, however, that the ACTH receptor gene (MC2R) is expressed in AIMAH tissue, and most patients with AIMAH respond to ACTH with variable increases of baseline cortisol levels, depending on the size of the adrenocortical mass.[12,17,24] This is in contrast to PPNAD, sporadic large benign adenomas, and adrenocortical carcinomas, which are mostly ACTH unresponsive.[25,26]

ACTH responsiveness is not the only difference between adrenocortical cancer and AIMAH. The latter is a benign process that has never, to our knowledge, been shown to either metastasize or transform to cancer. Histology is also distinct: the nodules are composed of two types of cells, those with clear cytoplasm (lipid-rich) that form cordon nest-like structures, and those with a compact cytoplasm (lipid-poor) that form small nest or island-like structures.[2, 27]

Diagnosis of AIMAH is almost always based on imaging studies (computed tomography preferably, but also magnetic resonance imaging) in a patient with the biochemical demonstration of ACTH-independent CS. Final diagnosis is made with the histological examination of the tissue (see above) (FIG. 1). Surgical treatment is recommended for most patients with CS and AIMAH, although it is questionable whether the adrenalectomy should always be bilateral: unilateral or partial adrenalectomies have also been performed in some patients with, reportedly, good results (although almost never cure, since the disease is, by definition, bilateral). This treatment (unilateral adrenalectomy in bilateral adrenal hyperplasia) has been tried with some success in several patients with idiopathic hyperaldosteronism and bilateral enlargement of the adrenal glands.

STEROIDOGENESIS IN AIMAH

Overall, steroid hormone synthesis in AIMAH is an inefficient process; the total increased gluco- or mineralocorticoid hormone secretion is the result of a significant increase in adrenocortical mass; consequently, it is the increased number of adrenocortical cells that accounts for hormonal overproduction, and not augmented synthesis per cell.[24,27,28] Inefficient hormonosynthesis in AIMAH is, perhaps, the result of altered steroidogenic enzymatic pathways: immunohistochemical studies showed

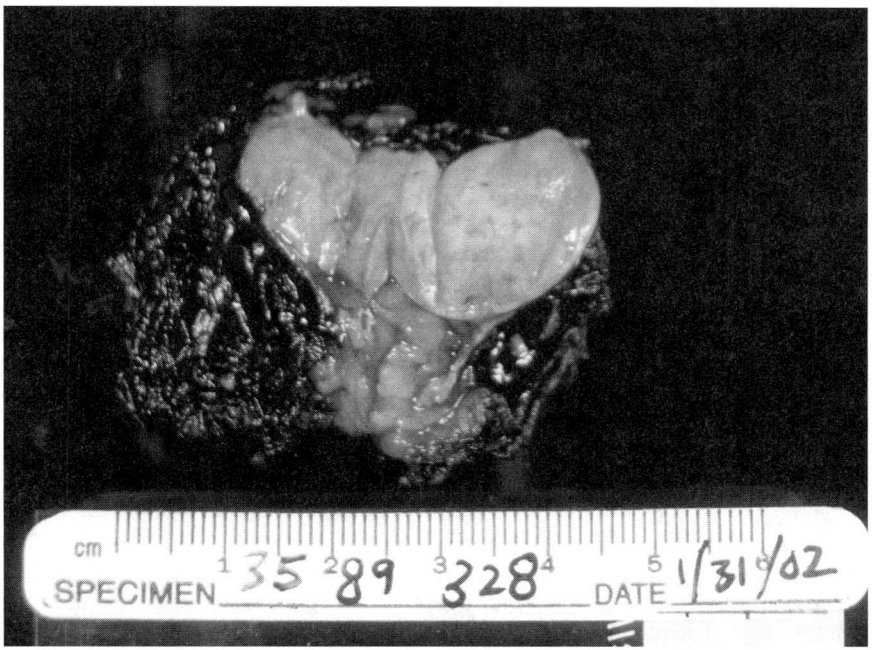

FIGURE 1. Macronodular adrenocortical hyperplasia of the right adrenal gland of a patient with Cushing's syndrome secondary to AIMAH. The gland is enlarged and consists of multiple nodules of different sizes.

that 3β-HSD2 was expressed only in large clear cells, whereas CYP17 (P-450C17) was seen predominantly in small compact cells. The differential expression of these two steroidogenic enzymes has been observed only in AIMAH and not in other adrenocortical diseases.[28] Immunoreactivities for CYP11A1 (P-450scc), CYP21A2 (P-450c21), and CYP11B2 (P-450C11) were present in both cell types.[27,29,30]

GENETICS

Most cases of AIMAH appear to be sporadic; however, the number of reported patients remains small, and for most of them family screening has not been performed. There are few reports of familial cases of AIMAH; their presentation suggests an autosomal dominant transmission.[11,31–33]

AIMAH has also been described as part of multiple endocrine neoplasia syndrome type 1 (MEN 1), which is caused by the tumor suppressor gene *menin*.[34] In 1992, Skogseid *et al.* investigated 33 MEN 1 patients and demonstrated bilateral adrenal enlargement in 7 of them (21%). More than half of the remaining patients also had unilateral adrenal disease.[35] More recently, Burgess *et al.* reviewed the prevalence of adrenal lesions occurring in 33 patients from a single kindred with MEN 1.[36] The overall prevalence of adrenal disease was 36%; bilateral macronodular cortical hyperplasia was present in 6%. The authors reported that an additional member of the pedigree was found at autopsy to have bilateral macronodular cortical hyperplasia. More recently, a 30% prevalence of adrenal disease was found in an Italian cohort of MEN 1 patients,[37] although a much lower incidence was reported in Swiss kindreds.[38] All cases were hormonally silent; this appears to be the case in the vast majority of patients with MEN 1; a few patients with MEN 1 and ACTH-independent CS are known to us (Stratakis *et al.*, unpublished observation), but they seem to be the exception rather than the rule. AIMAH has also been described in McCune-Albright syndrome (MAS). MAS may present in infancy or early childhood with CS that is characterized by the presence of Gs_α-activating mutations in affected tissue.[39]

In 1994, C. Gicquel *et al.* studied clonal composition of different nodules in adrenal tissue of a patient with CS secondary to AIMAH.[40] Different clonal patterns were found both between glands and within the same gland; in addition, clonality differed between nodules, too. Monoclonal adenomas from the same adrenal gland did not arise from the same progenitor cell. The authors proposed that in AIMAH different stages of a common multistep tumoral process may be present in different locations at the same time, as suggested by the heterogeneous clonal pattern.[41] F. Beuschlein *et al.* further supported this hypothesis by reporting polyclonal and monoclonal areas in adrenocortical tissue from a 32-year-old woman with AIMAH, whereas patients with ACTH-dependent macronodular adrenal disease had an exclusively polyclonal pattern.[42]

The product of the tumor suppressor gene *TP53*, p53, is known to be widely involved in adrenal tumorigenesis.[43,44] Recently, Latronico *et al.* studied 55 patients with benign and malignant adrenocortical tumors from Brazil and found that 35% of them had a *TP53* mutation (Arg337His) that had been previously identified to be highly prevalent in pediatric patients with adrenal cortical carcinoma from southern Brazil.[45,46] One 45-year-old woman with bilateral nonfunctioning adrenal tumor was included in the study and did not show the mutation.[45] In addition, Wachenfeld *et al.* studied three cases of AIMAH and found no aberrant p53 expression by immunohistochemistry; interestingly, however, one of these patients showed loss-of-heterozygosity (LOH) of the 17p-*TP53* locus.[47]

INVESTIGATION OF cAMP SIGNALING IN AIMAH

Normally, hypothalamic CRH (and vasopressin) stimulate pituitary ACTH secretion.[48] ACTH, in turn, activates cAMP synthesis in the adrenal cortex, through its binding to the ACTHR (a G protein–coupled receptor), the subsequent activation of Gs_α subunit of heterotrimeric Gs protein, and stimulation of adenylyl cyclase isoforms 2/4 and 5/6.[49] cAMP activates protein kinase A (PKA), which phosphorylates nuclear transcription factors, including cAMP response element modulator (CREM), cAMP response element binding protein (CREB), and activating transcription factor-1 (ATF-1).[50,51] These transcription factors bind to the cAMP response element (CRE) sequence and modulate the expression of target genes.

Investigation of the *MC2R* gene has not shown any mutations in three patients with nodular hyperplasia.[52,53] There is one patient in whom an activating mutation

TABLE 1.

	Sex	Age	Clinical Picture	McCune-Albright manifestations	Adrenal disease	ACTH levels	Treatment	$G_s\alpha$ mutations
Aaskog, 1968[107]	F	4 months	Cushingoid	SP, FD, sexual precocity	bilateral nodular hyperplasia	NR	bilateral adrenalectomy	NR
Macmahon, 1971[108]	F	10 years	Reported at autopsy	SP, FD, sexual precocity, pituitary nodular hyperplasia	bilateral nodular hyperplasia	NR	diagnosis at autopsy	NR
Benjamin, 1973[55]	M	17 years	Cushingoid	SP, FD, nodular goiter, pituitary chromophobe adenoma, Leydig cell hyperplasia	L: adenoma R: nodular hyperplasia	NR	unilateral left adrenalectomy	A201C
Danon, 1975[57]	F	6 months	Cushingoid	SP, FD, sexual precocity	bilateral nodular hyperplasia	suppressed	bilateral adrenalectomy	A201H
Mauras, 1986[56]	M	3.7 years	poor growth	SP, FD, sexual precocity, acromegaly, hyperthyroidism	bilateral nodular hyperplasia	suppressed	bilateral adrenalectomy	A201H
Yoshimoto, 1991[109]	F	2.3 months	Cushingoid	SP, FD, hyperthyroidism	bilateral nodular hyperplasia	suppressed	dead at 4.4 months from cardiac failure	NR
Schenker, 1993[110]	M	10 months	Cushingoid	SP, FD, hyperthyroidism	bilateral hyperplasia	suppressed	bilateral adrenalectomy	A201H
	F	2 months	growth failure	SP, FD, sexual precocity, hyperthyroidism	bilateral nodular hyperplasia	NR	bilateral adrenalectomy	A201C
	M	17 years	reported at autopsy	SP, FD, pituitary chromophobe adenoma	adrenal nodular hyperplasia	NR	diagnosis at autopsy	A201H
Kirk, 1999[18]	F	3 months	Cushingoid	SP, FD, sexual precocity, thyrotoxicosis	bilateral nodular hyperplasia	suppressed	bilateral adrenalectomy	NR
	F	5 months	Cushingoid	SP, FD, sexual precocity, thyrotoxicosis	bilateral nodular hyperplasia	NR	bilateral adrenalectomy	NR
	F	3 months	Cushingoid	SP, FD		suppressed	spontaneous resolution[a]	NR
	M	4 months	Cushingoid	SP, FD	bilateral nodular hyperplasia	suppressed	bilateral adrenalectomy	NR
	F	3 months		SP, FD, sexual precocity	scan: bilateral enlarged adrenals	normal limits	spontaneous resolution[b]	NR

ABBREVIATIONS: ACTH: adrenocorticotropin hormone; NR: not reported; SP: skin pigmentation; FD: fibrous dysplasia.
[a] Lack of suppression with dexamethasone and lack of diurnal variation in cortisol production at 6 months, but Cushingoid appearance disappeared.
[b] Paradoxical rise of serum and urine cortisol levels after dexamethasone administration.

TABLE 2. Cases of macronodular adrenal hyperplasia not associated with McCune-Albright syndrome screened for G protein mutations

	Patients screened (n)	Gs_α Exon 8	Gs_α Exon 9	$Gi2_\alpha$ Exon 5	$Gi2_\alpha$ Exon 6
Boston, 1994[58]	1	A201C (n:1)	0	0	0
Lieberman, 1994[9]	1	0	0	ND	ND
Fragoso, 1999[111]	2	A201S (n:1)	0	0	0
Kobayashi, 2000[112]	1	0	0	ND	ND
Fragoso, 2001[113]	4	A201S (n:1) A201H (n:1)	0	0	0

ABBREVIATIONS: ND: not done; A: arginine; C: cysteine; H: histidine; S: serine.

of this gene may be present; his case was presented in an abstract format, and he had intermittent ACTH-independent CS and relatively small enlargement of the adrenal glands upon imaging.[54]

In 1991, two activating mutations were detected in four patients with MAS. Histidine was substituted for arginine at position 201 of Gs_α, and cysteine was substituted for the same arginine residue.[39] Mutations of Gs_α subunit are localized in its GTPase domain and cause constitutive activation of adenylyl cyclase by inhibition of the intrinsic GTPase activity. Adrenal tumors were present in three patients, and G-protein mutations were shown in two of them (1 adenoma and 2 adrenocortical nodular hyperplasia).[55–57] The association of MAS with adrenal hyperplasia is probably more frequent than what has been recognized previously. At least 12 cases confirmed by histology have been reported in the literature; these are summarized in TABLE 1. In 1994, Boston et al. described an infant with MAS, ACTH-independent CS, and bilateral adrenal hyperplasia.[58] This infant was found to have an activating mutation in Gs_α (arginine201cysteine). Asymptomatic lesions of fibrous dysplasia were detected later on bone scan, which may be consistent with diagnosis of McCune-Albright syndrome. However, there is only a small number of patients with macronodular adenal hyperplasia unrelated to McCune-Albright who have been tested for protein G mutation; mutations were found in more than 40% of them (TABLE 2).

Lyons et al. identified mutations in adrenal tumors in another alpha-subunit subfamily protein, the Gi_α. The mutation was present in codon 179 of the Gi_α subunit in 2 out of 5 adrenal adenomas and 1 out of 6 adrenal carcinomas.[59] However, furthers studies failed to identify this mutation in AIMAH or adrenal adenoma or carcinomas.[60–63] Although Gs_α or Gi_α mutations are infrequent in other adrenal lesions, further studies are needed to evaluate its prevalence in AIMAH.

Other oncogenes may be involved in adrenal tumorigenesis. In 1998, a group from Taiwan screened 15 functional unilateral tumors for mutation in the H-ras and K-ras gene. Mutations of the K-ras gene were detected in 7 tumors.[64] Further studies showed that normal adrenocortical cells transfected with K-ras mutants significantly increased cortisol secretion,[65] and K-ras mutation decreased GTPase activity.[66]

PATHWAYS INVOLVED IN ADRENAL TUMORIGENESIS

The previous discussion highlighted the involvement of the cAMP-pathway in MMAD pathogenesis. There is some evidence that cAMP-dependent pathways may be involved in tumor formation. For example, mutationally activated G proteins have been identified in pituitary adenomas and thyroid tumors (see other chapters in this book). Persistent adenylyl cyclase activation secondary to Gs_α mutation induces cAMP, production and it has been shown that cAMP stimulates the growth and hormone secretion of pituitary somatrotrophs and thyroid cells.[67] In the adrenal cortex, ACTH regulates steroid synthesis;[68] however its role in cell proliferation is less clear. *In vivo* ACTH seems to have proliferative effects. For example, as mentioned earlier, in patients with increased ACTH production—for example, Cushing's disease or paraneoplastic ectopic ACTH secretion—adrenal glands may become hyperplastic and significantly enlarged.[5] Meanwhile, *in vitro* studies demonstrated that ACTH promotes a growth-inhibiting effect on cell proliferation.[69] In 1997, Lofti *et al.* reported for the first time a weak *in vitro* mitogenic response secondary to ACTH in cultures of the mouse Y1 adrenocortical tumor cell line.[70,71] The study suggested that mitogenicity was independent of cAMP/PKA, because it was also found in PKA-deficient mutants of the same cell line. Other studies showed that the mitogen-activated protein kinase (MAPK) pathway may be involved in cell proliferation. The ras-dependent pathway may activate signaling of the mitogen-activated protein kinase (MAPK). The protein kinase RAF is increased and stimulates ERK1 (p44MAPK) and ERK2 (p44MAP). Transcription factors *fos* and *jun* are also increased and modify the cell cycle from the G0/G1 phase to S phase.[72–74] Recently, Le and Schimmer suggested that PKA has a negative influence on MEK activation from the MAP kinase pathway.[75]

All these data taken together provide evidence that the G protein–coupled receptor signaling pathway is connected to nuclear gene expression by the MAPK cascade, at least in Y1 mouse adrenocortical tumor cells. PKA may phosphorylate *Raf* kinase, which, in turn, decreases its downstream effectors and results in a decrease of cellular proliferation.[76]

ABERRANT RECEPTORS IN ADRENAL TUMORIGENESIS

What leads to excess steroidogenesis in primary adrenal CS (in the absence of detectable levels of ACTH)? There is increasing evidence that adrenocortical steroidogenesis may be mediated by non-ACTH circulating hormones for which their respective receptors are expressed in the adrenal tumors. The two first *in vivo* well-characterized examples of aberrant cortisol regulation in AIMAH were reported in 1992. Two patients had food-dependent CS secondary to AIMAH, in which cortisol production was mediated by the postprandial physiological increase in plasma levels of GIP. GIP would then bind to GIP receptors ectopically localized in the adrenal cortex.[77,78] To date, food-dependent CS has been identified in at least 11 AIMAH patients and in 7 with adrenal adenomas.[7] Aberrant stimulation of cortisol in response to exogenous arginine-vasopressin or lysine-vasopressin,[79–82] catecholamines,[83,84] LH/hCG,[17,85] serotonin receptor agonists,[17,85] angiotensin II,[86] and leptin[87] have also been described in AIMAH.

The presence of abnormal hormone receptors was found to be more frequently present in AIMAH than in adrenal adenomas or carcinomas.[84,88] Three recent studies screened 15 patients with bilateral AIMAH *in vivo*, 11 with CS and 4 with subclinical CS for the presence of aberrant response to different non-ACTH stimuli. They demonstrated that all patients showed at least one aberrant cortisol regulation.[17,84,88] *In vitro* evidence of the involvement of aberrant receptors in adrenal tumorigenesis and of specific studies in AIMAH will be discussed next.

The first indication of tumor stimulation by aberrant hormone receptor was in a corticosterone-producing rat adrenocortical carcinoma.[89] Adenyl cyclase activity was stimulated by ACTH, epinephrine, norepinephrine, thyroid-stimulating hormone (TSH), FSH, and LH. The tumor cyclase was not responsive to angiotensin II, vasopressin, glucagon, insulin, growth hormone, parathyroid hormone, or thyrocalcitonin. Only ACTH stimulated normal control adrenal tissue. A noteworthy finding was that there were no additive effects of different hormones that would all stimulate adenylyl cyclase in neoplastic cells, in addition to ACTH. This observation suggested that the different hormone receptors act through a common cyclase catalytic unit.[90] In 1980, Matsukura *et al.* reported *in vitro* adenylate cyclase responses to different hormones in human adrenocortical tumors. Of five adrenocortical adenomas (all functional), three were stimulated by norepinephrine, two by epinephrine, one by TSH, and one by LH in addition to ACTH and NaF. An adrenal carcinoma was stimulated by PGE and NaF but not by ACTH. ACTH, glucagon, and NaF activated an ACTH-independent primary adrenocortical nodular hyperplasia tumor. An interesting part of the study was the absence of abnormal hormone receptor responses in two ACTH-dependent hyperplastic adrenocortical tissues and in two normal human adrenals. This suggested that the presence of altered hormone receptors in adrenal tumors is probably not secondary to ACTH secretion or increased glucocorticoid production itself.

Aberrant adrenal receptors may reflect functional cell surface alterations secondary to the tumoral transformation with loss of the hormonal specificity of cell surface receptors.[25] K. Nomata *et al.* studied the expression of proliferating cell nuclear antigen (PCNA) by immunohistochemical staining in a 59-year-old man with AIMAH. PCNA is a nuclear protein that is involved in DNA synthesis. PCNA was found mostly in the epithelial cells of the adrenocortical region, but not in the interstitial cells where relatively little staining was observed in normal adrenal gland cells. These data confirmed that some hormonal factor that is specific for adrenal cortical cell growth might be involved in AIMAH.[91]

MOLECULAR STUDIES IN ADRENAL TUMORS WITH ABERRANT HORMONAL RECEPTORS

Most receptors aberrantly coupled to adrenal steroidogenesis in adrenal adenomas or macronodular hyperplasia belong to the serpentine receptor family of G protein–coupled receptors.[7] In food-dependent CS, clinical observations suggested that an abnormal responsiveness to GIP may be involved.[77,78,92–98] Serum cortisol concentration increased after meals and after infusion of GIP, whereas there was no response in normal healthy persons. Overexpression of GIP receptors has been demonstrated in adrenals of GIP-dependent CS patients compared to normal adrenal tissues[93,95,97] and non-GIP-dependent adrenal CS.[92,93,95] Analysis of the coding sequence of GIP receptor in GIP-dependent CS identified no mutations. However, alternatively spliced isoforms of the receptor were found in GIP-dependent and normal adrenal tissues.[93] Lebrethon et al. showed that cultured GIP-dependent tumor adrenal cells responded to GIP with cortisol secretion in a dose-dependent manner to GIP (as is the case with ACTH). ACTH and GIP did not show an additive effect when combined.[95] When these two hormones were applied together in tumor cells of a food-dependent, cortisol-secreting adenoma, they also stimulated cortisol production (in addition to cAMP) but did not lead to inositol 1,4,5-triphosphate IP3 generation. GIP and ACTH also stimulated DNA synthesis and p42–p44 MAP kinase activity; levels of ACTH receptor mRNA, on the other hand, were low.[96,97]

In a case of AIMAH and CS, cortisol and aldosterone were responsive to elevation of plasma catecholamines. High-affinity binding sites compatible with β_1- or β_2-adrenergic receptors were found in the adrenal tissue from this case but not in control adrenals. They were efficiently coupled to steroidogenesis as shown by adenylyl cyclase stimulation with isoproterenol.[83] In vitro studies from patients with AIMAH and vasopressin responsiveness showed that the cortisol response was mediated by V1-vasopressin receptor. However, V1-vasopressin receptors are normally present in adrenal cortex; thus, the mechanism involved here remains to be elucidated.[7,79,80]

In another patient with CS secondary to AIMAH, cortisol production was regulated by LH/hCG and serotonin 5-HT_4 agonists. At least three splice variants of the LH/hCG receptor and three isoforms of the 5-HT_4 receptors were found to be expressed in her adrenal tissue.[99] Recently, Bugalho et al. described a 37-year-old woman with CS secondary to an adrenal adenoma in which cortisol secretion was increased by hCG injection.[100] LH/hCG receptors were revealed by immunohistochemistry in the adrenal's tissue, and a heterozygous Gsp mutation at codon 201 (A201C) was found.[100] It is noteworthy that mice transgenic for bLHβ-CTP do get CS secondary to bilateral adrenal hyperplasia. These mice express ectopic adrenal LH/hCG receptors, which are not found in control animals.[101]

However, even if there is in vivo and in vitro evidence that these different receptors are coupled to the adrenal steroidogenesis in AIMAH, little is known about their effect on cell proliferation. Their role in the process of adrenal tumorigenesis remains to be elucidated.

cAMP-DEPENDENT PATHWAY IN NON-AIMAH ADRENAL TUMORS

The involvement of the cAMP-dependent pathway in adrenal tumorigenesis has recently been demonstrated by the identification of the gene for Carney complex.[102] Carney complex is characterized by spotty skin pigmentation, myxomas, schwannomas, and tumors of two or more endocrine glands, including primary pigmented nodular adrenocortical disease, a rare form of benign adrenal tumor (reviewed elsewhere in this book).[103] Mutations of the gene encoding the protein kinase type I-α regulatory subunit have been detected in 44% of familial and 35% of sporadic cases of Carney complex.[104] All the reported *PRKAR1A* mutations lead to the production of a truncated protein product, which, in turn, leads to enhanced intracellular signaling by PKA and a greater response to cAMP.[102] Recently, two teams have been interested in characterizing the role of PKA-dependent transcription factor in adrenocortical tumor cells. Groussin *et al.* demonstrated that even if CREB is not expressed in the H295R cell line (human adrenocortical carcinoma), intracellular transcriptional regulation of CRE is preserved by the compensatory activity of CREM.[105] More recently, the study of 8 human adrenocortical adenomas and 8 adrenocortical carcinomas demonstrated that CREB was not transcribed in 4 adrenal carcinomas, but was detectable in all normal adrenal tissues and adenomas. The authors suggested that the lack of expression of CREB in adrenal carcinomas may reflect a loss of cell differentiation.[106]

CONCLUSIONS

In summary, AIMAH is a rare form of benign, bilateral adrenal tumor characterized by adrenocortical hyperplasia with nonpigmented nodules. Its most frequent clinical presentation is CS. Molecular studies are limited in AIMAH. G protein mutations that cause ACTH-independent adenylate cyclase activation have been shown in macronodular adrenal hyperplasia associated with MAS and in a few isolated cases of AIMAH. Abnormal cortisol regulation by non-ACTH factors has been shown *in vivo* and *in vitro* in a subgroup of patients with AIMAH. The role of these receptors in the process of tumorigenesis has not yet been elucidated. Further studies are needed to determine if other components of the cAMP-dependent pathway and other signaling pathways are involved in macronodular adrenal hyperplasia.

REFERENCES

1. ORTH, D.N. 1995. Cushing's syndrome. N. Engl. J. Med. **332:** 791–803.
2. AIBA, M. *et al.* 1991. Adrenocorticotropic hormone-independent bilateral adrenocortical macronodular hyperplasia as a distinct subtype of Cushing's syndrome. Enzyme histochemical and ultrastructural study of four cases with a review of the literature. Am. J. Clin. Pathol. **96:** 334–340.
3. SMALS, A.G. *et al.* 1984. Macronodular adrenocortical hyperplasia in long-standing Cushing's disease. J. Clin. Endocrinol. Metab. **58:** 25–31.
4. LAMBERTS, S.W., E.G. BONS & H.A. BRUINING. 1984. Different sensitivity to adrenocorticotropin of dispersed adrenocortical cells from patients with Cushing's disease with macronodular and diffuse adrenal hyperplasia. J. Clin. Endocrinol. Metab. **58:** 1106–1110.
5. DOPPMAN, J.L. *et al.* 1988. Macronodular adrenal hyperplasia in Cushing disease. Radiology **166:** 347–352.
6. STRATAKIS, C.A. & L.S. KIRSCHNER. 1998. Clinical and genetic analysis of primary bilateral adrenal diseases (micro- and macronodular disease) leading to Cushing syndrome. Horm. Metab. Res. **30:** 456–463.
7. LACROIX, A. *et al.* 2001. Ectopic and abnormal hormone receptors in adrenal Cushing's syndrome. Endocr. Rev. **22:** 75–110.
8. KIRSCHNER, M.A *et al.* 1964. Nodular cortical hyperplasia of adrenal glands with clinical and pathological features suggesting adrenocortical tumor. J. Clin. Endocrinol. Metab. **24:** 947–955.
9. LIEBERMAN, S.A., T.R. ECCLESHALL & D. FELDMAN. 1994. ACTH-independent massive bilateral adrenal disease (AIMBAD): a subtype of Cushing's syndrome with major diagnostic and therapeutic implications. Eur. J. Endocrinol. **131:** 67–73.

10. SWAIN, J.M. *et al.* 1998. Corticotropin-independent macronodular adrenal hyperplasia: a clinicopathologic correlation. Arch. Surg. **133:** 541–545; discussion: 545–546.
11. DOPPMAN, J.L. *et al.* 2000. Adrenocorticotropin-independent macronodular adrenal hyperplasia: an uncommon cause of primary adrenal hypercortisolism. Radiology **216:** 797–802.
12. SHINOJIMA, H. *et al.* 2001. Clinical and endocrinological features of adrenocorticotropic hormone-independent bilateral macronodular adrenocortical hyperplasia. J. Urol. **166:** 1639–1642.
13. KLOOS, R.T. *et al.* 1995. Incidentally discovered adrenal masses. Endocr. Rev. **16:** 460–484.
14. YOUNG, W.F., JR. 2000. Management approaches to adrenal incidentalomas. A view from Rochester, Minnesota. Endocrinol. Metab. Clin. North Am. **29:** 159–185, x.
15. ANGELI, A. *et al.* 1997. Adrenal incidentaloma: an overview of clinical and epidemiological data from the National Italian Study Group. Horm Res. **47:** 279–283.
16. BARZON, L. *et al.* 1999. Risk factors and long-term follow-up of adrenal incidentalomas. J. Clin. Endocrinol. Metab. **84:** 520–526.
17. BOURDEAU, I. *et al.* 2001. Aberrant membrane hormone receptors in incidentally discovered bilateral macronodular adrenal hyperplasia with subclinical Cushing's syndrome. J. Clin. Endocrinol. Metab. **86:** 5534–5540.
18. KIRK, J.M. *et al.* 1999. Cushing's syndrome caused by nodular adrenal hyperplasia in children with McCune-Albright syndrome. J. Pediatr. **134:** 789–792.
19. HAYASHI, Y. *et al.* 1998. A case of Cushing's syndrome due to ACTH-independent bilateral macronodular hyperplasia associated with excessive secretion of mineralocorticoids. Endocr. J. **45:** 485–491.
20. MALCHOFF, C.D. *et al.* 1989. Adrenocorticotropin-independent bilateral macronodular adrenal hyperplasia: an unusual cause of Cushing's syndrome. J. Clin. Endocrinol. Metab. **68:** 855–860.
21. IZUMI, T. *et al.* 1997. Adrenocorticotropin-independent bilateral macronodular adrenocortical hyperplasia presenting as pre-Cushing's syndrome. Urol. Int. **58:** 262–265.
22. YAMADA, Y. *et al.* 1997. Preclinical Cushing's syndrome due to adrenocorticotropin-independent bilateral adrenocortical macronodular hyperplasia with concurrent excess of gluco- and mineralocorticoids. Intern. Med. **36:** 628–632.
23. CHEITLIN, R.A. *et al.* 1988. Cushing's syndrome due to bilateral adrenal macronodular hyperplasia with undetectable ACTH: cell culture of adenoma cells on extracellular matrix. Horm .Res. **29:** 162–167.
24. MORIOKA, M. *et al.* 1997. ACTH-independent macronodular adrenocortical hyperplasia (AIMAH): report of two cases and the analysis of steroidogenic activity in adrenal nodules. Endocr. J. **44:** 65–72.
25. MATSUKURA, S. *et al.* 1980. Multiple hormone receptors in the adenylate cyclase of human adrenocortical tumors. Cancer Res. **40:** 3768–3771.
26. LAMBERTS, S.W. *et al.* 1990. Characterization of adrenal autonomy in Cushing's syndrome: a comparison between in vivo and in vitro responsiveness of the adrenal gland. J. Clin. Endocrinol. Metab. **70:** 192–199.
27. SASANO, H., T. SUZUKI & H. NAGURA. 1994. ACTH-independent macronodular adrenocortical hyperplasia: immunohistochemical and in situ hybridization studies of steroidogenic enzymes. Mod. Pathol. **7:** 215–219.
28. SASANO, H. 1994. Localization of steroidogenic enzymes in adrenal cortex and its disorders. Endocr. J. **41:** 471–482.
29. KOIZUMI, S. *et al.* 1994. Adrenocorticotropic hormone-independent bilateral adrenocortical macronodular hyperplasia: a case report and immunohistochemical studies. Endocr. J. **41:** 429–435.
30. WADA, N. *et al.* 1996. Adrenocorticotropin-independent bilateral macronodular adrenocortical hyperplasia: immunohistochemical studies of steroidogenic enzymes and post-operative course in two men. Eur. J. Endocrinol. **134:** 583–587.
31. FINDLAY, J.C. *et al.* 1993. Familial adrenocorticotropin-independent Cushing's syndrome with bilateral macronodular adrenal hyperplasia. J. Clin. Endocrinol. Metab. **76:** 189–191.
32. MINAMI, S. *et al.* 1996. ACTH independent Cushing's syndrome occurring in siblings. Clin. Endocrinol. **44:** 483–488.

33. COOPER, R.J et al. 1998. Familial Cushing's syndrome secondary to bilateral macronodular hyperplasia. The 80th Annual Meeting of the Endocrine Society, p. 333.
34. CHANDRASEKHARAPPA, S.C. et al. 1997. Positional cloning of the gene for multiple endocrine neoplasia-type 1. Science. **276:** 404–407.
35. SKOGSEID, B. et al. 1992. Clinical and genetic features of adrenocortical lesions in multiple endocrine neoplasia type 1. J. Clin. Endocrinol. Metab. **75:** 76–81.
36. BURGESS, J.R. et al. 1996. Adrenal lesions in a large kindred with multiple endocrine neoplasia type 1. Arch. Surg. **131:** 699–702.
37. BARZON, L. et al. 2001. Multiple endocrine neoplasia type 1 and adrenal lesions. J. Urol. **166:** 24–27.
38. CLERICI, T. et al. 2001. 10 Swiss kindreds with multiple endocrine neoplasia type 1: assessment of screening methods. Swiss Med. Wkly. **131:** 381–386.
39. WEINSTEIN, L.S. et al. 1991. Activating mutations of the stimulatory G protein in the McCune-Albright syndrome. N. Engl. J. Med. **325:** 1688–1695.
40. GICQUEL, C. et al. 1994. Clonal analysis of human adrenocortical carcinomas and secreting adenomas. Clin. Endocrinol. **40:** 465–477.
41. GICQUEL, C., X. BERTAGNA & Y. LE BOUC. 1995. Recent advances in the pathogenesis of adrenocortical tumours. Eur. J. Endocrinol. **133:** 133–144.
42. BEUSCHLEIN, F. et al. 1994. Clonal composition of human adrenocortical neoplasms. Cancer Res. **54:** 4927–4932.
43. REINCKE, M. et al. 1994. p53 mutations in human adrenocortical neoplasms: immunohistochemical and molecular studies. J. Clin. Endocrinol. Metab. **78:** 790–794.
44. LIN, S.R., Y.J. LEE & J.H. TSAI. 1994. Mutations of the p53 gene in human functional adrenal neoplasms. J. Clin. Endocrinol. Metab. **78:** 483–491.
45. LATRONICO, A.C. et al. 2001. An inherited mutation outside the highly conserved DNA-binding domain of the p53 tumor suppressor protein in children and adults with sporadic adrenocortical tumors. J. Clin. Endocrinol. Metab. **86:** 4970–4973.
46. RIBEIRO, R.C. et al. 2001. An inherited p53 mutation that contributes in a tissue-specific manner to pediatric adrenal cortical carcinoma. Proc. Natl. Acad. Sci. USA **98:** 9330–9335.
47. WACHENFELD, C. et al. 2001. Discerning malignancy in adrenocortical tumors: are molecular markers useful? Eur. J. Endocrinol. **145:** 335–341.
48. BORNSTEIN, S.R. & G.P. CHROUSOS. 1999. Clinical review 104: Adrenocorticotropin (ACTH)- and non-ACTH-mediated regulation of the adrenal cortex: neural and immune inputs. J. Clin. Endocrinol. Metab. **84:** 1729–1736.
49. COTE, M. et al. 2001. Expression and regulation of adenylyl cyclase isoforms in the human adrenal gland. J. Clin. Endocrinol. Metab. **86:** 4495–4503.
50. DE CESARE, D. & P. SASSONE-CORSI. 2000. Transcriptional regulation by cyclic AMP-responsive factors. Prog. Nucleic Acid Res. Mol. Biol. **64:** 343–369.
51. MAYR, B. & M. MONTMINY. 2001. Transcriptional regulation by the phosphorylation-dependent factor CREB. Nat. Rev. Mol. Cell. Biol. **2:** 599–609.
52. LATRONICO, A.C. et al. 1995. No evidence for oncogenic mutations in the adrenocorticotropin receptor gene in human adrenocortical neoplasms. J. Clin. Endocrinol. Metab. **80:** 875–877.
53. LIGHT, K. et al. 1995. Are activating mutations of the adrenocorticotropin receptor involved in adrenal cortical neoplasia? Life Sci. **56:** 1523–1527.
54. ALOI, J.A. et al. 1995. Episodic ACTH-independent Cushing's syndrome associated with a point mutation of the ACTH receptor. Endocrine Society, 99.
55. BENJAMIN, D.R. & J.W. MCROBERTS. 1973. Polyostotic fibrous dysplasia associated with Cushing syndrome. Arch. Pathol. **96:** 175–178.
56. MAURAS, N. & R.M. BLIZZARD. 1986. The McCune-Albright syndrome. Acta Endocrinol. (Suppl.) **279:** 207–217.
57. DANON, M. et al. 1975. Cushing syndrome, sexual precocity, and polyostotic fibrous dysplasia (Albright syndrome) in infancy. J. Pediatr. **87:** 917–921.
58. BOSTON, B.A. et al. 1994. Activating mutation in the stimulatory guanine nucleotide-binding protein in an infant with Cushing's syndrome and nodular adrenal hyperplasia. J. Clin. Endocrinol. Metab. **79:** 890–893.

59. LYONS, J. et al. 1990. Two G protein oncogenes in human endocrine tumors. Science 249: 655–659.
60. REINCKE, M. et al. 1993. No evidence for oncogenic mutations in guanine nucleotide-binding proteins of human adrenocortical neoplasms. J. Clin. Endocrinol. Metab. 77: 1419–1422.
61. YASHIRO, T. et al. 1994. Point mutations of ras genes in human adrenal cortical tumors: absence in adrenocortical hyperplasia. World J. Surg. 18: 455–460; discussion: 460–461.
62. GICQUEL, C. et al. 1995. Oncogenic mutations of alpha-Gi2 protein are not determinant for human adrenocortical tumourigenesis. Eur. J. Endocrinol. 133: 166–172.
63. DEMEURE, M.J. et al. 1996. Gip-2 codon 179 oncogene mutations: absent in adrenal cortical tumors. World J. Surg. 20: 928–931; discussion: 931–932.
64. LIN, S.R. et al. 1998. Mutations of K-ras oncogene in human adrenal tumours in Taiwan. Br. J. Cancer 77: 1060–1065.
65. HSU, C.H. et al. 2001. Significantly increased cortisol secretion in normal adrenocortical cells transfected with K-ras mutants derived from human functional adrenocortical tumors. DNA Cell Biol. 20: 231–238.
66. LIN, S.R. et al. 2000. Decreased GTPase activity of K-ras mutants deriving from human functional adrenocortical tumours. Br. J. Cancer 82: 1035–1040.
67. DHANASEKARAN, N., L.E. HEASLEY & G.L. JOHNSON. 1995. G protein-coupled receptor systems involved in cell growth and oncogenesis. Endocr. Rev. 16: 259–270.
68. SIMPSON, E.R. & M.R. WATERMAN. 1988. Regulation of the synthesis of steroidogenic enzymes in adrenal cortical cells by ACTH. Annu. Rev. Physiol. 50: 427–440.
69. RAMACHANDRAN, J. & A.T. SUYAMA. 1975. Inhibition of replication of normal adrenocortical cells in culture by adrenocorticotropin. Proc. Natl. Acad. Sci. USA 72: 113–117.
70. LOTFI, C.F. et al. 1997. Unmasking a growth-promoting effect of the adrenocorticotropic hormone in Y1 mouse adrenocortical tumor cells. J. Biol. Chem. 272: 29886–29891.
71. LOTFI, C.F. et al. 2000. Proliferative signaling initiated in ACTH receptors. Braz. J. Med. Biol. Res. 33: 1133–1140.
72. GUTKIND, J.S. 1998. The pathways connecting G protein-coupled receptors to the nucleus through divergent mitogen-activated protein kinase cascades. J. Biol. Chem. 273: 1839–1842.
73. LOTFI, C.F. et al. 2000. Role of ERK/MAP kinase in mitogenic interaction between ACTH and FGF2 in mouse Y1 adrenocortical tumor cells. Endocr. Res. 26: 873–877.
74. LOTFI, C.F. & H.A. ARMELIN. 2001. cfos and cjun antisense oligonucleotides block mitogenesis triggered by fibroblast growth factor-2 and ACTH in mouse Y1 adrenocortical cells. J. Endocrinol. 168: 381–389.
75. LE, T. & B.P. SCHIMMER. 2001. The regulation of MAPKs in Y1 mouse adrenocortical tumor cells. Endocrinology 142: 4282–4287.
76. JORDAN, J.D. & R. IYENGAR. 1998. Modes of interactions between signaling pathways. Biochem. Pharmacol. 55: 1347–1352.
77. LACROIX, A. et al. 1992. Gastric inhibitory polypeptide-dependent cortisol hypersecretion—a new cause of Cushing's syndrome. N. Engl. J. Med. 327: 974–980.
78. REZNIK, Y. et al. 1992. Food-dependent Cushing's syndrome mediated by aberrant adrenal sensitivity to gastric inhibitory polypeptide. N. Engl. J. Med. 327: 981–986.
79. LACROIX, A. et al. 1997. Abnormal adrenal and vascular responses to vasopressin mediated by a V1-vasopressin receptor in a patient with adrenocorticotropin-independent macronodular adrenal hyperplasia, Cushing's syndrome, and orthostatic hypotension. J. Clin. Endocrinol. Metab. 82: 2414–2422.
80. ARNALDI, G. et al. 1998. Variable expression of the V1 vasopressin receptor modulates the phenotypic response of steroid-secreting adrenocortical tumors. J. Clin. Endocrinol. Metab. 83: 2029–2035.
81. DAIDOH, H. et al. 1998. In vivo and in vitro effects of AVP and V1a receptor antagonist on Cushing's syndrome due to ACTH-independent bilateral macronodular adrenocortical hyperplasia. Clin. Endocrinol. 49: 403–409.
82. PERRAUDIN, V. et al. 1995. Vasopressin-responsive adrenocortical tumor in a mild Cushing's syndrome: in vivo and in vitro studies. J. Clin. Endocrinol. Metab. 80: 2661–2667.

83. LACROIX, A. et al. 1997. Propranolol therapy for ectopic beta-adrenergic receptors in adrenal Cushing's syndrome. N. Engl. J. Med. **337:** 1429–1434.
84. MIRCESCU, H. et al. 2000. Are ectopic or abnormal membrane hormone receptors frequently present in adrenal Cushing's syndrome? J. Clin. Endocrinol. Metab. **85:** 3531–3536.
85. LACROIX, A., P. HAMET & J.M. BOUTIN. 1999. Leuprolide acetate therapy in luteinizing hormone–dependent Cushing's syndrome. N. Engl. J. Med. **341:** 1577–1581.
86. NAKAMURA, Y. et al. 2001. Case of adrenocorticotropic hormone-independent macronodular adrenal hyperplasia with possible adrenal hypersensitivity to angiotensin II. Endocrine **15:** 57–61.
87. PRALONG, F.P. et al. 1999. Food-dependent Cushing's syndrome: possible involvement of leptin in cortisol hypersecretion. J. Clin. Endocrinol. Metab. **84:** 3817–3822.
88. BERTHERAT, J. et al. 2001. Systematic screening confirms that illicit membrane receptors are frequent and often multiple in bilateral ACTH-independent macronodular adrenal hyperplasia. (AIMAH). The Endocrine Society's 83rd Annual Meeting, 233.
89. SCHORR, I. & R.L. NEY. 1971. Abnormal hormone responses of an adrenocortical cancer adenyl cyclase. J. Clin. Invest. **50:** 1295–1300.
90. SCHORR, I. et al. 1971. Multiple specific hormone receptors in the adenylate cyclase of an adrenocortical carcinoma. J. Biol. Chem. **246:** 5806–5811.
91. NOMATA, K. et al. 1995. Proliferating cell nuclear antigen in ACTH-independent bilateral macronodular adrenal hyperplasia. Int. J. Urol. **2:** 203–205.
92. DE HERDER, W.W. et al. 1996. Food-dependent Cushing's syndrome resulting from abundant expression of gastric inhibitory polypeptide receptors in adrenal adenoma cells. J. Clin. Endocrinol. Metab. **81:** 3168–3172.
93. N'DIAYE, N. et al. 1998. Adrenocortical overexpression of gastric inhibitory polypeptide receptor underlies food-dependent Cushing's syndrome. J. Clin. Endocrinol. Metab. **83:** 2781–2785.
94. LUTON, J.P. et al. 1998. Aberrant expression of the GIP (gastric inhibitory polypeptide) receptor in an adrenal cortical adenoma responsible for a case of food-dependent Cushing's syndrome. Bull. Acad. Natl. Med. **182:** 1839–1849; discussion: 1849–1850.
95. LEBRETHON, M.C., et al. 1998. Food-dependent Cushing's syndrome: characterization and functional role of gastric inhibitory polypeptide receptor in the adrenals of three patients. J. Clin. Endocrinol. Metab. **83:** 4514–4519.
96. CHABRE, O. et al. 1998. Gastric inhibitory polypeptide (GIP) stimulates cortisol secretion, cAMP production and DNA synthesis in an adrenal adenoma responsible for food-dependent Cushing's syndrome. Endocr. Res. **24:** 851–856.
97. CHABRE, O. et al. 1998. Cushing's syndrome due to a gastric inhibitory polypeptide-dependent adrenal adenoma: insights into hormonal control of adrenocortical tumorigenesis. J. Clin. Endocrinol. Metab. **83:** 3134–3143.
98. N'DIAYE, N. et al. 1999. Asynchronous development of bilateral nodular adrenal hyperplasia in gastric inhibitory polypeptide-dependent Cushing's syndrome. J. Clin. Endocrinol. Metab. **84:** 2616–2622.
99. N'DIAYE, N. et al. 2001. Characterization of aberrant LH/hCG and serotonin 5-HT4 receptors in adrenal Cushing's syndrome. The Endocrine Society's 83rd Annual Meeting, 235.
100. BUGALHO, M.J. et al. 2000. Presence of a Gs alpha mutation in an adrenal tumor expressing LH/hCG receptors and clinically associated with Cushing's syndrome. Gynecol. Endocrinol. **14:** 50–54.
101. KERO, J. et al. 2000. Elevated luteinizing hormone induces expression of its receptor and promotes steroidogenesis in the adrenal cortex. J. Clin. Invest. **105:** 633–641.
102. KIRSCHNER, L.S. et al. 2000. Mutations of the gene encoding the protein kinase A type I-alpha regulatory subunit in patients with the Carney complex. Nat. Genet. **26:** 89–92.
103. STRATAKIS, C.A., L.S. KIRSCHNER & J.A. CARNEY. 2001. Clinical and molecular features of the Carney complex: diagnostic criteria and recommendations for patient evaluation. J. Clin. Endocrinol. Metab. **86:** 4041–4046.

104. KIRSCHNER, L.S. et al. 2000. Genetic heterogeneity and spectrum of mutations of the PRKAR1A gene in patients with the carney complex. Hum. Mol. Genet. **9:** 3037–3046.
105. GROUSSIN, L. et al. 2000. Loss of expression of the ubiquitous transcription factor cAMP response element-binding protein (CREB) and compensatory overexpression of the activator CREMtau in the human adrenocortical cancer cell line H295R. J. Clin. Endocrinol. Metab. **85:** 345–354.
106. PERI, A. et al. 2001. Variable expression of the transcription factors cAMP response element-binding protein and inducible cAMP early repressor in the normal adrenal cortex and in adrenocortical adenomas and carcinomas. J. Clin. Endocrinol. Metab. **86:** 5443–5449.
107. AARSKOG, D. & E. TVETERAAS. 1968. McCune-Albright's syndrome following adrenalectomy for Cushing's syndrome in infancy. J. Pediatr. **73:** 89–96.
108. MACMAHON, H.E. 1971. Albright's syndrome—thirty years later. (Polyostotic fibrous dysplasia). Pathol. Annu. **6:** 81–146.
109. YOSHIMOTO, M. et al. 1991. A case of neonatal McCune-Albright syndrome with Cushing syndrome and hyperthyroidism. Acta Paediatr. Scand. **80:** 984–987.
110. SHENKER, A. et al. 1993. Severe endocrine and nonendocrine manifestations of the McCune-Albright syndrome associated with activating mutations of stimulatory G protein GS. J. Pediatr. **123:** 509–518.
111. FRAGOSO, M.C et al. 1999. Activating mutation in the stimulatory guanine nucleotide-binding protein in a woman with Cushing's syndrome by ACTH-independent macronodular adrenal hyperplasia. The Endocrine Society's 81st annual meeting.
112. KOBAYASHI, H. et al. 2000. Mutation analysis of Gsalpha, adrenocorticotropin receptor and p53 genes in Japanese patients with adrenocortical neoplasms: including a case of Gsalpha mutation. Endocr. J. **47:** 461–466.
113. FRAGOSO, M.C. et al. 2001. Activating mutations in the stimulatory guanine nucleotide-binding protein in Cushing's syndrome by ACTH independent macronodular adrenal hyperplasia. The Endocrine Society's 83rd annual meeting, 234–235.

Protein Kinase A Signaling

"Cross-Talk" with Other Pathways in Endocrine Cells

AUDREY ROBINSON-WHITE AND CONSTANTINE A. STRATAKIS

Unit on Genetics and Endocrinology (UGEN), Developmental Endocrinology Branch, National Institute of Child Health and Human Development, Bethesda, Maryland 20892-1862, USA

ABSTRACT: Protein kinase A (PKA) signaling, in "classic" endocrine cell functioning, is known to mediate cAMP effects, generated through adenylate cyclase as a response to the activation of G protein–coupled receptors (GPCRs). This signaling system is highly versatile; its flexibility is supported by a number of adenylate cyclases, four PKA regulatory and three catalytic subunits, and several phosphodiesterases that close the negative feedback loop of cAMP generation, most molecules that are expressed in a tissue-specific manner. A central question, however, remains: how do the hundreds of GPCRs mediate their specific effects? Tissue specificity of the expression of the various components of the PKA system, albeit necessary, cannot be the only answer. It helps more to view PKA as a central hub that interacts with a variety of other signaling pathways in endocrine cells, not only mediating but also communicating cAMP effects to the mitogen-activated protein kinase (MAPK), protein kinase C and B (PKC and PKB/Akt, respectively). The net result of these complex interactions, evidence for which is reviewed in this chapter, is what we know as "cAMP effects." It is, perhaps, because of this complexity that investigations of PKA signaling *in vivo* and *in vitro* often give contradictory results and are difficult to interpret.

KEYWORDS: cyclic AMP; endocrine cell signaling; integration; mitogen-activated protein kinase (MAPK)

INTRODUCTION

The cellular responses to extracellular stimuli are mediated by a host of intracellular signaling pathways. Frequently, however, a direct correlation between receptor activation and intracellular events such as nuclear transcription and the final biological response cannot be made. This lack of correlation suggests that the signaling pathways interact (cross-talk) to modify and influence the outcome of a specific extracellular signal. These interactions increase the cellular capacity for processing and integrating diverse information. Interactions between receptors and pathways

---

Address for correspondence: Constantine A. Stratakis, M.D., D.Sc., Chief, Unit on Genetics and Endocrinology, DEB, NICHD, NIH, Building 10, Room 10N262, 10 Center Dr. MSC1862, Bethesda, MD 20892-1862. Voice: 301-496-4686/402-1998; fax: 301-435-4358.
stratakc@cc1.nichd.nih.gov

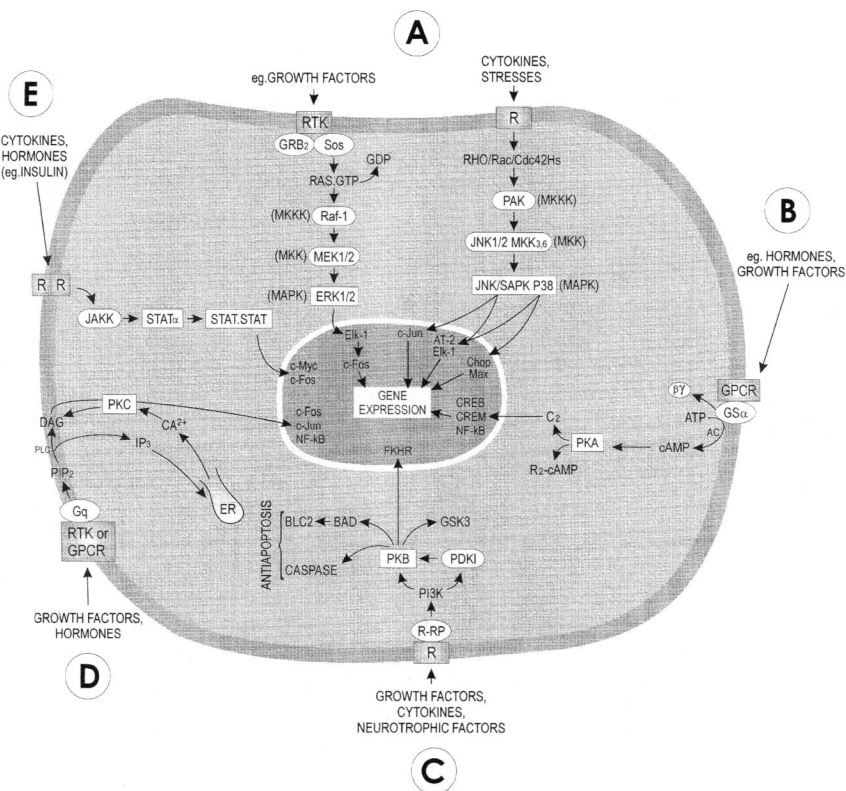

FIGURE 1. A general overview of the major signal transduction pathways in mammalian cells: **(A)** mitogen-activated protein kinase (MAPK); **(B)** protein kinase A (PKA); **(C)** protein kinase B (PKB/Akt); **(D)** Janus kinase/signal transducers and activators of transcription (JAK/STAT) pathways.

also frequently depend on the developmental stage, and may be species, tissue, or cell specific.[1]

The protein kinase A (PKA) pathway plays a part in many endocrine and nonendocrine cell responses. It modifies and influences the outcome of receptor activation in these tissues, and interacts with a variety of other signal transduction mechanisms in a complicated, multileveled manner. In this review, we will attempt to provide an outline of these pathways and their interactions and emphasize the role of PKA as the central hub of endocrine receptor signaling.

MAJOR SIGNALING PATHWAYS IN MAMMALIAN CELLS

MAP Kinase Pathway

The MAP kinases (mitogen-activated protein kinases, MAPK), a family of protein kinases, mediate diverse cellular processes including cell growth, differentiation, cytokine production, and survival. The family includes over one dozen MAPK members arranged in separate but interacting pathways on the basis of sequence homology and function. These include the ERK (extracellular signal–regulated kinase), JNK/SAPK (c-Jun N terminal kinase/stress-activated protein kinase), and p38 MAPK cascades. Each cascade consists of a three-member kinase core, MKKK (MAPKK kinase), MKK (MAPK kinase), and MAPK.[2] Upon cell stimulation, the initial core member phosphorylates and activates the succeeding member in a serial manner until the final kinase is activated. Activation occurs in the absence of a regulatory subunit, usually at both a tyrosine and a threonine phosphorylation site. The final activated kinase then proceeds to regulate cellular responses through the activation of transcription factors. Two of the many characteristics of the MAP kinases stand out: (i) the various members of this family of enzymes employ distinct mechanisms for detecting input from other signaling pathways to enhance or suppress their function; and (ii) the signal transmitted through the MAPKs is amplified, since each successive protein in the cascade is in greater abundance than its regulator.[3]

The best studied of the MAP kinases are the 44- and 42-kDA proteins (FIG. 1A), ERK 1 and ERK 2 kinases (ERK1/2), although other less known isoforms (ERK3, 4, 5, and 7) exist in mammalian cells. ERK1/2 are stimulated by a vast number of ligands (e.g., serum, growth factors, cytokine stresses, and other mitogens). Upon ligand-induced stimulation of receptor tyrosine kinase (RTK) activity, adapter molecules (Grb_2) containing Sh2 and Sh3 motifs bind to the receptors. The Sh3 domain of the adapter links the receptor to a guanine nucleotide exchange factor (GEF), son of sevenless (Sos), that enhances activation of the Ras GTP-binding protein. GDP is released and, in turn, GTP binds to and activates Ras. Activated Ras then binds to the ubiquitous Raf-1 (a MKK kinase, MKKK), the first core member in the cascade. It possibly binds to all three Raf isoforms, A-Raf, B-Raf, and Raf-1, and brings the isoform to the plasma membrane for activation by other protein kinases, such as Src, PKC, and PAK.[2–4] Active Raf dissociates from Ras and transmits a signal through phosphorylation of its substrate MEK1/2 (MKK). MEK1/2 is the second core member of the cascade and is found in the cytoplasm. This activity is regulated by the protein 14-3-3, which functions as a scaffold to stabilize both active and inactive forms of Raf, or by Ksr (kinase suppresser of Ras), which binds all three kinases, c-Raf 1, MEK1/2, and ERK1/2 (MAPK).[3,5] Activated MEK1/2 phosphorylates and activates the final core member ERK1/2, whose substrates include downstream kinases (e.g., $p90^{rsk}$) and various transcription factors, and most notably the ternary complex factor Elk-1 for the expression of early response genes leading to cell growth and differentiation.[2,4]

The three-kinase cascade that defines ERK1/2 is not as well characterized in the other MAP kinases subfamilies, due to the tendency of the intermediary members to phosphorylate and activate many MEKs and MAP kinases.[3] While the ERK1/2 proteins are strongly activated by mitogens and growth factors, the MAP kinases, JNK/SAPK and p38, are often, but not exclusively, activated by inflammatory cytokines and cellular stresses (e.g., ionizing radiation, heat shock, oxidative and osmotic stresses, and other stresses), and are associated with the promotion of apoptosis and cytokine production.[6] Therefore, these cascades will be discussed together.

The JNK/SAPK kinases consist of 46- and 54-kDa proteins (JNK1, JNK2, and JNK3), while p38 kinase has four isoforms (p38α, -β, -δ, -γ).[3,7] Upon stimulation of both pathways by stress conditions, activation may occur through members of the

Rho subfamily of small GTP-binding proteins (Rac 1, Cdc42Hs). Although Ras may interact with the JNK/SAPK pathway through the activation of MEK1, JNK/SAPK activity is predominately Ras independent.[6] These proteins stimulate PAK (p21-activated kinase, MKKK) in the p38 pathway, and MKKK in the JNK/SAPK pathway. Both MKK kinases then stimulate the MEK kinases, JNK kinase 1 and 2 for JNK/SAPK and MKK3 and 6 for the p38s.[7,8] Activated MEKs then proceed to stimulate MAP JNK/SAPK and p38. Once activated, the MAPKs can phosphorylate the same transcription factors (e.g., AT-2 and Elk-1), different transcription factors (e.g., c-Jun by JNK/SAPK, and Chop and Max by p38), and other kinases (MAPKAP, p90^{rsk}S6 kinase).[2] These downstream targets control a variety of cellular responses, such as growth, differentiation, and apoptosis.

Protein Kinase A Pathway

The protein kinase A (PKA) holoenzyme is a tetramer consisting of a dimeric regulatory subunit (R2) and two monomeric catalytic subunits (C2). There are at least two PKA isoforms, type I and type II. Four genes have been identified that code the regulatory subunits: RIα, RIβ, RIIα, and RIIβ. The RI subunit has a ubiquitous tissue distribution, while RIβ is localized primarily in brain, testis, and B and T lymphocytes. RIIα is also ubiquitous in its distribution, whereas RIIβ is found mainly in brain, adipose, and some endocrine tissues.

R subunits may determine the intracellular location of PKA. Studies on the cellular location of PKA indicate that PKA type I is found predominately in the cytoplasm and PKA type II is associated primarily with membranes, subcellular organelles, and the cytoskeleton. Anchoring proteins (with specific binding sites mainly for the RII subunits) determine the intracellular localization of the PKA tetramer; compartmentalization may be the basis for many of the specific functions of the PKA holoenzyme.[9]

The PKA pathway (FIG. 1B) may be stimulated upon ligand binding to a G protein–coupled receptor (GPRC). The activated receptor stimulates the production of cAMP from adenylate cyclase (AC) and ATP, through the Gs$_\alpha$ subunit of the heterotrimeric G protein. However, adenylate cyclase can also integrate signals from Gi G protein subunits—for example, from the βγ and the inhibitory G1 and G0—as well as from protein kinase C. cAMP binds to the R subunits of the inactive holoenzyme (PKA). The catalytic subunits then dissociate and become active. The free catalytic subunit can phosphorylate serine and threonine residues; its entrance into the cell nucleus and subsequent phosphorylation of transcription factors—for example, CREB, CREM, NF-kB and nuclear receptors—form the basis of PKA regulation of transcriptional activation.[10] As reviewed elsewhere in this book, PKA plays a large role in cellular differentiation and has complex effects on cell proliferation.[9,11]

Protein Kinase B Pathway

Protein kinase B (c-Akt/PKB), a serine/threonine kinase, is so named because of its structural similarities to the protein kinases A and C. Therefore, it has also been called RAC-PK (related to the A and C protein kinases).[12] Activation may occur at the plasma membrane (FIG. 1C) by the receptor binding of growth factors (e.g., insulin and insulin-like growth factors), cytokines, and neurotrophic factors. The activated receptor (or nonreceptor; i.e., p60-src) phosphorylates a key regulatory protein (through monomeric Ras or Ras-related protein, R-RP) phopsphatidylinositol-3 kinase (PI3K), which is involved in several different signaling pathways. PI3K is a heterodimer of two subunits, both catalytic and regulatory, having molecular masses of 110 kDa (p110) and 85 kDa (p85), respectively. There are also at least five isoforms of each subunit. PI3K is activated by two main processes.[13] First, p85 and p110 must assemble into a heterodimer; and second, the heterodimer must interact with activator proteins (AP). The main activator proteins contain tyrosine-phosphorylated amino acid sequences and include receptors (e.g., for platelet, epidermal, or insulin-like growth factors) as well as nonreceptor (p60-src) tyrosine kinases. Binding of the activator protein causes a conformational change in the heterodimer of PI3K, leading to its activation. PI3K has double enzymatic activity (both lipid kinase and protein kinase) and the ability to activate several signal proteins such as PKB, PKC, phosphoinositide-dependent kinase (PDK), Ras, and Rac Cd42. PKB is activated in two main stages: first, the PH (pleckstrin homology)-domain of PKB binds with the main products of a lipid kinase reaction catalyzed by PI3K, PtdIns P, and PtdIns P2; second, it is phosphorylated in its threonine 308 position by PDK-1 kinase.[13]

The role of the activated PKB is the preferential control of cell antiapoptotic mechanisms. PKB can block apoptosis through several independent mechanisms. It can (i) phosphorylate BAD to prevent its binding to the proapoptotic Bcl-2 protein; (ii) inhibit the activities of proteases of the caspase family that are activated during apoptosis; and (iii) directly phosphorylate and activate a p70 S6 kinase, which has a distinct antiapoptotic property (and increases the activity of PKB). PKB also directly phosphorylates glycogen synthetase kinase-3 (GSK-3) to decrease its activity, and to prevent the induction of programmed cell death.[13] Last, PKB phosphorylates the winged-helix family of transcription factors, *forkhead* (FKHR).[14,15] These actions of PKB, however, must be cautiously interpreted since many of its cellular activities are cell cycle and differentiation stage specific.[14,16]

Protein Kinase C Pathway

PKC is a large superfamily of protein kinases comprising at least 10 unequivocal members or isozymes. Each has a distinct pattern of tissue distribution and function, but they are closely related in structure. They can be grouped according to their requirements for activation. Conventional PKCs, PKC-α, -$β_I$, -$β_{II}$, and -γ, are activated by phosphatidyl serine, Ca^{2+}, and diacerylglycerol (DAG). Novel PKCs, PKC-δ, -ε, -η, and -φ, do not require Ca^{2+}; and atypical PKCs, PKC -ζ and -λ, require only phosphatidyl serine. The most recently discovered members of the PKC superfamily are the protein kinase C–related kinases (PRKs) that are insensitive to Ca^{2+}, DAG, and phorbol esters.[17] Upon cell stimulation (FIG. 1D) by various agents (mitogens, growth factors, phorbol ester) and oncogenic Ras, the plasma membrane phospholipid, phosphotidylinositol 4-5 bisphosphate (PIP_2), is broken down by phospholipase C (PLC). The resultant hydrolysis generates the production of inositol triphosphate (IP_3) and membrane diacylglycerol (DAG). IP_3 causes the release of Ca^{2+} from the endoplasmic reticulum (ER). The DAG produced from PIP_2 is transient, however; it is frequently followed by a more sustained increase following the hydrolysis of phosphatydylcholine (PC) by phospholipase D (PLD).[18] Mobilized

Ca^{2+} causes PKC to bind to the cytosolic leaf of the plasma membrane, where it is activated by DAG. This activation results in an array of cell responses through the activation of cell-specific transcription factors (e.g., c-Fos, c-Jun, NF-kB).[19]

JAK/STAT Protein Kinase Pathway

The Janus family of kinases (JAKs) play a large part in signaling for cell survival. They mediate survival signals from cytokine receptors to a family of signal transducers and activators of transcription (STAT) factors. However, the JAK/STAT pathway is poorly understood (FIG. 1E). In some systems, hormones, (e.g., interferon) bind to and activate receptors without intrinsic kinase activity. The activated receptor dimerizes (R.R) and activates JAK kinase. JAK kinase phosphorylates a STATα monomer. The monomer dimerizes, and the dimer inters the nucleus to activate transcription factors (c-Myc and c-Fos)[19] for a cell survival response. A contradiction exists, however, in some cell systems, whereby the activation of STAT may lead to cell cycle arrest or cell death[20,21] through the expression of caspase-1. This is a consequence of the activation by several cytokines of at least four JAK isoforms and seven STAT molecules. These molecules have distinct roles in different cell systems. For example, STAT3 may function in cell survival since it is essential for the early development of the mouse embryo.[22]

CROSS-TALK IN ENDOCRINE CELLS

In endocrine signaling, the linear model of the PKA and other pathways described above gives way to a mosaic of multiple interactions. Although other cAMP-regulated pathways exist in endocrine tissues (perhaps without the need for PKA mediation[16,23]), we will present examples of cAMP-regulated PKA interactions.

The Hypothalamic-Pituitary Axis

The integration and coordination of hormonal signaling at the hypothalamic-pituitary level is the keystone of the regulation of some of the most important endocrine secretory functions. An example is that of the gonadal axis, required for normal sexual maturation and reproductive functions in mammals. A key regulator of this system is hypothalamic gonadotropin-releasing hormone (GnRH). This peptide, synthesized in hypothalamic neurosecretory cells, is released into the hypothalamo-hypophyseal portal circulation, in a pulsatile fashion, and binds to specific membrane receptors (GnRHRs). It stimulates the anterior pituitary to synthesize and release the gonadotropins lutenizing hormone (LH) and follicle-stimulating hormone (FSH). These hormones enter the circulation and regulate gonadal steroidogenesis and gamete formation.[24]

Signal transduction in pituitary gonadotropes requires GPCR and the activation of multiple G proteins. GnRHRs can couple to heterotrimeric G proteins at the second and third loops of their cytoplasmic domain and activate more than one signal pathway, including those of the PKA, PKC, and MAPK. The work of Cheng and Leung[24] illustrates this mechanism (FIG. 2A). GnRHR-stimulated gonadotropin release is dependent on intra- and extracellular–derived Ca^{2+} and PKC via the G protein $Gq/11\alpha$ subunit (the major G protein subunit employed) as well as cAMP generation (after G protein Gs_α stimulation) and PKA activation.[24] However, GnRHR has also been shown to couple with Gi_α in several cancer cell types to reduce cAMP levels and to turn off adenyl cyclase (activated by Gs_α). On the other hand, the stimulation of Gi_α releases $G\beta\gamma$ subunits that can stimulate IP_3 through PLC to produce PKC for the activation of MAPK and, perhaps, the direct stimulation of gonadotropin release.[25] Several groups have also demonstrated that GnRHR activates the MAPK pathway, including ERK, JNK, and p38 MAPK. The activation of PLC by Gq stimulates PKC to activate the MAPK pathway. MAPK is also activated independently of PKC, through PKA as stimulated by Gs_α, and by $G\beta\gamma$-mediated PI3K and Src stimulation of Ras.[24] The $\beta\gamma$ subunits released by activation of Gi_α may also stimulate adenylate cyclase subunit II to produce cAMP. Thus, MAPK activation by GnRHR is mediated by Gp/11, Gs, Gi, and $G\beta\gamma$; in addition, PKA is involved in gonadotropin secretion, not only directly through cAMP activation, but also indirectly through its interactions with MAPK.

Thyroid Signaling

The role of the major signal transduction pathways in the control of human thyroid function (i.e., iodide uptake and organization, T3 secretion, differentiation, and proliferation) has been extensively explored (FIG. 2B). In normal thyroid follicular cells, two signal transduction pathways, PKA and PKC, and a protein tyrosine kinase (PTK)–mediated mechanism appear to be involved. Thyrotropin (TSH) stimulates thyroid cell function via a GPCR (TSHR) that elevates cAMP levels to activate PKA. Epidermal growth factor (EGF), acting through a tyrosine kinase–mediated mechanism (RTK); and the phorbol ester, 12-O tetradecanoylphorbol 13-acetate (TPA), both activate cell proliferation. A high degree of antagonism, as well as coordination, exists between these diverse pathways. TPA and EGF appear to inhibit thyroid cell differentiation mediated by PKA. TPA also inhibited forskolin and 8-bromo-cAMP effects. On the other hand, TSH (acting via PKA) attenuated the mitogenic action of PKC and cell proliferation by EGF.[26]

In the normal rat thyroid cell, Ras activity appears to be necessary for TSH/cAMP-dependent early-stage cell cycle progression and mitogenesis (e.g., transition from Go to G1).[16,27] Upon TSH stimulation, cAMP simultaneously activates PKA and influences the selection of Ras effectors (i.e., PI3K vs. Raf-1).[27] TSH/cAMP/PKA causes the phosphorylation and stabilization of the p85-p110 complex of PI3K, leading to the preferential accumulation of Ras/PI3K over Ras/Raf-1. Simultaneously, cAMP inhibits Raf-1/ERK1/2 signaling by decreasing the availability of Raf-1 to Ras. These ineractions may be present in other cell types and illustrate how cAMP/PKA can selectively influence Ras effector pathways.[16]

Signal transduction in the control of human thyroid carcinoma presents a different picture (FIG. 2B). In one study, four new human thyroid papillary carcinoma cell lines were treated with TSH, forskolin, 8-BrcAMP, and β-adrenergic receptor agonists (norepinephrine, epinephrine, and isoproterenol).[28] TSH (0.01–1 mU/mL) had no effect on cell proliferation or on cAMP accumulation, and cells did not contain TSH receptor mRNA. Forskolin (0.1–10 µmol/L) reduced cell growth but had no effect on cAMP accumulation. The β-adrenergic agonist induced cAMP accumulation

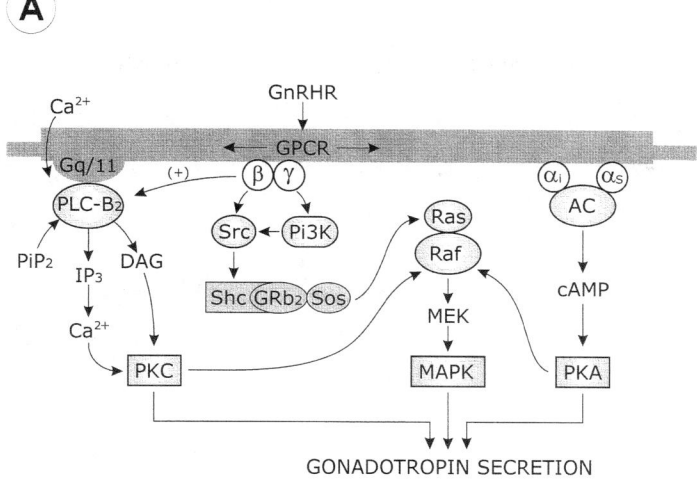

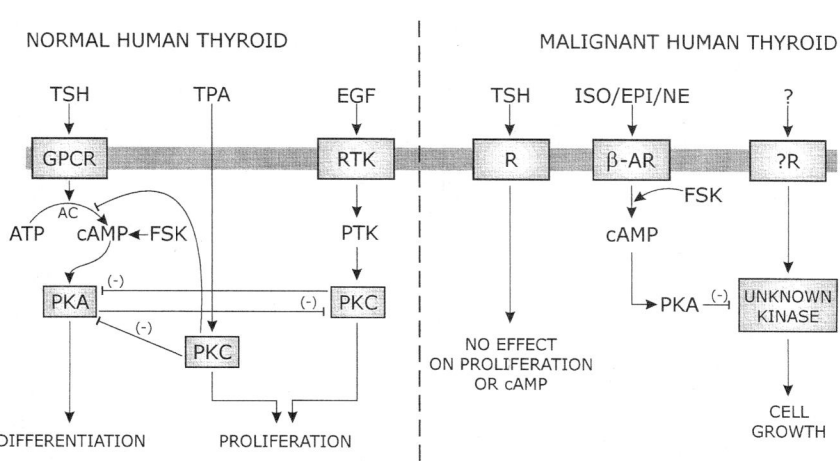

FIGURE 2. Interactions of PKA in endocrine cell signal transduction: examples in (**A**) hypothalamic-pituitary gonadal axis; (**B**) thyroid; (**C**) adrenal; (**D**) ovary; and (**E**) pancreas.

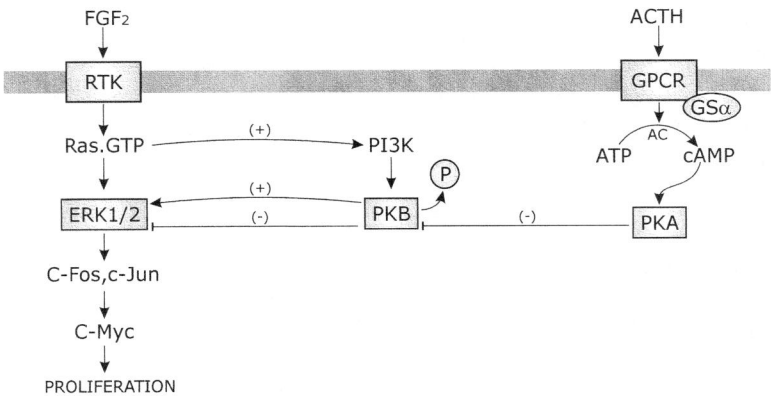

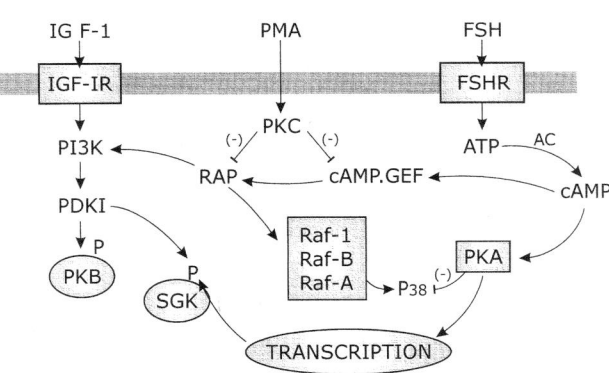

FIGURE 2. *Continued.*

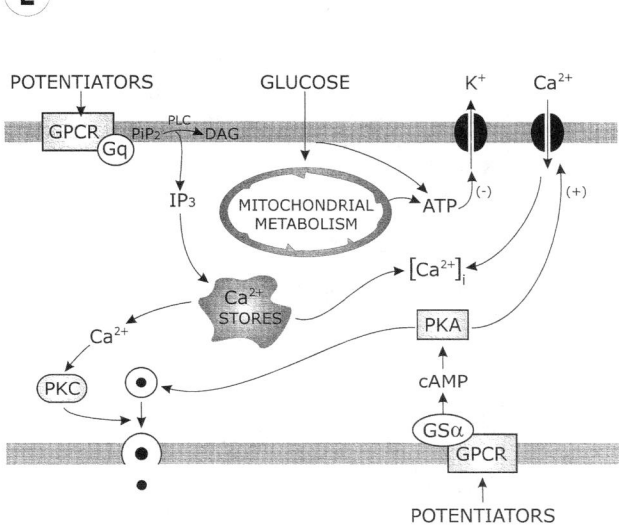

FIGURE 2. *Continued.*

and reduced cell growth. The rank order of potency was isoproterenol (ISO) > epinephrine (EPI) > norepinephrine (NE). Binding studies with [^{3}H] CGP-12177, a specific β-adrenergic agonist, revealed that the thyroid carcinoma cells had β-adrenergic receptors (β-AR) and were sensitive to the growth inhibitory effects of cAMP.[28] These results support the possibility (as shown in rat FRTL-Tc cells) that expression of β–adrenergic receptors generally results in inhibition of thyroid carcinoma cell growth.[29] However, the mechanisms of induction of β-adrenergic receptor expression in these thyroid cell lines remain as yet unknown.[28,29]

Adrenocortical Signaling

The cellular mechanisms underlying the effects of ACTH on adrenal cortex tumor cell growth have been studied in the mouse. A schematic view of the possible mechanisms is seen in FIGURE 2C. In G_o/G_1 cell cycle–arrested mouse Y1 adrenocortical tumor cells, fibroblast growth factor-2 (FGF-2) elicits a strong mitogenic response, a rapid and transient activation of ERK1/2 MAP kinases, induction of *c-Fos* and *c-Jun* genes and their protein products, and the stimulation of DNA synthesis. FGF-2 induces the *c-myc* gene and stabilizes its protein by an ERK1/2-dependent process. ACTH$_{39}$ has been shown to have a small positive (through a transient effect on ERK early in G_1) and a large negative mitogenic effect on this process. It blocks FGF-2 mitogenic activity at mid G_1, while ERK1/2 activation and c-Fos protein induction remain intact. It is suggested that the growth-inhibitory effect of ACTH at mid-G_1 represents an ACTH-regulated checkpoint that limits cell cycle progress.[30] Y1 cells possess an amplified and overexpressed *c-Ki-Ras* protooncogene and exhibit high constitutive levels of Ras.GTP. FGF_2 up-regulates Ras.GTP, activates PI3K, and promotes Akt/PKB phosphorylation. Phosphorylated Akt/PKB has an antiapoptotic and mitogenic effect in Y1 cells. Rather than inhibition of phosphorylation, $ACTH_{39}$, acting through Gs_α/adenylate cyclase, activates cAMP/PKA and causes a rapid dephosphorylation of Akt/PKB and a posttranscriptional down-regulation of the c-Myc protein.[31,32] Although other reports show that inhibition of Akt/PKB activity leads to cell cycle arrest due to $p27^{kip}$ protein expression mediated by AFX/forkhead transcription factor,[33,34] the dephosphorylation of Akt/PKB appears to be at least part of the mechanism by which $ACTH_{39}$ blocks G_1-S transition in Y1 adrenocortical cancer cells, at least in the mouse.

Ovarian Signal Transduction

The growth of ovarian follicles and the differentiation of ovarian granulosa cells are complex processes that depend on the sequential stimulation by FSH and LH. In small follicles, FSH regulates granulosa cell proliferation. As follicles mature to a preovulatory stage, the expression of genes that encode P450 aromatase and the LH receptor are induced by FSH. LH, in a surge, rapidly decreases proliferation and stimulates the terminal differentiation of the granulosa cell to a luteal cell. During granulosa cell development and differentiation, FSH and LH bind to GPCR and stimulate adenylate cyclase and cAMP production. At this point, cAMP may act as a molecular "switchboard" to control several cellular signaling pathways in granulosa cells. The model of Gonzalez-Robayna *et al.*[35] in rat granulosa cells (FIG. 2D) illustrates this mechanism: FSH, in addition to PKA, also activates PI3K through a cAMP/GEF/Ras-Rap/Raf pathway to cause the phosphorylation of the Akt/PKB and sgk proteins. cAMP, independently of PKA and PI3K, also causes the phosphorylation of the MAP kinase p38, probably through a cAMP-GEF/Ras-Rap/Raf mechanism. Insulin-like growth factor–1 (IGF-1) can also mediate the phosphorylation of PKB, independently of cAMP, through PI3K. This process can be partially abolished through the action of PKC, stimulated by the phorbol ester phorbol myristate acetate (PMA). PMA abolishes FSH, but not IGF-1R–mediated PKB phosphorylation, at a level upstream of PI3K. Therefore, cAMP acts in rat granulosa cells by PKA-dependent and -independent mechanisms to control several cell-signaling cascades.

In another study,[36] signal transduction in OVCAR-8 human ovarian cancer cells was analyzed. The expression of the RIα subunit of the cAMP-dependent PKA is increased in human cancers that display autocrine and paracrine pathways for epidermal growth factor (EGF)–related proteins. The two regulatory subunits of PKA, RI and RII, are differentially expressed. RI is preferentially expressed in transformed cells during early stages of ontogenesis, and in normal cells transiently exposed to mitogens. RII is expressed in normal tissues and in growth-arrested cancer cells following treatment with cAMP analogues (e.g., 8-cl-cAMP) or differentiating agents. EGF-like growth factors (e.g., transforming growth factor, TGF2) bind to EGF receptors and activate intracellular tyrosine kinase domains to stimulate MAP kinase activity for OVCAR-8 cancer cell growth and differentiation. These receptors are found in the majority of human cancers examined (including breast, lung, and colorectal cancers).[36,37] The interaction between the EGF pathway and RI occurs through the direct binding of the RI subunit to the Grb2 adapter protein upstream of the MAPK pathway.[37]

Testicular Signal Transduction

In testicular Sertoli cells, FSH stimulates the production of cAMP and the synthesis and expression of P450 aromatase. In rat Sertoli cells, while all RIα, RIIα, RIIβ and Cα subunits have been found,[38] most cAMP-dependent activity appears to be mediated by the RIIβ subunit. In these cells, TPA-induced PKC stimulation inhibited both FSH-stimulated cAMP formation and androgen aromatization.[39] It has also been shown that TPA can transiently stimulate both RIα and RIIβ mRNA levels, while it has no effects on other PKA subunits. When cells were treated with both 8-CPTcAMP and TPA, there was additive effect on the stimulation of RIα mRNA, and inhibition of RIIβ levels; treatment of cells with cycloheximide (a protein synthesis inhibitor) completely blocked the effect of TPA on RIIβ mRNA, but not on RIα. Therefore, the inhibitory effect of PKC on RIIβ was dependent on ongoing protein synthesis. Although the precise sites of PKC-PKA interactions in the testis have yet to be determined, these data indicate that multiple and distinct mechanisms are involved in the stimulation and inhibition of the two main PKA subunits (RIα and RIIβ).[40]

Pancreatic Signal Transduction

Normal blood glucose levels depend on the secretion of insulin from the pancreatic islet β cells, as well as on the biological action of this hormone on its target tissues. Insulin secretion (FIG. 2E) is potentiated by hormones and neurotransmitters that activate PKA and PKC, through cAMP and PLC, respectively. Glucose in high concentrations generates an elevated intracellular ATP/ADP ratio that leads to closure of K^+ channels, and the subsequent opening of Ca^{2+} channels, which leads to an increased intracellular Ca^{2+} $[Ca^{2+}]_i$ load.[42] One possible mechanism of Ca^{2+} entry is the phosphorylation of L-type Ca^{2+} channels by PKA.[43] Other potentiators of insulin secretion cause IP_3 to release Ca^{2+} from intracellular stores for the activation of PKC. The release of insulin from secretory granules is orchestrated by PKA, PKC, and myosin light-chain (MLC) protein phosphorylation by myosin light-chain kinase (MLCK).[44] The coordinated activity of PKC and PKA on islet β-cells that results in insulin secretion is a subject of intense investigation.

CONCLUSION

The various signaling pathways talk to each other in endocrine and other cells, forming a complicated network of interactions to simultaneously amplify some and dampen the activation of other signals, both extracellular and intracellular. The role of PKA in endocrine signal transduction is that of a main hub: depending on the species, tissue, cell type, and ligand involved, the hub responds with amazing speed, accuracy, and versatility to achieve the intended goal by interacting with the MAPK, PKB, PKC, and other pathways. Knowledge of the precise nature of these interactions in both normal and abnormal cells, albeit an immensely difficult task, is vital to our understanding of endocrine cell function and the design of effective therapeutic approaches to the treatment of endocrine and other diseases, including cancer.

ACKNOWLEDGMENT

We thank Mrs. Caroline Sandrini (Indaial, Santa Catarina, Brazil) for preparing the artwork for this article.

REFERENCES

1. BARRETT, J.C. 1995. Mechanisms for species differences in receptor-mediated carcinogenesis. Mutat. Res. **333:** 189–202.
2. GARRINGTON, T.P. & G.L. JOHNSON. 1999. Organization and regulation of mitogen-activated protein kinase signaling pathways. Curr. Opin. Cell. Biol. **11:** 211–218.
3. PEARSON, G., F. ROBINSON, T.B. GIBSON, et al. 2001. Mitogen-activated protein (MAP) kinase pathways: regulation and physiological functions. Endocr. Rev. **22:** 153–183.
4. CHAUDHARY A., W.G. KING, M.D. MATTALIANO, et al. 2000. Phosphatidylinositol 3-kinase regulates Raf1 through PAK phosphorylation of serine 338. Curr. Biol. **10:** 551–554.
5. YIP-SCHNEIDER, M.T., W. MIAO, A. LIN, et al. 2000. Regulation of Raf-1 kinase domain by phosphorylation and 14-3-3 association. Biochem. J. **351** (Pt.1): 151–159.
6. VERHEIJ, M., G.A. RUITER, S.F. ZERP, et al. 1998. The role of the stress-activated protein kinase (SAPK/JKN) signaling pathway in radiation-induced apoptosis. Radiol. Oncol. **47:** 225–232.
7. MINDEN, A. & M. KARIN. 1997. Regulation and function of the JNK subgroup of MAP-kinases. Biochem. Biophys. Acta **1333:** F85–F104
8. IP, Y.T. & R.J. DAVIS. 1998. Signal transduction by the c-Jun N-terminal kinase (JNK)—from inflammation to development. Curr. Opin. Cell Biol. **10:** 205–219.
9. TASKEN, K., R. SOLBERG, F.B. FOSS, et al. 1995. Protein serine kinases. In The Protein Kinase Facts Book. G. Hardie & S. Hanks, Eds.: 58–63. Academic Press. London.
10. DANIEL, P.B., W.H. WALKER & J.F. HABENER. 1998. Cyclic AMP signaling and gene regulation. Ann. Rev. Nutr. **18:** 353–383.
11. BERTHERAT, J. 1997. Nuclear effects of the cAMP pathway activation in somatotrophs. Horm. Res. **47:** 245–250.
12. COFFER, P.J. & J.R. WOODGETT. 1991. Molecular cloning and characterization of a novel putative protein-serine kinase related to the cAMP-dependent and protein kinase C families. Eur. J. Biochem. **201:** 475–481.
13. KRASSILNIKOV, M.A. 2000. Phosphatidylinositol-3 kinase dependent pathways: the role in control of cell growth, survival and malignant transformation. Protein Biochem. **65**(1): 59–67.
14. GUO, S., G. RENA, S. CICHY, et al. 1999. Phosphorylation of serine 256 by protein kinases disrupts transactivation of FKHR and mediates effects of insulin or insulin like growth factor-binding protein-1 promotes activity through a conserved insulin response sequence. J. Biol. Chem. **274:** 17184–17191.
15. BRUNET, A., A. BONNI, M.J. ZIGMOND, et al. 1999. Akt promotes cell survival by phosphorylating and inhibiting a forkhead transcription factor. Cell **96:** 857–868.
16. CIULLO I., G. DIEZ-ROUX, M. DI DOMENICO, et al. 2001. cAMP signaling selectively influences Ras effector pathways. Oncogene **20:** 1186–1192.
17. WEBB, B.L..J, S.J. HIRST & M.A. GIEMBYCZ. 2000. Protein kinase C isozymes: a review of their structure, regulation and role in regulating airway smooth muscle tone and mitogenesis. Brit. J. Pharmacol. **130:** 1433–1452.
18. NISCHIZUKA, Y. 1995. Protein kinase C and lipid signaling for sustained cellular responses. FASEB J. **9:** 484–496.
19. LODISH, H., A. BERK, S.L. ZIPURSKY, et al. 2000. Cell-to-cell signaling: hormones and receptors. In Molecular Cell Biology: 848–909. W.H. Freeman. New York.
20. CHIN, Y.E., M. KITAGAWA, K. KUIDA, et al. 1997. Activation of the STAT signaling pathway can cause expression of caspase 1 and apoptosis. Mol. Cell. Biol. **17:** 5328–5337.

21. CHIN, Y.E., M. KITAYAWA, W-C. SU, *et al.* 1996. Cell growth arrest and induction of cyclin-dependent kinase inhibitor p21$^{WAF/CIP1}$ mediated by STAT1. Science **272:** 719–722.
22. TAKEDA, K., K. NOGUCHI, W. SHI, *et al.* 1997. Targeted description of the mouse STAT 3 gene leads to early embryonic lethality. Proc. Natl. Acad. Sci. USA **94:** 3801–3804.
23. RICHARD, J.S. 2001. New signaling pathways for hormones and cyclic adenosine 3,'5'-monophosphate action in endocrine cells. Mol. Endocrinol. **15:** 209–218.
24. CHENG, K.W. & P.C.K. LEUNG. 2000. The expression, regulation and signal transduction pathways of the mammalian gonadotropin-releasing receptor. Can. J. Physiol. Pharmacol. **78:** 1029–1052.
25. CAMPS, M., A. CAROZZI, P. SCHNABEL, *et al.* 1992. Isozyme-selective stimulation of phospholipase C-β2 by G protein βγ-subunits. Nature **360:** 684–686.
26. KRAIEM, Z., O. SADEH, M. YOSEF & A. AHARON. 1995. Mutual antagonistic interactions between the thyrotropin adenosine 3',5'-monophosphate) and protein kinase C/epidermal growth factor (tyrosine kinase) pathways in cell proliferation and differentiation of cultured human thyroid follicles. Endocrinology **136:** 585–590.
27. CASS, L.A. & J.L. MEINKOTH. 2000. Ras signaling through PI3K confers hormone-independent proliferation that is compatible with differentiation. Oncogene **19:** 924–932.
28. OHTA. K., X-P. PANG, L. BERG & J.M. HERSHMAN. 1997. Growth inhibition of new human thyroid carcinoma cell lines by activation of adenylate cyclase through the β–adrenergic receptor. J. Clin. Endocrinol. Metab. **82:** 2633–2638.
29. ENDO, T., H. SHIMURA, T. SAITO & T. ONAYA. 1990. Cloning of malignantly transformed rat thyroid (FRTL) cells with thyrotropin receptors and their growth inhibition by 3',5'-cyclid adenosine monophosphate. Endocrinology **126:** 1492–1497.
30. LOFTI, C.F.P., Z. TODOROVIC, H.A. ARMELIN & B. SCHIMMER. 1997. Unmasking a growth-promoting effect of the adrenocorticotropic hormone in Y1 adrenocortical tumor cells. J. Biol. Chem. **272:** 29886–29891.
31. LEPIQUE, A.P., F.L. FORTI, M.S. MORAES & H.A. ARMELIN. 2000. Signal transduction in Go/G1-arrested mouse Y1 adrenocortical cells stimulated by ACTH and FGF2. Endocr. Res. **26:** 825–832.
32. FORTI, F.L. & H.A. ARMELIN. 2000. ACTH inhibits a Ras-dependent anti-apoptotic and mitogenic pathway in mouse Y1 adrenocortical cells. Endocr. Res. **26:** 911–914.
33. COLLADO, M., R.H. MEDEMA, I. GARCIA-CAO, *et al.* 2000. Inhibition of the phosphoinositide 3-kinase pathway induces a senescence-like arrest mediated by p27 Kip 1. J. Biol. Chem. **275:** 21960–21968.
34. MEDEMA, R.H., G.J. KOPS, J.L. BOSS, *et al.* 2000. AFX-like forkhead transcription factors mediate cell cycle regulation by Ras and PKB through p27. Nature **404:** 782–787.
35. GONZALEZ-ROBAYNA, I.J., A.E. FALENDER, S. OCHSNER, *et al.* 2000. Follicle-stimulating hormone (FSH) stimulates phosphorylation and activation of protein kinase B (PKB/Akt) and serum and glucocorticoid-induced kinase (Sgk): evidence for A kinase-independent signaling FSH in granulosa cells. Mol. Endocrinol. **14:** 1283–1300.
36. ALPER, O., N.F. HACKER & Y.S. CHO-CHUNG. 1999. Protein kinase A-1a subunit-directed antisense inhibition of ovarian cancer cell growth: crosstalk with tyrosine kinase signaling pathway. Oncogene **18:** 4999–5004.
37. CIARDIELLO, F. & G. TORTORA. 1998. Interactions between the epidermal growth factor receptor and type I protein kinase A: biological significance and therapeutic implications. Clin. Cancer Res. **4:** 821–828.
38. OYEN, O., A. FRAYSA, M. SANDBERG, *et al.* 1987. Cellular localization and age-dependent changes in mRNA for cAMP-dependent protein kinase in rat testis. Biol. Reprod. **37:** 947–956.
39. CONTI, M. & L. MONACO. 1986. Modulatory mechanisms of the hormonal response of testicular cells. Endocrinology of the testis. Excerpta Med. Int. Congr. Ser. **716:** 89–100.
40. TASKEN, K.A., H.K. KNUTSEN, L. EIKVAR, *et al.* 1992. Protein kinase C activation by 12-o-tetradecanoylphorbol 13-acetate modulates messenger ribonucleic acid levels for two of the regulatory subunits of 3',5'-cyclic adenosine monophosphate-depen-

dent protein kinase (RIIβ and PIIα) via multiple and distinct mechanisms. Endocrinology **130:** 1271–1280.
41. WOLLHEIM, C.B., J. LANG & R. REGAZZI. 1996. The exocytotic process of insulin secretion and its regulation by Ca^{2+} and G-proteins. Diabetes Rev. **4:** 276–297.
42. MACFARLANE, W.M., S.B. SMITH, R.F. JAMES, *et al.* 1997. The p38/reactivating kinase mitogen-activated protein kinase cascade mediates the activation of the transcription factor insulin upstream factor 1 and insulin gene transcription by high glucose in pancreatic beta-cells. J. Biol. Chem. **272:** 20936–20944.
43. LEISER, M. & N. FLEISHER. 1996. cAMP-dependent phosphorylation of the cardiac type α1 subunit of the voltage-dependent Ca^{2+} channel in a murine pancreatic β-cell line. Diabetes **45:** 1412–1418.
44. YU, W., T. NIWA, T. FUKASAWA, *et al.* 2000. Synergism of protein kinase A, protein kinase C, and myosin light-chain kinase in the secretory cascade of the pancreatic β-cell. Diabetes **49:** 945–952.